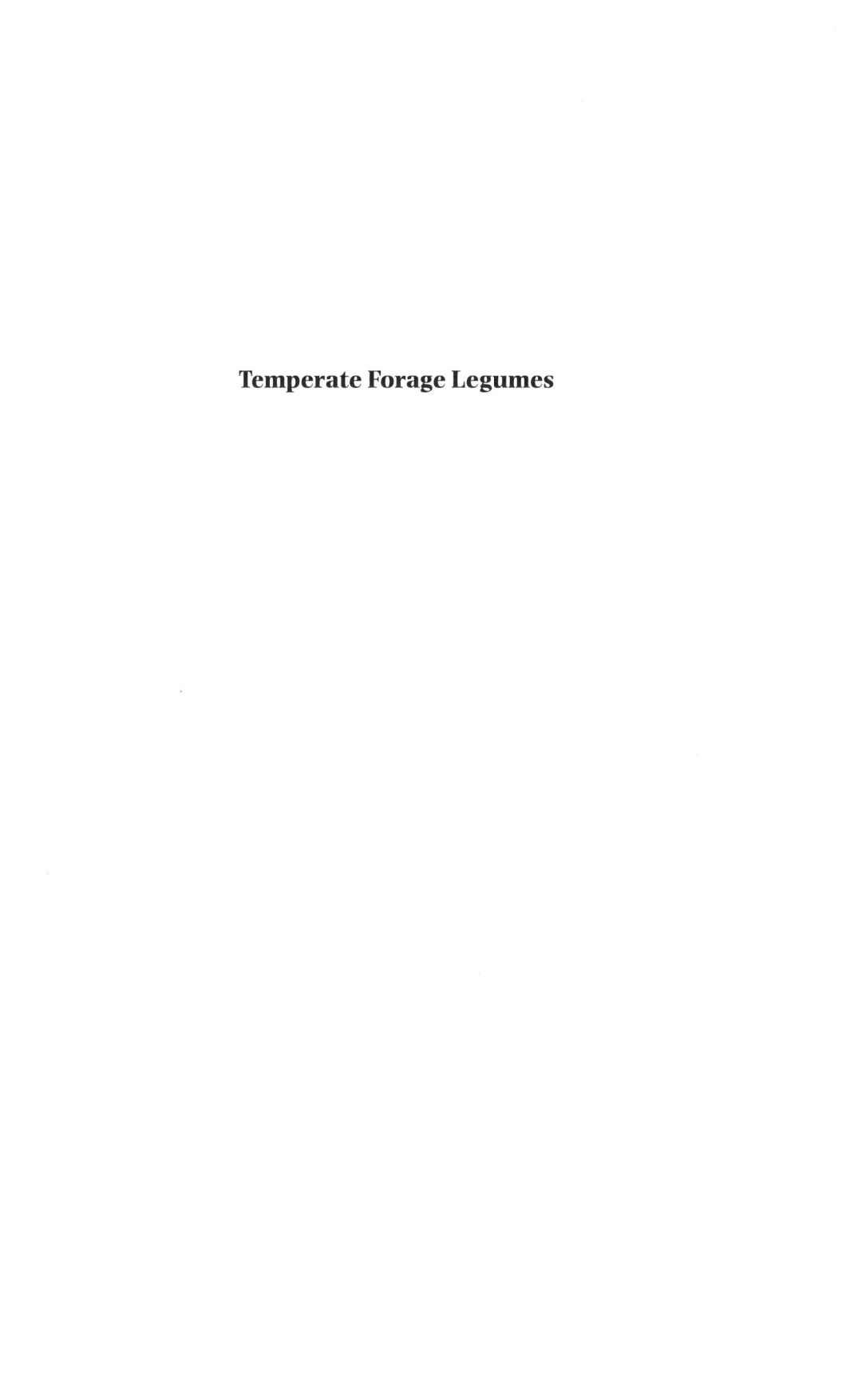

Temperate Forage Legumes

Temperate Forage Legumes

J. Frame
*formerly of the Scottish Agricultural College
Auchincruive, Ayr
UK*

J.F.L. Charlton
*AgResearch Grasslands
Palmerston North
New Zealand*

and

A.S. Laidlaw
*Department of Applied Plant Science
Queen's University of Belfast
UK*

CAB INTERNATIONAL

CAB INTERNATIONAL
Wallingford
Oxon OX10 8DE
UK

Tel: +44 (0)1491 832111
Fax: +44 (0)1491 833508
E-mail: cabi@cabi.org

CAB INTERNATIONAL
198 Madison Avenue
New York, NY 10016–4314
USA

Tel: +1 212 726 6490
Fax: +1 212 686 7993
E-mail: cabi-nao@cabi.org

©CAB INTERNATIONAL 1998. All rights reserved. No part of this publication may be reproduced in any form or by any means, electronically, mechanically, by photocopying, recording or otherwise, without the prior permission of the copyright owners.

A catalogue record for this book is available from the British Library, London, UK

ISBN 0 85199 214 5

Library of Congress Cataloging-in-Publication Data
Frame, John
    Temperate forage legumes / J. Frame, J.F.L. Charlton, A.S. Laidlaw.
      p.   cm.
    Includes index.
    ISBN 0–85199–214–5 (alk. paper)
    1. Legumes. 2. Forage plants. I. Charlton, J.F.L.
II. Laidlaw, A.S. (A. Scott) III. Title.
SB203.F735   1997
633'.3'0912--dc21   97–22229
                                                   CIP

Drawings by J.F.L. Charlton
Typeset in Photina 10/12 by Columns Design Limited, Reading
Printed and bound in the UK by Biddles, Guildford and King's Lynn

Contents

Acknowledgements		vi
Preface		vii
1.	Introduction	1
2.	White Clover	15
3.	Lucerne (syn. Alfalfa)	107
4.	Red Clover	181
5.	Subterranean Clover	225
6.	Birdsfoot Trefoil and Greater Lotus	245
7.	Alsike Clover and Sainfoin	273
8.	Serradella, Sulla and Tagasaste	291
9.	Prospects for Forage Legumes	313
Index		319

Acknowledgements

We are grateful to the following organizations and individuals for permission to use material for the figures and tables listed.

Academic Press, Orlando (Table 2.1 and Fig. 2.5); Dr A.J. Parsons, New Zealand (Fig. 2.3); Agronomy Society of New Zealand, Christchurch, and New Zealand Grassland Association, Palmerston North (Tables 2.4 and 2.7); L'institut Technique d'Elevage Bovin, Paris (Table 2.4); Kluwers Publishers, Dordrecht (Fig. 2.4); Dr R.D. Harkness (Fig. 2.7); British Grassland Society, Reading (Table 2.8, and Figs 2.6 and 2.8); European Grassland Federation and Irish Grassland and Animal Production Association (Table 2.9); Continuing Committee of the International Grassland Congress (Tables 3.2 and 5.2); Oxford University Press (Fig. 3.2); Food and Agriculture Organisation, Rome (Fig. 3.5); American Society of Agronomy, Madison (Fig. 3.6); Blackwell Science Ltd, Oxford (Fig. 4.3); Lincoln University Library, Christchurch (Table 5.1); *Australian Journal of Agricultural Research* (Fig. 5.2); Agronomy Society of New Zealand (Tables 8.1 and 8.2); Department of Agriculture, Western Australia (Table 8.4).

Preface

Temperate forage legumes play an invaluable part in grassland farming in many regions of the world, encompassing cool to warm temperate climes. The well documented advantages of forage legumes are their contributions to the nitrogen economy of swards through nitrogen fixation, and their superior feeding values to those of grasses. Another key feature is the adaptation to different environmental and management conditions exhibited by specific legume species, and even some essentially 'Mediterranean' or 'subtropical' legumes are being successfully exploited in temperate grassland farming. In addition, a large reservoir of genetic variability is available to draw upon for breeding improved cultivars, whether by conventional methods or by innovative biotechnology. These benefits have been increasingly examined in recent times, not only in regions traditionally reliant on the contribution from forage legumes (e.g. Australasia), but in countries (e.g. those in north-west Europe) in which high usage of fertilizer nitrogen on grass swards is the norm. Sustainability, environmental friendliness and energy input efficiency have all become grassland farming watchwords. As a consequence, the mainly beneficial role and potential of forage legumes in these respects have been increasingly appreciated over the past 15–20 years, whether from scientific, practical or economic viewpoints. Thus, this book, dealing with each species comprehensively, is timely.

In a series of sections including growth, development, physiology, nitrogen fixation, agronomy, management, utilization, herbage quality and animal productivity, the individual species are overviewed and the existing body of knowledge assessed and presented. Where relevant, specific disadvantages and disbenefits such as bloat, cyogenic potential and other antiquality factors, are also identified and evaluated. Information has been drawn selectively from the worldwide research and development database, including the numerous publications of the authors themselves. The detail and length of the overviews for each species are a reflection of their relative worldwide importance and use in grassland farming. The 'big three' are white clover, lucerne and red clover, followed by subterranean clover, the *Lotus* species and a selected number of lesser-used or niche species, namely, alsike clover, sainfoin, serradella, sulla and tagasaste.

In the preparation of this book we have drawn heavily on our wide experience in grassland research, development, advisory work and lecturing which includes experience in several countries of the world. In this respect we are grateful for the encouragement received and wisdom imparted over the years by past and present colleagues, farmers and ranchers.

We fully acknowledge the many diverse and valuable research and advisory publications which were consulted and/or cited. However, due to the constraints of space many references were omitted which would otherwise have been included in detailed literature reviews of the different forage legume species; any oversight of germane references is unintentional. Library staff at our associated institutes are thanked for their unstinting efforts in response to requests for published information.

Constructive comment and advice on specific chapters were given by G.E.D. Tiley and R.F. Gooding (Scotland), G. Parente (Italy), J.M. Prosperi (France), B. Frankow-Lindberg (Sweden), G.A. Jung, D.P. Belesky, R.L. McGraw and P.R. Beuselinck (USA), H.T. Kunelius (Canada) and M.J. Blumenthal (Australia). None the less, any shortcomings or omissions in the text are our own and comments or suggestions for improvements in future editions will be welcomed.

Dedication is made to our families who were towers of strength and support during the writing of the book, and in some instances contributed directly by word processing. Two of the co-authors (JF, ASL) acknowledge the splendid plant drawings by the third (JFLC). Finally, thanks and appreciation are extended to staff in the publishing division of CAB International for their encouragement, help and meticulous attention to detail in producing the finished product.

J. Frame, J.F.L. Charlton and A.S. Laidlaw

Introduction 1

The legume plant family, *Leguminosae*, is one of the largest in the world and, with almost cosmopolitan distribution, has played an essential part in human development and civilizations (Isely, 1982). Greek and Roman historians were the first to record the benefits of legumes as green manures for subsequent crops (Rogers, 1976), but the explanation in terms of molecular nitrogen (N_2)-fixation capability was not forthcoming until the mid-nineteenth century and that of the legume–*Rhizobium* symbiosis until the mid-twentieth century (Taylor, 1985a). Within a forage context, the *Leguminosae* subfamily *Papilionoideae* has occupied the key position, with the genera *Trifolium* and *Medicago* particularly prominent in sustainable farming systems in temperate climes.

In modern times, the benefits of forage legumes have been propounded in terms of their

- contribution to the nitrogen (N) economy of grassland, especially in extensive low-input, low-cost systems of animal production, and of subsequent crops, from N_2 fixation;
- superior feeding value – a function of both nutritive value and voluntary intake – in relation to grasses.

These benefits have been increasingly examined over the past 15–20 years, with the ensuing result of an ever-expanding volume of published literature from the research and development work.

In many countries and regions of the world, such as Western Europe, recognition that N was the most important nutrient limiting grassland production led to increasing rates of application from the 1950s onwards and the consequent intensification of stocking. Other sources of N, such as soil, excreta from grazing livestock and organic manure from winter housing, and biological N from legumes, became completely subsidiary to fertilizer N. The usage of forage legumes declined generally and, when sown in mixed grass/legume swards, the adverse effects of N fertilizer on their growth and development markedly reduced the impact of biological N. The Netherlands, with its virtually total reliance on nitrogenous fertilizer, was a classic example of this ethos. However, in the early 1970s there was a sharp upturn in the cost of fossil fuel and, as a consequence of the high energy consumption in the manufacture of inorganic nitrogenous fertilizer, its cost was expected to escalate. This reawakened interest in legumes and stimulated scientific and economic appraisal of the role and potential of legumes in grassland production (Frame and Newbould, 1986; Sprent and 't Mannetje, 1996). Despite forage legumes being favourably comparable to grass receiving high applications of nitrogenous fertilizer in many animal production systems, use of legumes in grassland has not yet increased markedly.

In other temperate countries with well-developed grassland-based animal production systems, forage legumes were, and still are, heavily relied upon, e.g. white clover (*Trifolium repens*) in New Zealand, lucerne (*Medicago sativa*) in the USA, subterranean clover (*Trifolium subterraneum*) in Australia. On agricultural soils worldwide, the estimate for annual N_2 fixation is about 90 Mt of N, of which *c.* 56% is fixed by forage legumes, including the temperate forage legumes, while the estimate of fertilizer N applied annually is 60 Mt (Hauck, 1988).

New Zealand and the UK offer an interesting comparison. Although similar areas of grassland are farmed, over 1 Mt of N is biologically fixed annually in the former country and only 10,000–20,000 t of fertilizer N is used, compared with *c.* 80,000 t of N fixed and approximately 750,000 t of fertilizer N used annually in the latter (Ball and Crush, 1985; Whitehead, 1986), although a more recent estimate of N_2 fixed in New Zealand is 1.57 Mt of N (Caradus *et al.*, 1996). From these estimates, UK grassland receives annually less than 10 kg N ha^{-1} from N_2 fixation, demonstrating the meagre contribution which legumes are currently making to the N economy of the grassland. In contrast to the potential contribution which legumes can make to N budgets, which are outlined later in this and in other chapters of this book, these estimates also emphasize the scope which exists to exploit N_2 fixation in the UK and other countries in a similar position.

In recent years, various points of concern have arisen *vis-à-vis* continued intensive use of fertilizer N, including its steadily increasing cost, the growing realization of the drain on finite fossil-fuel energy which its manufacture entails, and the environmental consequences, such as leaching of nitrate into watercourses and thence drinking supplies. As projected by Wittwer (1978), 'The current and projected natural gas dependency of chemically fixed nitrogen fertilizer remains as one of the most flagrant violations of good economics, use of a non-renewable resource. It is inconceivable for us to continue to go this route.' Comparison between the fossil-energy expenditure in a grass/N-fertilizer system and a grass/clover system, both receiving the equivalent of 200 kg N ha^{-1}, shows the latter to expend only 5% the fossil energy of the N-fertilizer system (Wood, 1996).

Environmental protection, the cost of which will be increasingly borne by grassland farmers, has assumed ever-increasing importance in those countries, e.g. in the European Union (EU), where there is also overproduction of ruminant and crop products. In many other countries, public and governmental concern has also raised the profile of environmental issues. Thus, more research is being focused on the N cycle in relation to application of inorganic fertilizer or organic manures – and also to forage-legume use, since it is likely to increase – especially on how to reduce the losses or leakages of N from animal and crop production systems ('t Mannetje and Jarvis, 1990; Van der Meer and Van der Putten, 1995).

There have to be practical and profitable reasons why grassland farmers should replace N-fertilizer use with forage-legume technology. Perhaps the case for forage legumes was not always portrayed as clearly as it could have been, nor were there sufficient clear-cut management guidelines available. Undoubtedly, the forage legumes were often unreliable and variable in persistent performance in practice. Even in New Zealand, white-clover contribution to pasture yield has varied markedly, from no more than 10–15% in many pastures on the flat and rolling country to only about 5% on the hill country (Lancashire, 1990). Nevertheless, in recent years a large number of the biological complexities of the major forage legumes, especially white clover and lucerne, have been unravelled, with respect both to their potential as forage in animal production systems and to the factors causing their variability in performance. Research on these complexities has always been ongoing in countries or regions of countries which traditionally relied on forage legumes, but has burgeoned in some other countries, such as in the UK, where systems based on N-fertilized grass have been prevalent (Laidlaw and Frame, 1988; Novoselova and Frame, 1992).

Apart from the consolidation of forage-legume use in regions such as Australasia, it is widely predicted that forage legumes will be increasingly utilized in many temperate regions of the world, within Europe for example. Changes in EU policy are resulting in reduced stocking rates and biological changes, such as an increase in the genetic merit of dairy cows, are encouraging a shift in emphasis towards individual animal performance. With many of the forage legumes having superior nutritive and feeding value over grass, their importance in the diet of ruminants will undoubtedly increase. Temperate legumes are also grown and utilized in high altitude areas with dry, cold winters in subtropical and tropical regions; notably, species of the genera *Trifolium*, *Medicago*, *Melilotus* and *Vicia* have a wide latitudinal range in both northern and southern subtropics (Skerman et al., 1988).

The major advantage that legumes possess is their ability to fix substantial amounts of atmospheric N in association with *Rhizobium* bacteria. As is well documented, legume N benefits grass growth in mixed swards, particularly where there is sufficient vigorous legume growth with effective N_2-fixing rhizobia, e.g. in grazed grass/white-clover swards, where the main transfer route of clover N is via ingested herbage and there is subsequent return to pasture of excretal N, particularly in the urine. In some other forage legumes, less N is 'lost' from the plant and so a high proportion of fixed N is utilized directly in legume herbage production, e.g. lucerne where it is grown as a monoculture or as the dominant component of a mixed species, where its N_2-fixing ability is put to good use in supplying most of the sward's N requirement, enhancing plant protein and promoting general productivity. In red clover, for example, N_2 fixation can contribute up to 80% of total N assimilated (Heichel et al., 1985).

The quantity of N fixed by forage legumes can differ widely, reflecting the influence of a range of factors, such as *Rhizobium*-strain effectiveness, quantity of available soil mineral N and applied N from fertilizers or organic manures, utilization regime, climatic influences and the proportion of legume present in the sward. Variation in the estimates of N_2 fixation between studies is also a consequence of the methodology used, as techniques may not be comparable. Nitrogenase is the enzyme responsible for the reduction of N_2 and an assay of its activity by acetylene reduction is not always calibrated against the more direct method using ^{15}N isotope (Marriott and Haystead, 1993), and questionable assumptions have to be made when interpreting isotope data in some studies (Witty, 1983). In a world context, Burton (1985) estimated that *Trifolium* species fix 50–350 kg N ha^{-1} annually, while Nutman (1976), reviewing international literature, quoted a range for white clover of 45–673 kg N ha^{-1} fixed, although the upper value in the latter range was for swards grown in N-impoverished subsoil in New Zealand. Other estimates of the amounts of N_2 fixed annually (kg N ha^{-1}) by white clover are: 85–265 from a series of coordinated trials in New Zealand (Hoglund et al., 1979), 74–280 in UK lowland swards (Cowling, 1982) and 100–150 in UK hill and upland swards (Newbould, 1982). The range cited for red clover (*Trifolium pratense*) is 125–220 (Rohweder et al., 1977; La Rue and Patterson, 1981) and for birdsfoot trefoil (*Lotus corniculatus*) 60–138 kg N ha^{-1} (Farnham and George, 1994), while N_2 fixation rates for lucerne can approach 300 kg N ha^{-1} under optimum conditions (Sheehy et al., 1984).

In addition to increasing the level of available soil N, forage legumes have been shown to have a beneficial effect on soil structure, aided by their association with mycorrhizal fungi (Miller and Jastrow, 1996). The hyphae and decomposing roots bind soil particles into stable macroaggregates and the relatively rapid turnover of roots provides substrate for microbes to produce polysaccharides, which also contribute to soil structure.

Renewed and increased appraisal has underlined the nutritional potential of forage legumes for animal production. In general, legumes are highly acceptable to ruminants, whether for grazing or as well-preserved silage and hay. The lower contents of structural fibre and higher protein contents of legumes compared with grasses result in improved voluntary intake and digestion processes and a more efficient absorption of nutrients (Ulyatt et al., 1977; Beever and Thorp, 1996).

In many comparative trials with grass species, legumes have proved superior not only in nutritive value, i.e. animal production response per unit of intake (Campling, 1984; Thomson, 1984), but also in feeding value, i.e. animal production response. Much of this advantage is attributable to enhanced forage intakes, although high concentrations of protein and minerals are also of major nutritional significance. Additionally, at similar stages of maturity the dicotyledonous legumes have faster rates of particle breakdown in the rumen and more rapid clearance of the particles from the rumen than the monocotyledonous grasses (Ulyatt et al., 1986; Waghorn et al., 1989). The arrangement of veins in the leaves is considered to contribute to these differing rates of breakdown, parallel venation of the grasses imposing more resistance than the reticulum of veins in legume leaves (Wilman et al., 1996).

The superior feeding value of forage legumes over grass at similar stages of digestibility, either fresh or conserved, has been demonstrated in many studies. For example, better dairy-cow performance, in terms of higher yields of milk, milk protein and lactose, was achieved from red-clover silage than from perennial-ryegrass silage compared at similar digestibility levels (Thomas et al., 1985). Improved intakes and performance from red clover silage were obtained even when its digestibility was lower than that of grass silage (Copeman and Younie, 1982).

Similar advantages from legume use have been obtained in grazing experiments, for example, with grass/white-clover swards versus N-fertilized grass swards (Wilkins et al., 1991). However, while such positive results have usually also been obtained in large-scale system trials with sheep, notably in individual lamb liveweight gains postweaning (Vipond et al., 1993), the advantages have been less consistent with dairy or beef cattle, especially when complete systems and not just components of systems are considered (Davies and Hopkins, 1996). *Inter alia*, a more precise matching of specific animal requirements to the seasonal pattern of grass/legume sward growth seems necessary (Bax and Schils, 1993).

As an alternative to comparing systems in reality, economic-feasibility studies have been undertaken, in which costs and values have been assigned to physical inputs and outputs in systems based on forage legumes and grass receiving N fertilizer, e.g. 18-month beef from grass/white-clover swards (Doyle and Morrison, 1983; Doyle and Bevan, 1996) and lucerne for conservation (Doyle and Thomson, 1985). While the profitability of legume-based systems has been repeatedly demonstrated, consistent year-to-year production is assumed, and at a relatively high level in the case of lucerne. Long-term system experiments have exposed the weakness of such assumptions (Stewart and Haycock, 1984). Ironically, in New Zealand, the leading protagonist of white-clover use, a recent dairy-farm simulation model predicts that maximum physical and economic performance from dairy pasture is best achieved with clover contents of 30–40% and N fertilizer rates of 100–200 kg N ha^{-1} annually (Clark and Harris, 1996). Yet exploitation of their joint benefits in mixed swards has proved elusive in the past, except with low inputs of fertilizer N applied tactically (Fothergill and Davies, 1993).

The antiquality factors in forage legumes can militate against the full expression of their feeding value. The most recognized antiquality factor is the potential risk of bloat (tympanites) in livestock grazing swards rich in white clover, lucerne or red clover. However, a number of preventive and control measures, including management and specific treatments, are available (Essig, 1985; Carruthers *et al.*, 1987). Phyto-oestrogens in *Trifolium* spp., especially red clover and subterranean clover, can reduce reproductive fertility, particularly in sheep, but can be minimized by cultivar choice and manipulation of grazing management. The cyanogenic potential of white clover is of concern in some countries, but there is considerable variation among cultivars (Lehmann *et al.*, 1991). Condensed tannins (CT), which are present in several forage legume species, e.g. birdsfoot trefoil and greater lotus, can inhibit digestion in the rumen if the CT content in the herbage is at high levels (D'Mello and Macdonald, 1996), but, as discussed below, the benefits of CT outweigh the disadvantages.

On a world basis, a wide array of animal production systems based on the judicious use of forage legumes, sown in monoculture or in legume/grass mixtures, is practised in grassland farming. Considering swards sown with white clover, estimates indicate 9 Mha in New Zealand, 6 Mha in Australia and 5 Mha in the USA (Marten *et al.*, 1989). Highlighting the New Zealand model, various types of cattle and sheep enterprises – spring calving for milk production and spring lambing for lamb production in particular – are based almost exclusively on grass/white-clover pastures, although lucerne swards play a significant supplementary role in some dryland areas, for both grazing and conservation (Wynn-Williams, 1982), and the products, among the most economically produced in the world, are thus highly competitive in world markets.

In parts of Europe, system studies have demonstrated the agronomic sustainability and economic viability of production systems with different animal species based on grass/white-clover swards rather than heavily N-fertilized grass swards (Vipond and Swift, 1992; Young, 1992), but the uptake of forage legumes has been slow, except in some former Eastern-bloc countries, where fertilizer N has been priced out of the market. Nevertheless, there are development schemes to promote white-clover use on commercial farms (Bax and Browne, 1995; Frankow-Lindberg *et al.*, 1996). One of the problems in transforming currently intensive systems from a fertilizer- to legume-N basis is that the proportion of total costs of the system due to fertilizer N may be relatively low, after costs of concentrate feed, silage making and other variable costs have been taken into account. The additional cost incurred by relying on fertilizer N may be considered by many farmers currently applying high N rates as an insurance against unpredictable and variable forage yield and quality and reduced risks to animal health.

Lucerne (syn. alfalfa) is widely grown throughout the world under both dry climatic and irrigated conditions. It is the major forage legume in the USA, with some 13 Mha (Marten *et al.*, 1989), and is a major forage in several southern and central European countries, more than 1 Mha being grown annually in Italy, for example (Guy, 1993). In comparative trials with other major temperate forage legumes, lucerne is usually the highest-yielding (e.g. Aldrich, 1984). Its chief uses are for hay, silage, artificially dried forage and pelleted meal. Due to the varying contribution of the grazing-resistant and winter-hardy yellow lucerne (*Medicago falcata*) to the ancestry of some cultivars, lucerne is also utilized by grazing animals, especially in Australia and New Zealand (Douglas, 1986). Its exploitation is not confined to the warmer temperate areas, being grown in some of the provinces of Canada.

Although it has the potential to be the highest-yielding of the common temperate forage legumes, lucerne is one of the most pest- and disease-prone species. Attack by pests, such as alfalfa weevil, a range of aphids and nematodes, and infection with fungi or bacteria causing wilt, leaf spots or stem and root diseases all contribute to lucerne stands falling short of their annual yield and persistence potential. Fortunately, there have been some notable successes over the past five decades or so, in which breeders have introduced resistance to some of the more troublesome pests and disease agents (Barnes, 1992; Barnes and Sheaffer, 1995). The application of biotechnology, for example by the production of transgenic plants, offers the prospect of further success in increasing pest resistance (Thomas *et al.*, 1994). In common with other forage legumes, lucerne has higher intake characteristics than most grasses at a comparable digestibility and a higher animal production response per unit of dry matter (DM) ingested (Campling, 1984).

Red clover, used mainly in a conservation role, is the most important legume in Scandinavia, where its winter hardiness is a valuable characteristic, and in some Mediterranean countries, northern Italy for example, where its drought resistance is of significant value. Its high production potential and suitability for silage in the UK have been demonstrated but not exploited in practice (Frame, 1990). In the USA, red clover is widely grown for hay and aftermath grazing in the humid north-east, but also, and particularly for seed production, in the north-west under irrigation, while in the south-east it is managed primarily as an annual (Taylor, 1985b); in the 1980s about 7 Mha of red-clover swards were grown in North America out of a world total of *c.* 20 Mha (Smith *et al.*, 1985), although its use in the USA has declined from 6 Mha in 1950 to 4.5 Mha in the 1990s (Taylor and Smith, 1995).

The winter annual, subterranean clover, is the most important forage legume in the drylands of Australia, with an estimated 17 Mha of pasture (Marten *et al.*, 1989). However, its use has spread to other regions, *inter alia* to drier areas of New Zealand, to its areas of origin in the Mediterranean basin and to other temperate regions, such as parts of the USA and southern Latin America. Having the capability to self-seed efficiently, it has the potential to fulfil the role of a perennial legume in pasture in areas too dry for white clover.

Birdsfoot trefoil (*L. corniculatus*) is of major importance for hay, silage and grazing in some regions of the world, for example, in north-east USA and south-east Canada (Beuselinck and Grant, 1995). It is in effect a pioneer legume for conditions which are less favourable for the more productive forage legumes, especially acid infertile areas. Greater lotus (*Lotus uliginosus*) is of lesser importance and is used in localized areas, in parts of eastern Australia and New Zealand for instance (Charlton, 1983; Blumenthal *et al.*, 1993).

A major characteristic of these two *Lotus* species, and also of sainfoin (*Onobrychis viciifolia*) and sulla (*Hedysarum coronarium*), is the presence in the herbage of CT. These constituents prevent bloat in ruminants and, nutritionally, they protect protein in the rumen and this protection improves amino acid supply and absorption in the small intestine (Waghorn *et al.*, 1990). Evidence is also mounting on the ability of CT to surmount the deleterious effects of intestinal parasites in grazing animals and reduce attack on sheep by flies, which lay their eggs in the fleece, resulting in maggots eating into animal flesh (Robertson *et al.*, 1995). These CT-containing species will also be particularly valuable for genetic manipulation with other species in order to create new types with advantageous characteristics.

Apart from the above-mentioned major forage legumes, discussed below in successive individual chapters, the minor perennial legumes – alsike clover (*Trifolium hybridum*), sainfoin, the annual pink (*Ornithopus sativus*) and yellow (*Ornithopus compressus*) serradella species, sulla and tagasaste (*Chamaecytisus palmensis*) – are also dealt with. These species are important in some areas, are used in niche situations or there is interest in their potential. In contrast, a number of other minor legumes have been omitted for various reasons, e.g. the small area sown in some cases or because they have been dealt with adequately elsewhere. For example, the winter annual *Trifolium* species, crimson clover (*T. incarnatum*) has been described by Knight (1985) and arrowleaf clover (*T. vesiculosum*) by Miller and Wells (1985), or the perennial *Trifolium* species, Caucasian clover (*T. ambiguum*), strawberry clover (*T. fragiferum*) and zigzag clover (*T. medium*), by Townsend (1985).

The value of forage legumes to the N economy of the world's grasslands and to ruminant production from them cannot be overemphasized. Yet there are still problems to be overcome, challenges to be met in research, development and technology transfer, and, above all, opportunities to achieve improved productivity by better use of forage legumes. The following chapters attempt to highlight these facets and thereby contribute to a better understanding of temperate forage-legume technology.

References

Aldrich, D.T.A. (1984) Lucerne, red clover and sainfoin: herbage production. In: Thomson, D.J. (ed.) *Forage Legumes*. Occasional Symposium No. 16, British Grassland Society, Hurley, pp. 121–126.

Ball, P.R. and Crush, J.R. (1985) Prospects for increasing symbiotic nitrogen fixation in temperate grasslands. In: *Proceedings of the XV International Grassland Congress, Kyoto, Japan*. The Science Council of Japan and the Japanese Society of Grassland Science, Tochigi-kem, pp. 56–62.

Barnes, D.K. (1992) Forage legume breeding past, present and future. In: *Proceedings of the 14th General Meeting of the European Grassland Federation, Lahti, Finland.* European Grassland Federation, pp. 78–86.

Barnes, D.K. and Sheaffer, C.C. (1995) Alfalfa. In: Barnes, R.F., Miller, D.A. and Nelson, C.J. (eds) *Forages.*, Vol. 1, *An Introduction to Grassland Agriculture*, 5th edn. Iowa State University Press, Ames, Iowa, pp. 205–216.

Bax, J.A. and Browne, I. (1995) *The Use of Clover on Dairy Farms.* Research Summary, Milk Development Council, London, 21 pp.

Bax, J.A. and Schils, R.L.M. (1993) Animal responses to white clover. *FAO/REUR Technical Series* 30, 7–16.

Beever, D.E. and Thorp, C. (1996) Advances in the understanding of factors influencing the nutritive value of legumes. In: Younie, D. (ed.) *Legumes in Sustainable Farming Systems.* Occasional Symposium No. 30, British Grassland Society, Reading, pp. 194–207.

Beuselinck, P.R. and Grant, W.F. (1995) Birdsfoot trefoil. In: Barnes, R.F., Miller, D.A. and Nelson, C.J. (eds) *Forages*, Vol. 1, *An Introduction to Grassland Agriculture*, 5th edn. Iowa State University Press, Ames, Iowa, pp. 237–248.

Blumenthal, M.J., Kelman, W.M. Lolicato, S., Hare, M.D. and Bowman, A.M. (1993) Agronomy and improvement of *Lotus*: a review. In: Michalk, D.L., Craig, A.D. and Collins, W.J. (eds) *Alternative Pasture Legumes 1993.* Technical Report 219, Department of Primary Industries, Adelaide South Australia, pp. 74–85.

Burton, J.C. (1985) *Rhizobium* relationships. In: Taylor, N.L. (ed.) *Clover Science and Technology.* ASA/CSSA/SSSA, Madison, Wisconsin, pp. 161–184.

Campling, R.C. (1984) Lucerne, red clover and other forage legumes: feeding value and animal production. In: Thomson, D.J. (ed.) *Forage Legumes.*Occasional Symposium No. 16, British Grassland Society, Hurley, pp. 140–146.

Caradus, J.R., Woodfield, D.R. and Stewart, A.V. (1996) Overview and vision for white clover. In: Woodfield, D.R. (ed.) *White Clover: New Zealand's Competitive Edge.* Grassland Research and Practice Series No. 6, New Zealand Grassland Association, Palmerston North, pp. 1–6.

Carruthers, V.R., O'Connor, M.B., Feyter, C., Upsell, M.P. and Ledgard, S.F. (1987) Results from the Ruakura bloat survey. In: Charlton, D. (ed.) *Proceedings of the Ruakura Farmers' Conference 1987.* Wellington, pp. 44–46.

Charlton, J.F.L. (1983) Lotus and other legumes. In: Wratt, G.S. and Smith, H.C. (eds) *Plant Breeding in New Zealand.* Butterworths/DSIR, Wellington, pp. 253–262.

Clark, D.A. and Harris, S.L. (1996) White clover or nitrogen fertilizer for dairying. In: Woodfield, D.R. (ed.) *White Clover: New Zealand's Competitive Edge.* Grassland Research and Practice Series No. 6, New Zealand Grassland Association, Palmerston North, pp. 107–114.

Copeman, G.J.F. and Younie, D. (1982) Feed quality and utilisation of red clover swards. In: Murray, R.B. (ed.) *Legumes in Grassland. Proceedings of the Fifth Study Conference of the Scottish Agricultural Colleges, Peebles.* The Scottish Agricultural Colleges, Aberdeen/Auchincruive/Edinburgh, pp. 53–58.

Cowling, D.W. (1982) Biological nitrogen fixation and grassland production in the United Kingdom. *Philosophical Transactions, Royal Society of London,* B 296, 397–404.

Davies, D.A. and Hopkins, A. (1996) Production benefits of legumes in grassland. In: Younie, D. (ed.) *Legumes in Sustainable Farming Systems.* Occasional Symposium No. 30, British Grassland Society, Reading, pp. 234–246.

D'Mello, J.P.F. and Macdonald, A.M.C. (1996) Anti-nutrient factors and mycotoxins in legumes. In: Younie, D. (ed.) *Legumes in Sustainable Farming Systems*. Occasional Symposium No. 30, British Grassland Society, Reading, pp. 208–216.

Douglas, J.A. (1986) The production and utilization of lucerne in New Zealand: review paper. *Grass and Forage Science* 41, 81–128.

Doyle, C.J. and Bevan, K. (1996) Economic aspects of legume-based grassland systems. In: Younie, D. (ed.) *Legumes in Sustainable Farming Systems*. Occasional Symposium No. 30, British Grassland Society, Reading, pp. 247–256.

Doyle, C.J. and Morrison, J. (1983) An economic assessment of the potential benefits of replacing grass by grass–clover mixtures for 18-month beef systems. *Grass and Forage Science* 38, 273–282.

Doyle, C.J. and Thomson, D.J. (1985) The future of lucerne in British agriculture: an economic assessment. *Grass and Forage Science* 40, 57–68.

Essig, H.W. (1985) Quality and anti-quality components. In: Taylor, N.L. (ed.) *Clover Science and Technology*. ASA/CSSA/SSSA, Madison, Wisconsin, pp. 309–324.

Farnham, D.E. and George, J.R. (1994) Harvest management effects on productivity, dinitrogen fixation and nitrogen transfer in birdsfoot trefoil–orchard grass communities. *Crop Science* 34, 1650–1653.

Fothergill, M. and Davies, D.A. (1993) White clover contribution to continuously stocked sheep pastures in association with contrasting perennial ryegrass. *Grass and Forage Science* 48, 369–379.

Frame, J. (1990) The role of red clover in United Kingdom pastures. *Outlook in Agriculture* 19, 49–55.

Frame, J. and Newbould, P. (1986) Agronomy of white clover. *Advances in Agronomy* 40, 1–88.

Frankow-Lindberg, B.E., Danielsson, D.A. and Moore, C. (1996) The uptake of white clover technology in farming practice. *FAO/REU Technical Series* 42, 37–43.

Guy, P. (1993) Lucerne in Europe: statistical elements. In: Rotili, P. and Zannone, L. (eds) *The Future of Lucerne Biotechnology, Breeding and Variety Constitution. Proceedings of the X International Conference of EUCARPIA*, Medicago *spp.* Group. Istituto Sperimentale per le Colture Foraggere, Lodi, pp. 13–17.

Hauck, R.D. (1988) A human ecosphere perspective of agriculture nitrogen cycling. In: Wilson, J.R. (ed.) *Advances in Nitrogen Cycling in Agricultural Ecosystems*. CAB International, Wallingford, pp. 3–19.

Heichel, G.H., Vance, C.P., Barnes, D.K. and Henjum, K.I. (1985) Dinitrogen fixation and N and DM distribution during four-year stands of birdsfoot trefoil and red clover. *Crop Science* 25, 101–105.

Hoglund, J.H., Crush, J.R., Brock, J.L., Ball, R. and Carran, R.A. (1979) Nitrogen fixation in pasture. XII. General discussion. *New Zealand Journal of Experimental Agriculture* 7, 45–51.

Isely, D. (1982) Leguminosae and *Homo sapiens*. *Economic Botany* 36, 46–70.

Knight, W.E. (1985) Crimson clover. In: Taylor, N.L. (ed.) *Clover Science and Technology*. ASA/CSSA/SSSA, Madison, Wisconsin, pp. 491–502.

Laidlaw, A.S. and Frame, J. (1988) Maximising the use of the legume in grassland systems. In: *Proceedings of the 12th General Meeting of the European Grassland Federation, Dublin, Ireland*. Irish Grassland Association, Belclare, pp. 34–46.

Lancashire, J. (1990) Special address: 150 years of grassland development in New Zealand. *Proceedings of the New Zealand Grassland Association* 52, 9–15.

La Rue, T.A. and Patterson, T.G. (1981) How much nitrogen do legumes fix? *Advances in Agronomy* 34, 15–38.

Lehmann, J., Meister, E., Gutzwiller, A., Jans, F., Charles, J.P. and Blum, J. (1991) Should one use white clover (*Trifolium repens* L.) varieties rich in hydrogen cyanide? *Revue Suisse d'Agriculture* 23, 107–112.

't Mannetje, L. and Jarvis, S. (1990) Nitrogen flows and losses in grazed grasslands. In: *Proceedings of the 13th General Meeting of the European Grassland Federation, Banská Bystrica, Czechoslovakia*, Vol. l, pp. 114–131.

Marriott, C.A. and Haystead, A. (1993) Nitrogen fixation and transfer. In: Davies, A., Baker, R.D., Grant, S.A. and Laidlaw, A.S. (eds) *Sward Measurement Handbook* 2nd edn. British Grassland Society, Reading, pp. 245–264.

Marten, G.C., Matches, A.G., Barnes, R.F., Brougham, R.W., Clements, R.J. and Sheath, G.W. (eds) (1989) *Persistence of Forage Legumes*. ASA/CSA/SSSA, Madison, Wisconsin, pp. 569–572.

Miller, J.D. and Wells, H.D. (1985) Arrowleaf clover. In: Taylor, N.L. (ed.) *Clover Science and Technology*, ASA/CSSA/SSSA, Madison, pp. 503–514.

Miller, R.M. and Jastrow, J.D. (1996) Contribution of legumes to the formation and maintenance of soil structure. In: Younie, D. (ed.) *Legumes in Sustainable Farming Systems*. Occasional Symposium No. 30, British Grassland Society, Reading, pp. 105–112.

Newbould, P. (1982) Biological nitrogen fixation in upland and marginal areas of the UK. *Philosophical Transactions, Royal Society of London*, B 296, 405–417.

Novoselova, A. and Frame, J. (1992) The role of legumes in European grassland. In: *Proceedings of the 14th General Meeting of the European Grassland Federation, Lahti, Finland*. European Grassland Federation, pp. 87–96.

Nutman, P.S. (1976) IBP field experiments on nitrogen fixation by nodulated legumes. In: Nutman, P.S. (ed.) *Symbiotic Nitrogen Fixation in Plants*. Cambridge University Press, Cambridge, pp. 211–237.

Robertson, H.A., Niezen, J.H., Waghorn, G.C., Charleston, W.A.G. and Jinlong, M. (1995) The effect of six herbages on liveweight gain, wool growth and faecal egg count of parasitised ewe lambs. *Proceedings of the New Zealand Society of Animal Production* 55, 199–201.

Rogers, H.H. (1976) Forage legumes (with particular reference to lucerne and red clover). In: *Report of the Plant Breeding Institute for 1975*. Plant Breeding Institute, Cambridge, pp. 22–57.

Rohweder, D.A., Shrader, W.D. and Templeton, W.C., Jr (1977) Legumes, what is their place in today's agriculture? *Crop Soils* 29, 11–15.

Sheehy, J.E., Minchin, F.R. and McNeill, A. (1984) Physiological principles governing the growth and development of lucerne, sainfoin and red clover. In: Thomson, D.J. (ed.) *Forage Legumes*. Occasional Symposium No. 16, British Grassland Society, Hurley, pp. 112–125.

Skerman, P.J., Cameron, D.G. and Riveros, F. (1988) *Tropical Forage Legumes*. FAO, Rome.

Smith, R.R., Taylor, N.L. and Bowley, S.R. (1985) Red clover. In: Taylor, N.L. (ed.) *Clover Science and Technology*. ASA/CSSA/SSSA, Madison, Wisconsin, pp. 457–470.

Sprent, J.I. and Mannetje, L. 't (1996) The role of legumes in sustainable farming systems: past, present and future. In: Younie, D. (ed.) *Legumes in Sustainable Farming Systems*. Occasional Symposium No. 30, British Grassland Society, Reading, pp. 2–14.

Stewart, T.A. and Haycock, R.E. (1984) Beef production from low N and high N S24 perennial ryegrass/Blanca white clover swards – a six year farmlet scale comparison. *Research and Development in Agriculture* 1, 103–113.

Taylor, N.L. (1985a) Clovers around the world. In: Taylor, N.L. (ed.) *Clover Science and Technology*. ASA/CSSA/SSSA, Madison, Wisconsin, pp. 1–6.

Taylor, N.L. (1985b) Red clover. In: Heath, M.E., Metcalfe, D.S. and Barnes, R.F. (eds) *Forages: The Science of Grassland Agriculture*, 4th edn. Iowa State University Press, Ames, Iowa, pp. 109–117.

Taylor, N.L. and Smith, R.R. (1995) Red clover. In: Barnes, R.F., Miller, D.A. and Nelson, C.J. (eds) *Forages*, Vol. 1, *An Introduction to Grassland Agriculture*, 5th edn. Iowa State University Press, Ames, pp. 217–226

Thomas, C., Aston, K. and Daley, S.R. (1985) Milk production from silage. 3. A comparison of red clover with grass silage. *Animal Production* 41, 23–31.

Thomas, J.C., Wasmann, C.C., Echt, C., Dunn, R.L., Bonnert, H.J. and McCoy, T.J. (1994) Introduction and expression of an insect proteinase-inhibitor in alfalfa. *Plant Cell Reports* 14, 31–36.

Thomson, D.J. (1984) The nutritive value of white clover. In: Thomson, D.J. (ed.) *Forage Legumes*. Occasional Symposium No. 16, British Grassland Society, Hurley, pp. 78–92.

Townsend, C.E. (1985) Miscellaneous perennial clovers. In: Taylor, N.L. (ed.) *Clover Science and Technology*. ASA/CSSA/SSSA, Madison, Wisconsin, pp. 563–578.

Ulyatt, M.J., Lancashire, J.A. and Jones, W.T. (1977) The nutritive value of legumes. *Proceedings of the New Zealand Grassland Association* 38, 107–118.

Ulyatt, M.J., Dellow, D.W., John, A., Reid, C.S.W. and Waghorn, G.C. (1986) Contribution of chewing during eating and rumination to the clearance of digesta from the reticulorumen. In: Milligan, L.P., Grovum, W.L. and Dobson, A. (eds) *Control of Digestion and Metabolism in Ruminants. Proceedings of the VIth International Symposium on Ruminant Physiology, Banff, Canada*. Prentice Hall, Englewood Cliffs, pp. 498–515.

Van der Meer, H. and Van der Putten, A.H.J. (1995) Reduction of nutrient emissions from ruminant livestock farms. In Pollott, G.E. (ed.) *Grassland into the 21st Century*. Occasional Symposium No. 29, British Grassland Society, Reading, pp. 118–134.

Vipond, J.E. and Swift, G. (1992) Developments in legume use on hills and uplands. In: Hopkins, A. (ed.) *Grass on the Move*. Occasional Symposium No. 26, British Grassland Society, Reading, pp. 54–65.

Vipond, J.E., Swift, G., McClelland, T.H., Fitzsimons, J., Milne, J.A. and Hunter, E.A. (1993) A comparison of diploid and tetraploid perennial ryegrass and tetraploid ryegrass/white clover swards under continuous sheep stocking at controlled sward heights. 2. Animal production. *Grass and Forage Science* 48, 290–300.

Waghorn, G.C., Shelton, I.D. and Thomas, V.J. (1989) Particle breakdown and rumen digestion of fresh ryegrass (*Lolium perenne* L.) and lucerne (*Medicago sativa* L.) fed to cows during a restricted feeding period. *British Journal of Nutrition* 61, 409–423.

Waghorn, G.C., Jones, W.T., Shelton, I.D. and McNabb, W.C. (1990) Condensed tannins and the nutritive value of pasture. *Proceedings of the New Zealand Grassland Association* 51, 171–176.

Whitehead, D.C. (1986) Nitrogen in UK grassland agriculture. *Journal of the Royal Agricultural Society* 147, 190–200.

Wilkins, R.J., Huckle, A. and Clements, A.J. (1991) Effects of concentrate supplementation and sward clover content on milk production by spring calving dairy cows. In: Mayne, C.S. (ed.) *Management Issues for the Grassland Farmer in the 1990s*. Occasional Symposium No. 25, British Grassland Society, Reading, pp. 218–220.

Wilman, D., Mtengeti, E.J. and Moseley, G. (1996) Physical structure of 12 forage species in relation to rate of intake by sheep. *Journal of Agricultural Science, Cambridge* 126, 277–285.

Wittwer, S.H. (1978) Nitrogen fixation and productivity. *Bioscience* 28, 555.

Witty, J.F. (1983) Estimating N_2-fixation in the field using ^{15}N-labelled fertilizer: some problems and solutions. *Soil Biology and Biochemistry* 15, 631–639.

Wood, M. (1996) Nitrogen fixation: how much and at what cost? In: Younie, D. (ed.) *Legumes in Sustainable Farming Systems*. Occasional Symposium No. 30, British Grassland Society, Reading, pp. 26–35.

Wynn-Williams, R.B. (ed.) (1982) *Lucerne for the 80s*. Special Publication No. 1, Agronomy Society of New Zealand, Christchurch.

Young, N.E. (1992) Developments in legume use for beef and sheep. In: Hopkins, A. (ed.) *Grass on the Move*. Occasional Symposium No. 26, British Grassland Society, Reading, pp. 29–39.

White Clover 2

Introduction

On a world basis, white clover (*Trifolium repens* L.) is the most important true clover species for grazed swards within the genus *Trifolium*. This genus, comprising about 240 species, is found in moist temperate regions, Mediterranean areas and some cool subtropical parts of the world. In spite of this widespread distribution, only a handful of species have attained pastoral significance There is some controversy over the centre of origin of the genus, but suffice it to say that the Mediterranean basin and the Californian region of the western USA have their proponents; however, the balance of opinion favours the Mediterranean area (Taylor *et al.*, 1980), which is the centre of origin of 110 of the 237 *Trifolium* species (Zohary and Heller, 1984). White clover can be crossed with several related *Trifolium* species and may well have evolved from crosses between these species (Williams, 1983).

White clover was introduced or recognized as a component of pastures in north-west Europe about the seventeenth century (Zeven, 1991), but it and other clovers, notably red clover, played an important part in the cultural life of early peoples. This included their role as a source of honey and even as a religious symbol, prior to their specific cultivation for use by domesticated ruminants (Taylor, 1985). Following colonization of the Americas and Australasia by Europeans, white clover became widely distributed in pastures in temperate areas of these regions, particularly during the nineteenth century, a distribution assisted by its adaptability to a range of environmental and management conditions. Estimates indicate some 15 Mha of pasture with white clover in Australasia and 5 Mha in the USA (Marten *et al.*, 1989).

The role and potential of white clover has always been recognized, though not always fully realized, in temperate Australasia. There has been continued research there into optimizing its contribution to swards, as a supplier of rhizobially fixed nitrogen (N) to the grass component and as a forage of high nutritive value in its own right, leading in turn to a mixed sward of high feeding value. In other regions, such as Western Europe, its role has steadily declined from the 1950s onwards in favour of intensively stocked grass swards heavily fertilized with N. However, the value of white clover has been increasingly reappraised and researched over the past 10–15 years, not least in light of its attractions in sustainable, low-input and low-cost systems of animal production (Frame and Newbould, 1986; Hopkins *et al.*, 1994). The species has been improved by breeding since the 1930s, and over 300 cultivars are available (Caradus and Woodfield, 1996) to fulfil a current world annual consumption of 8000–10,000 t seed; 55–60% of this amount is used in the northern hemisphere, particularly in Western Europe and North America, and the remainder in the southern hemisphere, especially in New Zealand, southern Latin America and Australia (Mather *et al.*, 1996).

The Plant

Roots

The structure, development, function and genetics of white-clover root systems have been adequately reviewed (Caradus, 1990). Initially, white-clover seedlings develop an extensively branched tap root system. Subsequently, adventitious roots, with numerous lateral branches, arise from the nodes of stolons, which ramify from the mother plant. Adventitious roots develop from two primordia at each node and the development of these primordia into roots is dependent on relative humidity (RH) around the nodes, the RH requiring to be in excess of 90% before root initials develop into roots (Stevenson and Laidlaw, 1985).

The tap-root system of small-leaved wild white cultivars is not so well developed as that of large-leaved types, but the former's strong stoloniferous habit ensures good survival under adverse conditions (Caradus, 1977). Roots of white clover grow to a similar depth to those of temperate grass species, such as perennial ryegrass (*Lolium perenne*), but, although 80% of clover roots are usually present in the top 20 cm of soil (Caradus, 1990), clover has shorter root hairs and a smaller root-hair cylinder, together with less root mass in the upper soil layers, 0–20 cm, the most important zone for plant nutrition (Evans, 1977, 1978). The tap-root system is short-lived, particularly under moist soil conditions, and clover persistence then ultimately depends on the rooted plantlets which develop from stolon nodes. It is this ability, together with its stolon extension habit, which makes white clover a successful guerilla-type species, capable of spreading and establishing itself in suitable niche situations in grazed pastures.

Individual adventitious roots serve specific parts of the plant, due to the vascular structure of the roots, stolons and branches. Isotopic tracer studies show that nutrients taken up by individual roots are transported acropetally, the largest apparent sinks being those with the strongest vascular connection with the root (Hay and Sackville Hamilton, 1996). Consequently, the organs most favourably served by a specific root are the branch on the same node, its secondary branches, branches distal to the node on the same side of the stolon and the leaf on the second node distal to the root (Lotscher and Nösberger, 1996).

Forms of white clover adapted to dryland conditions tend to be more taprooted and can penetrate over 0.5 m into the soil seeking moisture supply for survival (Caradus and Woodfield, 1986). Types with strong tap-root features, such as the South African cultivar Dusi, have been selected specifically for use in dryland pastures where frequent moisture deficit and high summer temperatures occur.

Large numbers of root nodules are developed, first on the finer, upper branches of the tap root and later on the nodal adventitious roots, following infection by molecular N (N_2)-fixing rhizobia present in the soil or added through rhizobial inoculation of the sown seed. These nodules may be club-shaped, ellipsoidal or sometimes palmate.

Leaves

White clover is a glabrous plant, with leaves that are normally trifoliate and ovate to circular, have conspicuous toothed edges and are borne on long petioles (Fig. 2.1). Leaflets usually have some form of white leaf marking and sometimes a degree of dark red leaf coloration. The plant's stipules are pale and translucent and have a short tip.

Because of its spreading, stoloniferous habit, white clover is generally low-growing in pasture, but the more upright large-leaved forms thrive in taller swards grown for conservation, or in rotational cattle-grazing systems, as a consequence of petiole extension. Italian ladino types, e.g. cv. Espanso, exemplify this more erect form, and are used for pastures grazed predominantly by cattle or cutting swards for conservation.

Clover plants exhibit phenotypic plasticity in response to severe defoliation and other growth-limiting pressures. For example, continuous sheep stocking at a high grazing pressure results in progressive dwarfing of the leaves and petioles and stolon ramification is reduced. Conversely, in rotational cattle-grazing systems, the rest intervals encourage development of larger laminae and petioles and a more vigorous expansion of the stolon network.

Stolons

After the seedling stage, stolons emerge from the axils of the leaves to form a network of branched stolons, radiating from a central rosette, and stolon development in the lower axils of the primary axis soon confers a prostrate growth habit on the plant. During this phase, in which stolons and then branches are developing, white clover has the capacity to spread widely, its behaviour being conditioned by the microenvironment around the stolons. As the centre of the rosette dies, individual stolons assume an independent existence.

These stolons are perennating organs which accumulate carbohydrates, chiefly in the form of starch, and their apices are the sites of leaf production.

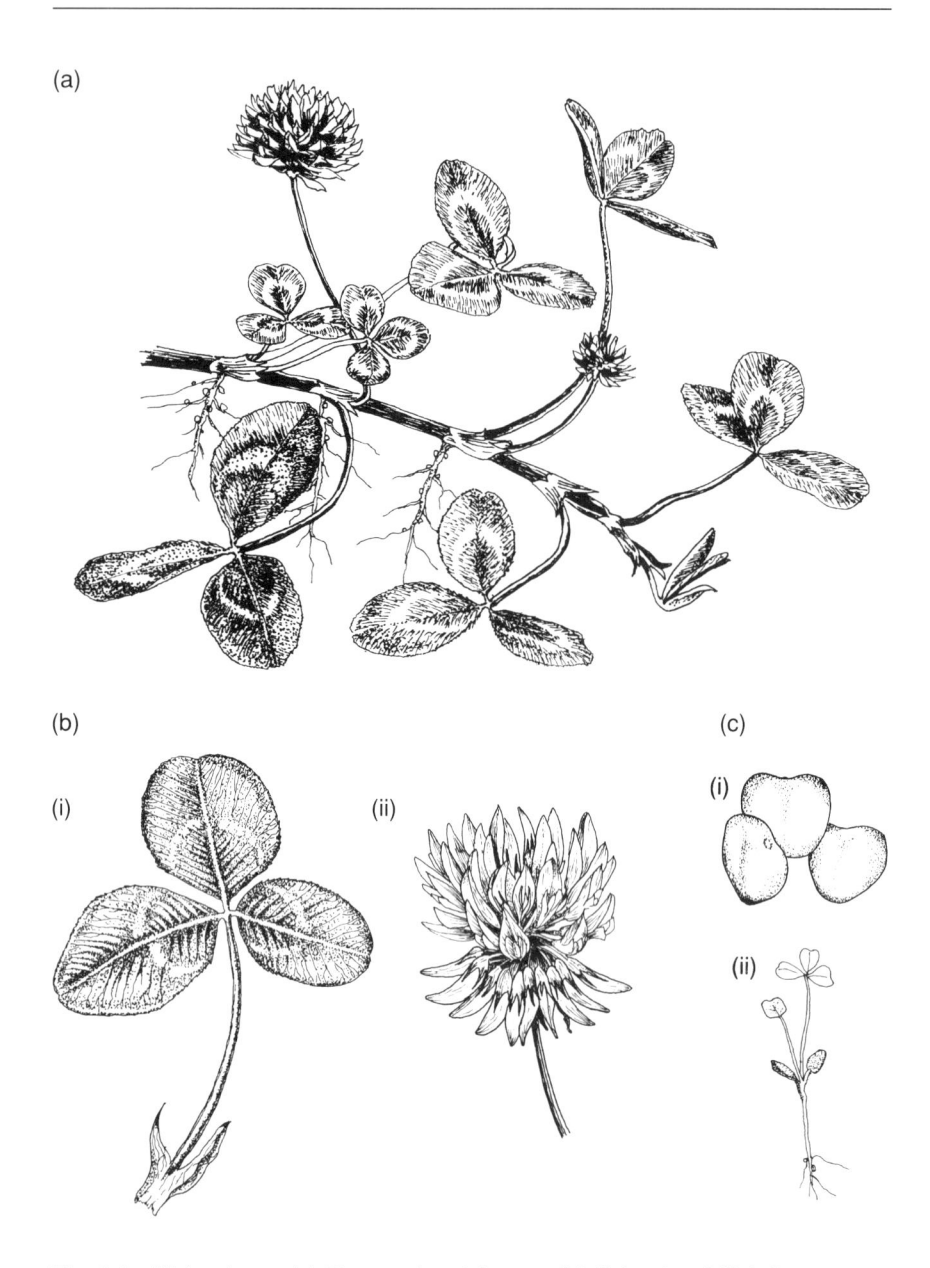

Fig. 2.1. White clover. (a) Plant and nodal roots. (b) (i) Leaf and (ii) inflorescence. (c) (i) Seeds and (ii) seedling.

The small-leaved, highly stoloniferous clover types tolerate severe defoliation better than other types, are more persistent, and have a superior ability to colonize bare spaces (Burdon, 1983). The importance of stolons in the variability in yield and persistence of white clover has now been recognized more clearly (Hay, 1983, 1985) and cultivars have been selected for particular stolon characteristics, such as degree of branching (Williams, 1983).

Clover persistence generally is therefore highly dependent on stolon development and continued replacement. Stolon densities from approximately 20 to 100 m m^{-2} in early spring are required if a grass/clover sward is to achieve a satisfactory clover yield later in the season, but, beyond 100 m of stolon m^{-2}, the response in yield declines rapidly (Rhodes, 1991; Rhodes et al., 1994). The buds that develop in the axils of the leaves on the stolon may develop into 'daughter' stolons or else inflorescences, according to the season. In the high temperatures, around 20°C optimum, and long days, 14–16 h, of summer, the production of inflorescences from the buds is favoured, rather than daughter stolons.

Stolon-based perennation is the principal means of clover persistence in most temperate climates (Harper, 1978), but regeneration from shed or buried seed is a major factor in the south-east of the USA (Evers, 1989) or subtropical eastern Australia (Jones, 1980, 1982), where white clover behaves more like an annual than a perennial species. Seed production from freely seeding types represents an advantageous survival mechanism in such situations and under dryland conditions generally (Macfarlane and Sheath, 1984). Stolon production is enhanced in summer, with its high temperatures and high irradiance in comparison with other seasons, but, conversely, is depressed by the low temperatures and low irradiance typical of cold springs; production of stolons is also depressed when the stolon buds are shaded by a tall, dense canopy of grass, such as occurs in response to applied fertilizer N.

Internode elongation on stolons is increased when the stolon is shaded (Thompson and Harper, 1988). Within patches, elongation is faster at the periphery than at the centre of the patch, and the probability of a branch and root being formed at nodes of comparable age at the edge of a patch is also increased relative to the centre, demonstrating the ecological significance of stolon sensitivity to its environment (Kemball et al., 1996).

There is an annual cycle of stolon burial in winter, re-emergence of growing points in spring and above-ground stolon development during summer (Hay, 1985). A small proportion of the surface stolon, classed as aerial, is close to ground level and is therefore susceptible to removal by grazing animals. In grazed swards, the proportion of stolon length on the surface in winter is higher in swards which have been rested during summer for a conservation cut than in those grazed all year or rested in spring prior to a cut and then grazed (Gooding, 1993).

Buried stolons, caused by stock trampling, wheel traffic and earthworm casting, have been identified as an important factor in clover persistence (Hay, 1983), since there is:

- initiation of subsurface branching in stolons capable of growing vertically to the soil surface;
- redirection of stolon apices to the surface;
- initiation of branch stolons where a vertical branch has reached the soil surface;
- production of bud-like structures, with several growing points, from vertically growing stolons.

However, if the leaves of the clover plant are grazed or destroyed by trampling, development of buried stolons is reduced and increased death of stolon branches occurs. This is attributable to acropetal translocation of energy reserves to new growth on the main stolons and a reduction in the partitioning of assimilates to support smaller buried branches (Grant and Marriott, 1989).

Buried stolons are thus a potentially beneficial factor in the regeneration of clover following decline or apparent disappearance in a sward. They supplement both the regeneration of plants by germination of buried seed, which are present in considerable quantities if the land has grown white clover over a long period, and the renewed development of plants that may have become stunted in survival mode, for example, in response to overintensive sheep grazing.

Flowers

The inflorescence is globular in shape, with 20–40 florets per head, and the peduncles are usually about twice the length of the petioles of the associated leaves. This flowering phase is less important as a determinant of herbage production than in grass. There is considerable variation among varieties in their temperature requirements for floral development, and flowering can be delayed by grazing or cutting the sward. White clover is usually cross-fertilized by honey-bee (*Apis mellifera*) and bumble-bee (*Bombus* spp.) pollination (Free, 1993), self-fertilization being prevented by the structure of the flowers, and the seed ripen 3–4 weeks after pollination. When self-pollinated, almost no seed is set, except in rare self-fertile plants or when temperatures are very high (Williams, 1983).

Seed

Seeds are relatively small, with 1000-seed weight of approximately 0.6–0.7 g (Charlton, 1992). The seed is irregularly heart-shaped to oval and has a smooth surface. In mass, seeds of white clover appear to be yellowish with a tinge of red, becoming brown with ageing.

White clover usually has a proportion of its seed with impermeable coats – so-called hard seed – which do not germinate immediately after sowing. They may do so over an extended period, following softening or abrasion of the seed coat, and therefore represent some insurance against an initial poor establishment. Hard-seededness is as high as 90% in hand-harvested seed samples, but is rarely greater than 10% in mechanically harvested seed (Scott and Hampton, 1985). Hard seed of white clover remain viable for long periods, the longest being for seed found under a 700-year-old demolished church in the UK, seed which germinated readily (Suckling and Charlton, 1978).

A proportion of white-clover seed ingested by grazing animals survives passage through the gut, and even increases in germination capacity as a result. In one assessment, the amount of seed in cow-dung samples averaged nearly 1 kg ha^{-1} from grazed unimproved pasture and nearly 5 kg ha^{-1} on grazed improved pasture (Suckling, 1951).

Physiology

Seedling physiology

White-clover seedlings show epigeal germination, i.e. elongation of the hypocotyl pushing the cotyledons above soil level. The first leaf is small, spade-like and unifoliate, but all subsequent leaves are trifoliate.

In the early stages of establishment the young seedling is dependent on cotyledonary reserves, so sowing depth has an impact on the probability of seedling survival. If sown too deeply, over 15–20 mm, the hypocotyl may not be capable of extending to the soil surface; if sown too shallowly, the seedling may be exposed to widely fluctuating moisture levels. Development of young seedlings of white clover that survive moisture stress is constrained, due to initial poor root production, and, for example, young plants subjected to 50% optimum soil moisture content had only 30% of the root biomass of those developed at optimum moisture (Engin and Sprent, 1973). In comparison with some other legumes, such as the medic *Medicago lupulina*, seedlings of white clover were more susceptible to drought (Foulds, 1978).

Although high temperatures inhibit germination, decreasing temperature levels have no effect, except on rate of emergence, and cultivars that reached 75% germination 2 days after sowing at 20°C took 4 days at 10°C and 8–9 days at 5°C (Hampton *et al.*, 1987). Seedling age affects white clover's ability to withstand low temperatures, cold-hardiness increasing approximately linearly with age, at least up to 3 months after sowing (Laidlaw and McBride, 1992).

Leaf initiation and development are strongly influenced by temperature and light. Under sward conditions, the life span of white-clover leaves, which determines their photosynthetic usefulness to the plant, is usually only 4–5 weeks from the time of emergence. In late season and throughout winter, the leaves on the older parts of the stolons senesce, leaving mainly underdeveloped leaves near the stolon apices, but a slow turnover of leaves continues during winter (Davies and Evans, 1982). When spring temperatures and radiation improve, there is a net increase in production. As the height of the canopy of a mixed sward increases, successive clover petioles extend higher and maintain their photosynthetic efficiency (Dennis and Woledge, 1982, 1983).

Seedlings which have recently nodulated grow more slowly than those depending on inorganic N, as the energy demand for N_2 fixation is high in relation to the amount of photosynthate, and so less is available for growth compared with those receiving inorganic N. Fixation of N can depress growth of seedlings by as much as 40% (Ryle *et al.*, 1979).

Temperature

The growth rate of white-clover shoots increases with temperature to about 25°C, in contrast to perennial ryegrass, which has an optimum of about 20°C (Davies, A., 1992). However, at temperatures lower than 10°C, the growth rate of perennial ryegrass is faster than that of white clover. This response of white-clover growth to a temperature range is reflected in the effect of temperature on net photosynthetic rate of single leaves of the two species (Fig. 2.2).

Leaf appearance rate is influenced strongly by temperature, a daytime temperature of 15.5°C resulting in double the leaf appearance rate of plants at 8.5°C. Petioles of plants at 23.5°C cease elongation 5 days after unfolding, compared with 15 days for those at 8.5°C. However, petiole extension rate at the higher temperature is more than twice that at the lower temperature (Boller and Nösberger, 1985). Young plants are more responsive to temperature than mature plants, and the optimum temperature for petiole extension changes in the field from spring to early summer, day/night temperatures of 20/15°C resulting in longer petioles than 15/10°C at the first harvest, whereas the reverse is the case at the second harvest (Frankow-Lindberg, 1987). In general, the optimum temperature for photosynthesis is increased as the temperature during leaf development is increased (Fig. 2.2). However, when some alpine ecotypes are grown at a low temperature, they have a greater photosynthetic

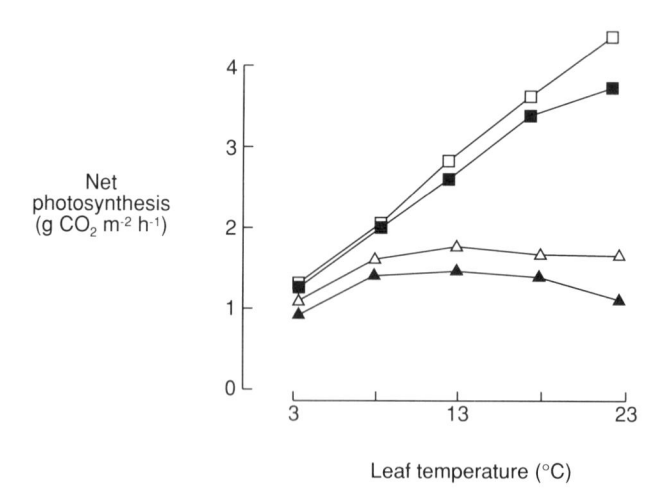

Fig. 2.2. Net photosynthesis of single white clover leaves from plants in a warm (□△) and cold (■▲) regime at 50 (△▲) and 250 (□■) J m^{-2} s^{-1} (from Woledge and Dennis, 1982).

capacity over a wide range of temperatures than those which develop at higher temperatures, and accumulate more carbohydrate in stolons than lowland ecotypes (Mächler and Nösberger, 1977). Source of ecotypes can influence the ability to grow at low spring and autumn temperatures (minimum 5°C), Mediterranean types having a higher capacity to grow at low temperatures than north European ecotypes (Eagles and Othman, 1988b).

Winter conditions result in petiole extension being more adversely affected than lamina and sheath extension in grass, and so clover laminae assume a less favourable position in the canopy than under conditions during the growing season. This results in a relatively lower *in situ* photosynthetic rate per unit of leaf area for clover than for grass during winter and a greater loss in leaf weight in clover than in grass (Woledge *et al.*, 1989).

Ability to withstand subzero temperatures depends on temperatures experienced by the ecotypes during hardening and the nutritional status of the plants, cold-hardiness being related directly to the duration over which the plant is maintained at low temperatures above 0°C, i.e. about 2°C (Collins and Rhodes, 1995). Ecotypes from northern Europe are generally more cold-hardy than those of Mediterranean origin and this is reflected in cultivars bred from such ecotypes (Eagles and Othman, 1981). Although the ability to grow at temperatures above zero is usually inversely related to ability to withstand subzero temperatures some ecotypes have the ability to do both (Rhodes *et al.*, 1989), and this has implications for the breeding of cold-hardy cultivars which can also grow early in the season (Collins *et al.*, 1991), a goal achieved in the UK-bred cultivars, small-leaved AberCrest and medium-leaved AberHerald (Collins *et al.*, 1996; Rhodes and Ortega, 1996).

Management can influence cold-hardiness of white-clover stolons, those which have had the opportunity to be photosynthetic and accumulate carbohydrate being more hardy (Harris *et al.*, 1983), and stolons of the most cold-hardy genotypes have the highest total non-structural carbohydrate (TNC) contents (Collins and Rhodes, 1995). However, imposing a management which should have increased TNC content in stolons in winter, i.e. increasing cutting interval from 2 to 8 weeks during the period late summer to winter, did not affect the ability of stolons to withstand low temperatures during the winter (Patterson *et al.*, 1995).

Low root temperature (5°C) causes a cold-induced water stress in white clover – more so than in grasses. This has been interpreted as the main reason for poor clover growth at low temperatures, rather than the reduction by low temperature of other processes, e.g. photosynthesis and N_2 fixation (Kessler and Nösberger, 1994). White-clover stolons have the potential to initiate branches during mild temperate winters, provided competition from grass is minimized (Patterson *et al.*, 1995)

Light

The effect of light, or, more accurately, irradiance, on the growth and development of white clover is twofold. It influences photosynthesis rate but it also

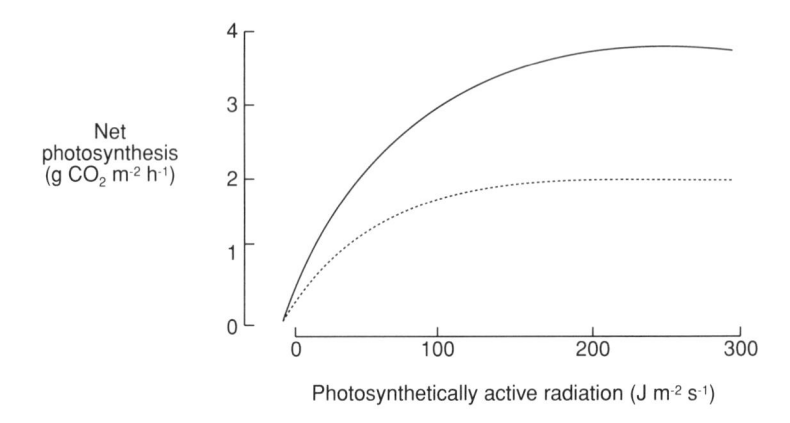

[Figure: Net photosynthesis (g CO₂ m⁻² h⁻¹) vs Photosynthetically active radiation (J m⁻² s⁻¹), showing two curves]

Fig. 2.3. Effect of irradiance on photosynthesis of single white clover leaves grown at 30 (------) and 120 (———) J m^{-2} s^{-1} (from Woledge and Parsons, 1986).

exerts a photomorphogenic effect. White-clover leaves developed under low irradiance levels have lower photosynthetic potential than those developed under high irradiance (Dennis and Woledge, 1983). As shown in Fig. 2.3, saturation for leaves grown at 135 µmol m^{-2} s^{-1} is 540 µmol m^{-2} s^{-1}, compared with about 900 µmol m^{-2} s^{-1} for those grown at 540 µmol m^{-2} s^{-1}. However, by extending petioles to ensure that the laminae are at the surface of the canopy before the leaflets are fully open, the photosynthetic potential may not be affected by shade conditions within the sward, unlike laminae of grasses (Woledge, 1971).

The response of net photosynthesis by single leaves to irradiance is higher at higher temperatures (see Fig. 2.2). As some leaves of white clover will be in shade conditions in the canopy when grown in a mixed sward, their contribution to canopy photosynthesis will decline with increasing temperature, reducing the photosynthetic efficiency of the canopy.

Although individual leaves are light-saturated at irradiances between 500 and 1000 µmol m^{-2} s^{-1}, canopies at optimum leaf area index (LAI) are saturated at about 1500 µmol m^{-2} s^{-1}. Owing to the horizontal arrangement of white-clover leaf laminae, estimates of the coefficient for the attenuation of light through the canopy, i.e. the extinction coefficient, are high (0.8–1), compared with those for grasses (0.2–0.5). Consequently, photosynthesis increases with additional increments in LAI to higher values of 7 or 8 in a mixed sward, compared with 3–4 for white clover.

Both the position of white-clover leaves within a canopy and leaf age are important in determining carbon dioxide (CO_2) assimilation rate. Young leaves have a higher photosynthesis rate than older leaves at the same height within the canopy, the decline in photosynthesis due to leaf age being accentuated at higher temperatures (Boller and Nösberger, 1985). In addition to differences in photosynthetic potential between leaves of different ages, younger leaves seem to assume positions in a canopy at which irradiance levels are higher than the average for all leaf positions at a given height in the canopy. Petioles elongate in dense canopies in response to low irradiance at their tip (Thompson, 1995).

Generally, white clover maintains a competitive position within a canopy with grasses during the growing season (Davidson and Robson, 1984), but there are limits, including genotypic limits, to the extension of petioles (Woledge *et al.*, 1992). During late autumn and winter, the laminae are relatively low in the canopy and so photosynthesis by white clover is severely affected (Woledge *et al.*, 1989). Sward defoliation in winter improves potential net photosynthesis of white-clover leaves in the following spring, presumably as they develop under higher irradiance conditions than they would within an undefoliated canopy dominated by grass (Laidlaw *et al.*, 1992).

Under controlled-environment conditions, daily irradiance can be manipulated by varying the radiant flux density (intensity) or varying the length of time a given irradiance level impinges on the plant (photoperiod). In this latter case, shortening the photoperiod from 16 to 12 h resulted in smaller leaves but a higher stolon branch population density, and reducing photoperiod from 16 to 8 h also reduced leaf appearance rate (Boller and Nösberger, 1983). Photoperiod effect on internode length varies but may be a function of genotype (Eagles and Othman, 1988a).

Development of white-clover plants is controlled by light quality (Solangaarachchi and Harper, 1987) and by short end-of-day exposure to red (R) and/or far red (FR) irradiance (Varlet-Grancher *et al.*, 1989). White-clover branching and petiole and internode lengths are strongly influenced by light quality at the base of the sward, which in turn is determined by the quality of incident radiation and interception and transmission of irradiance by the canopy (Thompson and Harper, 1988), although petiole length is also influenced by irradiance level impinging on the petiole tip (Thompson, 1995). The role of the relative proportion of radiation in the R waveband (660 nm) and FR waveband (730 nm), usually expressed as the R : FR ratio, in influencing stolon development has been confirmed, with branching rates being increased by enriching irradiance around the stolon with R light (Thompson, 1993). The inhibition of branching by low R : FR ratios appears to be perceived by the subtending leaf of the branch bud at the stolon apex, when the leaf is at an early stage of development (Robin *et al.*, 1994). However, this inhibition can be reversed by increasing the content of R light around the fully developed node associated with the suppressed bud, the R light only being effective if it is applied directly to the region of the node (Teuber and Laidlaw, 1996). Moreover, the petiole may be implicated in photoperception, as removal of the petiole strengthens the inhibitory effect on branching of shaded stolons (Davies and Evans, 1990a).

Flowering is also under control of light, particularly day length, and long days in excess of 14 h have a promotive effect on inflorescence induction and development. However, temperature and photoperiod interact, high temperature substituting for long days (Norris, 1989). There are a variety of flowering responses to day-length and temperature combinations in different types and cultivars of white clover, and the most widely grown cultivar, Grasslands Huia, responds as a short-day plant when grown at 15–25°C (Kendall and Stringer, 1985).

Moisture

White clover is adapted to a wide range of soils, but it does not thrive and is often absent in poorly drained soils (McAdam, 1983), shallow, drought-prone soils (Foulds, 1978; Thomas, 1984) or saturated, unamended peat (Burdon, 1983). The marked response of white clover to irrigation has been interpreted as white-clover sensitivity to drought (Cowling, 1982), but it can tolerate moderate drought conditions. However, irrigation with saline water, even with 20 mmol l^{-1} salt (NaCl), causes marked yield reduction, lower leaf water potential and lower canopy photosynthesis rates (Rogers *et al.*, 1993, 1994). Seasonal deficits of rainfall, soil type, slope and aspect affect clover demography more than annual rainfall *per se* (Bircham and Gillingham, 1986), although no direct relationship between white-clover content in swards and summer rainfall in permanent pasture was found in a study in The Netherlands (Kleter, 1968). Similarly, no consistent relationship between several weather parameters, including rainfall, and total herbage or white-clover dry matter (DM) production has been found, since production is affected by a complex of interacting factors (Frame and Boyd, 1984). Nevertheless, in the more extreme seasonal climate of southern Australia, summer drought is a major factor limiting clover performance and persistence (Hutchinson *et al.*, 1995).

White clover does not control water loss as efficiently as some other forage legumes, such as lucerne, the clover stomata being less responsive to drought conditions, and cuticular and mesophyll resistance may also be lower (Sanchez-Diaz and Sanchez-Marin, 1974). Leaf diffusive resistance increases with water stress but not enough to compensate for the loss.

Water stress affects leaf production in three ways, i.e. by reducing the rate of branching and so reducing the rate at which sites for leaf production are produced, by reducing leaf appearance rate and, thirdly, by reducing leaf size. Of the three, branching is the most sensitive and leaf appearance rate the least sensitive to water stress (Belaygne *et al.*, 1996).

Drought results in leaf senescence, leading to reduced transpiration, and so senescence is considered as a means of reducing water loss (Burch and Johns, 1978). White clover appears to have the ability to recover quickly from drought stress, a leaf water potential of −1.1 MPa after 17 days of drought returning to the original −0.5 MPa after 3 days of watering to alleviate the drought (Aparicio-Tejo *et al.*, 1980). Stolons adjust osmotically to drought stress and survive longer than leaves (Turner, 1991), and so plant survival following a severe drought is related to plant density or stolon growing-point number (Brock and Kim, 1994). In drought situations, grass tillers can provide cover and shade for stolons against direct solar radiation, and this, together with concomitantly lower soil-surface temperatures, aids clover survival in this respect; dense pastures resulting from set stocking were superior to the less dense swards under rotational grazing (Brock and Hay, 1993).

In response to moisture stress, cultivar differences in stolon axillary-bud development have been found in spaced plants in the field. Axillary buds of medium-leaved cv. Aberystwyth S100 developed more slowly than those of large-leaved cv. Olwen and small-leaved cv. Aberystwyth S184 at a maximum potential soil deficit of 60 mm (Thomas, 1984).

Nitrogen fixation in white clover is sensitive to drought, but the system can recover, although at a slow rate (Engin and Sprent, 1973). The moisture level at which nodules can lose water and still recover is not precisely known for white clover but is considered to be 20% of full moisture status, and the nodules seem to be more sensitive to drought than those of subterranean clover (*Trifolium subterraneum*), but less so than those of greater lotus (*Lotus uliginosus*) (Crush, 1987).

Perennation

The ability of white clover to perennate is dependent on the continued production of stolons. The network of stolons is built up by development of successive orders of branching, with primary branches being borne at nodes on the main stolon, secondary branches arising from the nodes of primary branches and so on. In instances where whole plants are being described, 'primary' plants are those with only one axis, i.e. a main stolon, and so this terminology should not be confused with the order to which a specific branch belongs. In the absence of limitations imposed by pests, diseases and environmental stress, and under constant management, stolon-growing point density declines during the winter, reaching a minimum in spring, after which it increases until autumn. The potential of axillary buds to develop into branches is also at its lowest in spring (Newton *et al.*, 1990).

The organization of the various orders of branches within a clover plant and the organs borne on these axes are well coordinated. Generally, the apex of the main axis has the highest priority for recently assimilated carbon (C) and only when this is satisfied do the younger leaves become exporters of C to axillary branch buds or young branches. Due to this well-coordinated structure, C can pass in both directions between the main axis and branches (Kemball and Marshall, 1995) and also between branches. For example, although young branches import C from the main axis, they are generally able to supply sufficient C for themselves or may receive some supplementary C from other branches, even if the main stolon is defoliated (Chapman *et al.*, 1992).

Conditions around the stolon determine its behaviour. The base of a dense sward will receive irradiance with a low R (660 nm) : FR (730 nm) ratio, and so the developing stolons will produce long internodes, although branch bud development may be suppressed (Thompson and Harper, 1988).

Young branches may also die if they are too young to produce petioles sufficiently long to put the laminae into positions in the canopy so as to intercept sufficient light to photosynthesize. Suppressed buds may develop into branches if conditions around the node improve, e.g. if herbage is cut and removed (Teuber and Laidlaw, 1996). However, if the suppressed bud is at a node older than the ninth from the apex, it is unlikely to produce a branch, an effect which may be related to starch reserves in the stolon, since starch content in the stolon declines basipetally from the apex and is relatively low at this, and older, nodes (Hay *et al.*, 1989).

Rooting at a node and production of a stolon branch at that node are correlated. Although it has been speculated that a branch will only develop if a root exists, some more recent evidence suggests that the probability of a root surviving when resources are limiting is higher if a branch has already developed at that node (Newton and Hay, 1994). Nevertheless, in experiments in which roots have been severed, the probability of an axillary bud developing into a branch is inversely related to the distance from the nearest rooted node proximal, i.e. closer to the base, to the bud, conditional on the root having developed from the primordium on the node at the same side of the stolon as the bud (Lotscher and Hay, 1996).

For an axillary bud to develop into a branch, the bud has to be viable, vegetative and non-dormant and the conditions have to be conducive for bud outgrowth. Although the majority of buds were non-viable under field conditions, only defoliation reduced viability from a range of stresses imposed on white-clover plants under controlled conditions, the effect being restricted to severe defoliation, and even then the reduced viability was temporary and confined to buds developing within the apical bud while the treatments were imposed (Newton and Hay, 1996). Conversely, outgrowth of buds into branches is much more sensitive to environmental and management conditions, with defoliation, phosphorus (P) deficiency, moisture stress, companion grass presence and treading all being implicated (Hay and Newton, 1996).

During periods of short-term clover decline, e.g. when the interval between defoliations is excessively long or N fertilizer has been applied, clover DM decline is associated with a decrease in the growing-point population density. The younger stolons and those belonging to the lower orders in the branching hierarchy suffer disproportionately to their number (Soussana *et al.*, 1995) and their contribution to total clover LAI declines (Laidlaw *et al.*, 1995a).

The longevity of a segment of stolon depends on the microenvironmental conditions, the state of other stolons associated with it and management. Stolon death rate is highest during winter, when 80% or more of the stolon tissue may be buried (Hay *et al.*, 1987). The interconnecting network of stolons is broken up and so, by the following spring, a new population of individual plants is established from surviving surface stolon and from some of the buried stolon.

Prolonged periods of dense herbage cover, frequent and close defoliation, drought, subzero temperatures – especially for unhardened plants – and pests or diseases are all factors that can hasten senescence of stolons. As clover plant numbers build up, due to stolon fragmentation in continuously stocked sheep swards, plant size declines and, immediately prior to a so-called clover 'crash', i.e. a sudden and severe decline in clover contribution to the sward, plants reach their minimum size and their maximum population density (Fothergill et al., 1996).

While stolon persistence is related to the stolon growing-point density, high stolon density is associated with small leaves. Attempts to break this association by selecting for variation in branching ability within different leaf-size populations have shown that selecting for short internodes, rather than for high branching frequency, offers a better prospect for success in breeding for types which have high growing point densities and large leaves (Caradus and Chapman, 1996).

Defoliation

White clover responds to defoliation by producing shorter petioles, smaller laminae and shorter stolon internodes, but cutting interval does not affect leaf appearance rate, unless it is extremely short or long (King et al., 1978). Branching of white-clover stolons is reduced by frequent close defoliation, a response associated with a reduction in total available carbohydrate (TAC) content (Jones and Davies, 1988). However, if mixed swards are defoliated infrequently, stolon growing-point density declines, due to irradiance at the base of the sward being very low or having a low R : FR ratio, which inhibits bud development. However, lengthening the defoliation interval causes a reduction in the density of the grass component, and because of the resultant improved light conditions at the base of the sward, white-clover petiole length, leaf weight, stolon length and stolon diameter all increase (Wilman and Asiegbu, 1982).

When leaf appearance and branching rates have been reduced due to frequent defoliation, e.g. weekly cuts simulating intense continuous grazing, they can be increased by introduction of a longer defoliation interval. This not only allows the number of leaves per stolon to accumulate, thus improving the build-up of TAC, but also reduces competitiveness by associated grass since its tiller number and leaf appearance rate decline (Grant and Barthram, 1991). The benefit to clover continues for a short time after frequent defoliation resumes, since the stolons continue to produce large leaves (Davies, A., 1992). Management that controls grass growth during winter results in increased stolon population density and branching, thereby improving potential clover persistence in the following season (Patterson et al., 1995).

Elevated carbon dioxide and global warming

Predictions that atmospheric CO_2 concentrations may double during the next century has stimulated studies on the effect of elevated CO_2 levels on white clover, both as single plants and in mixed swards. As temperature is predicted to increase with elevated CO_2, due to the greenhouse effect, such a scenario has been studied in white-clover plants. Those in an atmosphere of double the CO_2 concentration of current ambient temperature and 3°C higher grew faster and had 30% higher leaf photosynthesis (Ryle et al., 1992). However, the differences in growth rate were not maintained, due possibly to excessive accumulation of carbohydrate and restricted root development. Elevated CO_2 also results in lower specific leaf and specific root weights and N concentration in the herbage (Jongen et al., 1996).

In mixed swards, elevated CO_2 increases total yield, that of white clover increasing more than perennial ryegrass, with grass benefiting in the first year by 6%, compared with clover yield increasing by 18% (Nösberger et al., 1995), although drought has a greater detrimental effect on the clover than on grass in high atmospheric CO_2 conditions, compared with the sward under normal CO_2 conditions (Newton et al., 1996). White clover profits more from elevated CO_2 than grass in mixtures, at least in the early stages of sward development, due to N being the limiting factor for grass growth (Schenk et al., 1996) and elevated CO_2 having a positive effect on N_2 fixation (Crush, 1993).

From simulation studies in Scotland, it has been predicted that white clover in swards with a reasonable clover content will increase its contribution in the total herbage DM from 32 to 46% for an assumed 2°C rise in temperature and that the direct positive effect of elevated CO_2 on clover will be additional (Topp and Doyle, 1996a). Mixed swards cut for conservation are also predicted to have higher first-cut silage yields and higher contents of white clover should the anticipated changes in CO_2 levels, rainfall and temperature materialize (Topp and Doyle, 1996b).

Nitrogen fixation

Rhizobial populations of the strains *Rhizobium leguminosarum* bv. *trifolii* which infect the roots of white clover are highest in soils in which *Trifolium* species have been or are currently prevalent. However, ineffective strains may be the only types present in low-fertility soils, such as peat or peaty podzols. Introduction of white clover into such soils requires inoculation with effective and competitive strains of rhizobia (Newbould et al., 1982). When seed is rhizobially inoculated, it should be sown as soon as possible afterwards since survival of rhizobia on the seed is adversely affected by drying, and also by a water-soluble toxin which diffuses from the clover seed coat during inoculation (Hale, 1977).

Soil acidity is also known to have major effects on the rhizobium–whiteclover relationship, potentially adversely affecting rhizobia survival, multiplication prior to infection, infection of the root or nodule development or directly affecting white-clover growth; for example, liming an acid soil led to root-nodule development and subsequent white-clover growth, whereas no rhizobia were present when the pH level was 4.5 (Greenwood, 1961).

Although capable of withstanding slightly acid conditions, *R. leguminosarum* bv. *trifolii* grows better at pH 5.0 or above, as this ensures low soil aluminium (Al) and manganese (Mn), as well as the benefit of higher calcium (Ca) concentrations (Cooper et al., 1983). White-clover growth is affected less by low pH if it is dependent on inorganic N, as the fixation process is more sensitive to acidity than inorganic N uptake.

Heavy metals are more toxic to R. leguminosarum bv. trifolii and the N_2-fixation process than to the growth of white clover per se. Rhizobium may adapt to high heavy-metal contents without losing N_2-fixing ability (Smith and Giller, 1992). However, at levels in the soil in excess of the maximum permitted in the UK for sewage-sludge disposal, Rhizobium numbers and N_2-fixing effectiveness are reduced. Cadmium is particularly lethal to Rhizobium (Chaudri et al., 1992).

Ineffective strains of rhizobia are known to be widespread in clover-growing areas. Only 25% of nearly 500 Rhizobium isolates collected from 48 different sites in Scotland were effective in newly introduced cultivars of white clover, although more were effective in indigenous white clovers, and there was some evidence of specificity between cultivars and R. leguminosarum bv. trifolii strains (Holding and King, 1963).

Nodulation takes 1–3 weeks from infection, although under optimum conditions more than half the nodules can be formed after 1 week. High concentrations of soil mineral N delay infection, as well as reducing N_2 fixation in developed nodules, the effect depending on several factors, including concentration of the N, environmental conditions and physiological and morphological state of the clover, as well as amount, timing and form of applied inorganic N (Sprent and Minchin, 1983). Application of low levels of fertilizer N to white-clover seedlings delays infection but increases the eventual number of plants, but, when applied to mature plants, fertilizer N reduces N_2 fixation more or less in relation to the quantity applied to nodules (Nutman, 1962). Nevertheless, clover retains some N_2-fixing activity – c. 15% of total legume N – even at high soil N status (Davidson and Robson, 1985).

Subsequent to nodule development by an effective strain and conditional on nutrient requirements being met, including components of nitrogenase and the minerals iron (Fe), molybdenum (Mo) and cobalt (Co), the N_2-fixation rate is determined by the plant's environment and the supply of photosynthesis products. Conditional on sufficient photosynthetic reserves being available, such as in well-developed plants with stolons, N_2 fixation may be unaffected by photoperiod and so is not immediately dependent on current photosynthate (Haystead et al., 1979). It has been estimated that 4–6 mg C are required to fix 1 mg N, equivalent to 12–16% of the daily photosynthetic rate, and so this process is considerably more energy-demanding than uptake of a similar quantity of inorganic N (Ryle et al., 1989). However, net costs for N_2 fixation are met directly from solar energy, via photosynthesis, and not from non-renewable fossil energy, which is required for the manufacture of fertilizer N.

Rhizobial N_2-fixation rates in relation to clover production (N_2-fixation efficiency) vary widely. Crush (1987) calculated efficiency to range from 27 to 112 kg N t^{-1} of clover DM, from 30 published studies carried out in New Zealand, with absolute rates ranging from 17 to 680 kg N ha^{-1} year^{-1}; more than half of the estimates of fixation were within the range 35–80 kg N t^{-1} of DM, with annual N$_2$-fixation rates ranging from about 100 to 400 kg N ha^{-1} year^{-1}. This contrasts with UK estimates, which ranged from 50 kg N ha^{-1} year^{-1} in upland reseeds (Newbould, 1982) to 270 kg N ha^{-1} year^{-1} in a sward in Northern Ireland cut for silage (Halliday and Pate, 1976). In the latter study, N$_2$ fixation in white clover was found to be relatively insensitive to temperature in the range from 13 to 26°C but was adversely affected above and below that range. However, cultivar origin can influence the activity, since cultivars derived from Mediterranean germplasm, with cool-season growth, have peak activity at lower temperatures than cultivars from cold-climate germplasm (Crush *et al.*, 1993).

Nitrogen fixation is reduced by defoliation or prolonged shading and, when measured by nitrogenase activity, it is reduced very quickly, i.e. within a few hours of defoliation (Hartwig *et al.*, 1987); depending on the intensity of defoliation, recovery to predefoliation levels takes from 10 days (Chu and Robertson, 1974) to 26 days (Yoshida and Yatazawa, 1977), although the latter period is exceptional. Recovery to predefoliation levels usually takes about 2 weeks (Gordon *et al.*, 1990). The effect of defoliation or low photosynthetically active radiation (PAR) on N$_2$ fixation may be a consequence of changes in nodule-oxygen diffusion resistance acting as a feedback mechanism, rather than being due to the direct effect of reduced assimilate (Hartwig and Nösberger, 1994).

The DM growth of white-clover plants declines faster in response to declining PAR when relying on N$_2$ fixation than when supplied with inorganic N, and the amount of nodulated root declines more than the fixation rate per nodule (Kessler and Nösberger, 1994).

Nitrogen transfer

Nitrogen contained within white-clover herbage is only a portion of total N fixed. Other components of clover-derived N to the N economy of a sward include N in the decaying subterranean parts of the clover plant (roots, nodules and stolons) and litter, which eventually may be mineralized, and recycled excretal N from grazed herbage.

In both cases, much of this N is taken up by grass, as it utilizes inorganic N more efficiently than clover, and so the clover-derived N is usually referred to as 'transferred' N. Estimates of transferred N as a percentage of total N fixed vary. For example, in a cut upland sward in Wales transferred N accounted for almost 50% of N fixed during the year of measurement (Goodman, 1988), while in a grazed sward in New Zealand just over 20% of fixed N was transferred via the grazing animal and 25% underground (Ledgard, 1991). In cut swards, the proportion of clover N transferred is higher in swards containing a small-leaved than a large-leaved cultivar although it is not clear whether this is due to less N being available for transfer from the large-leaved type or to the large-leaved cultivar being more competitive with grass, resulting in grass being less able to take up the available N (Laidlaw *et al.*, 1996).

Subterranean N transfer is encouraged by stress conditions, e.g. drought or defoliation, which increase the rate of senescene of stolons, roots and nodules, turnover rate in clover roots being much faster than in grass roots (Laidlaw et al., 1996). Subsequent mineralization of organic N in decaying plant parts makes the N readily available to the accompanying grass. Although direct transfer of N from clover to perennial ryegrass has been demonstrated via mycelia of vesicular arbuscular mycorrhizae (VAM) in pot experiments (Haystead et al., 1988), this is unlikely to be a major mode of transfer within a grassland ecosystem (Rogers, 1993).

Transfer of clover N via the grazing animal has a long-term ecological impact on the contribution of white clover to pasture. The role of transferred N in the N cycle of a grazing system is presented diagrammatically in Fig. 2.4. The clover contribution to a mixed sward is cyclical in some circumstances, since the clover N encourages grass growth at the expense of clover. As clover contribution declines, N within the system also declines, due to the losses (volatilization, denitrification, leaching and offtake) exceeding inputs. Grass growth then becomes less aggressive, due to inadequate N supply, and there is increased opportunity for white clover to improve its contribution to the sward, and so the cyclical pattern of coexistence is repeated (Brougham et al., 1978). This cycle of contribution is expected when the sward is subjected to consistent management and conditions that are not severe enough to suppress white-clover growth independently of competition from grass. Models of the process predict that the two components will oscillate in dominance over a long period before they attain an equilibrium mixture (Thornley et al., 1995;

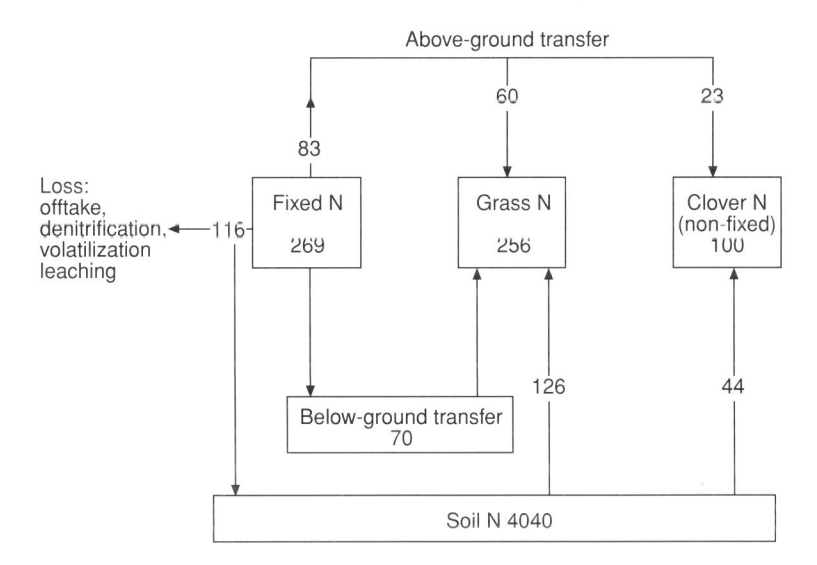

Fig. 2.4. Example of N cycled (kg N ha^{-1} year^{-1}) in a grass/white-clover sward grazed with dairy cows (from Ledgard, 1991).

Schwinning and Parsons, 1996). Experiments are often too short-term for the full cycling effect to be observed, but declines or crashes in clover contribution, followed by recoveries, have been observed in several long-term trials (Stewart, 1988; Tyson et al., 1990).

Although many estimates of N transfer from white clover to grass have been made, caution has to be exercised when interpreting the data, particularly from older studies, which did not involve the use of ^{15}N. As transfer is basically calculated from the difference in N in grass herbage in the presence and absence of white clover, assumptions have to be made about the proportion of clover N derived from N_2 fixation (Marriott and Haystead, 1993). Also, the soil N content is likely to be higher where grass is growing with the N_2-fixing clover and this may enhance mineralization of soil organic N. In turn, this increases the amount of mineral N available for uptake, resulting in an overestimate in transfer. This mineralized N cannot be identified separately from atmospheric N and so the contribution of fixed N to the total amount transferred is also overestimated.

Mineral nutrition

White clover is usually grown in association with grass, so it has to compete for mineral nutrients under sward conditions. Clover roots have a lower density of root hairs than perennial ryegrass, with less profusely branched roots, and a higher cation-exchange capacity (Jackman and Mouat, 1972). Consequently, white clover is less competitive than grasses in taking up mineral nutrients. Ranges in the mineral composition of white-clover DM from many studies are presented in Table 2.1.

The critical level of a nutrient, i.e. the minimum concentration within the plant that is required to produce 90 or 95% optimum growth, may differ from study to study, depending on availability of other nutrients, stage of development, part of the plant analysed, edaphic conditions and the ecotype or cultivar. Therefore, although critical concentrations of the main nutrients for white clover seem to be higher generally in grass/clover associations than in white-clover monocultures (Dunlop and Hart, 1987), differences due to some of the factors outlined above could account for some of the differences in critical concentrations for white clover between monocultures and mixtures.

While nitrate (NO_3^-) in soil solution reduces nitrogenase activity, NO_3^--uptake and N_2 fixation can occur together (Copeland and Pate, 1970). Comparison between yields produced by clover relying on fixed N versus inorganic N shows that a yield penalty is incurred by the plants fixing N biologically, mainly because of the higher respiratory cost of N_2 fixation, compared with that of inorganic N uptake.

Nitrogen content in white clover is usually in the range 35–40 g kg^{-1} DM. Nitrogen nutrition of white clover is not only influenced by availability of soil mineral N but also by other nutrients which are directly involved in nitrogenase activity, namely, Mo, Co, Ca, Fe, boron (B) and copper (Cu) (Dunlop and Hart, 1987).

Table 2.1. Mineral composition of white clover (from Frame and Newbould, 1986).

Constituent	Content range
(g kg⁻¹ DM)	
P	1.9–4.7
K	15.4–38.0
Ca	12.0–23.1
Mg	1.5–2.9
S	2.4–3.6
Na	0.5–2.0
Cl	3.4–15.6
(mg kg⁻¹ DM)	
Fe	102–448
Mn	40–87
Zn	22–32
Cu	5.4–9.7
Co	0.10–0.38
I	0.14–0.44
Mo	1.3–14.2
B	26–50
Se	0.005–153

Estimates of critical concentrations for P vary from 2 g P kg^{-1} DM (Rangeley and Newbould, 1985) and 2.3 g P kg^{-1} (Andrew, 1976) to almost 4 g kg^{-1} (Evans *et al.*, 1986b). Symptoms of P deficiency include the leaves assuming a dark green colour and the plant adopting a dwarf, prostrate habit. The P moves acropetally, with the plant sinks closest to each root involved in uptake having the highest priority (Kemball and Marshall, 1994; Lotscher and Hay, 1996). As a leaf ages, P can be remobilized towards younger tissue. This reutilization allows the clover plant to behave as if P were at an adequate level when the growing medium is becoming P-deficient, shoot growth being affected more than the roots (Dunlop and Hart, 1987). Phosphorus also accumulates in lateral roots and nodules. Some genotypes tolerate low P levels better and respond less to high P levels than less tolerant types (Scott, 1976).

Higher critical concentrations of P generally apply to white clover grown in association with perennial ryegrass (McNaught, 1970). Phosphorus applied to grass/white clover boosts white-clover growth more than grass, but, if soil N is in adequate supply, grass will profit more from P than white clover (Wolfe and Lazenby, 1973). None the less, the surface rooting habit of white clover results in higher P uptakes on P deficient soils after fertilizer application (Caradus, 1990). Vesicular arbuscular mycorrhizae increase P uptake by white clover in P-deficient soils (Powell, 1976), although in P-rich soils the association may reduce P concentration in white clover, the fungus possibly behaving as a parasite (Crush, 1976). Phosphorus availability is associated with pH, becoming less available as lime application is increased (Rangeley and Newbould, 1985).

However, white clover has the capacity to acidify its rhizosphere (although not as effective as greater lotus), thus solubilizing some forms of P, e.g. inorganic P in North Carolina phosphate rock (Trolove et al., 1996).

Estimated critical concentrations for potassium (K) are usually in the range 15–19 g K kg^{-1} DM, the level for leaf laminae being higher than for petioles (Kelling and Matocha, 1990), although for clover growing with perennial ryegrass the critical concentration may be over 20 g kg^{-1} DM. Potassium deficiency is characterized by the formation of necrotic spots or necrotic areas close to the margins, initially of the older leaves. Potassium only moves acropetally, petioles containing particularly high K levels (Wilkinson and Gross, 1967). In mixtures with perennial ryegrass, white clover will respond to K application more than the grass, because of the preferential uptake by the grass of existing available K to meet its requirement.

Estimates of critical Ca concentration vary from about 7 g kg^{-1} DM (Andrew and Norris, 1961) to 20 g kg^{-1} DM (Rangeley and Newbould, 1985). Some ecotypes adapted to low soil Ca levels are more efficient in Ca uptake than other non-adapted types (Snaydon and Bradshaw, 1969). Calcium is not mobile in phloem tissue and so tends not to be readily remobilized to younger tissues, especially during rapid growth. Deficiency symptoms include the petiole of young, fully expanded leaves collapsing when the lamina seems to be normal, followed by older leaves becoming necrotic and chlorotic between the veins.

Calcium is applied as lime ($CaCO_3$ or CaO), but, as pH is also raised, the extent to which the response of the plant is due to increase in Ca or to increase in pH is not always clear, the two having different effects. For example, Ca interacts with soil pH with regard to nodulation, high soil concentrations of Ca overcoming the inhibitory effect of low pH on the infection of roots and formation of nodules by *Rhizobium*.

Sulphur (S) critical concentrations range widely, from 1 to 3 g S kg^{-1} DM (Dunlop and Hart, 1987). Deficiency symptoms include yellowing of the older leaves and veins becoming light in colour. Phosphorus and S interact in grass/white-clover swards, P concentration in the herbage declining with increase in S, and S in herbage declining with increase in P. Recommended grass/white-clover sward balance for S and P is an S : P ratio of 0.7–0.8 (Sinclair et al., 1996a, b).

Magnesium (Mg) critical concentrations are around 1.2–1.4 g Mg kg^{-1} DM, older leaves on plants deficient in Mg having reddish-pink margins and chlorosis between the veins, with the symptoms affecting young leaves if severe. In some soils, Ca is excessively higher than Mg. If the exchangeable Ca : Mg ratio is greater than 20, Mg deficiency in white clover is induced when lime is applied (Carran, 1991). The problem can be circumvented by using Mg limestone.

Sodium (Na) can compensate in part for low K levels. However, Na may be a problem in soils which have been irrigated, due to accumulation of Na and chlorine (Cl), causing toxicity to white clover. Selections of white clover that are tolerant of high NaCl concentrations in soil have higher yields and leaf appearance rates at high salt concentrations and accumulate Cl more slowly than susceptible genotypes (Rogers and Noble, 1992).

From a survey of 51 cultivars of white clover in common use, iodine (I) concentration ranged from 0.08 to 0.21 mg I kg^{-1} DM (Crush and Caradus, 1995), concentration of I being more important for animals fed white clover than for the growth of the plant.

Data on critical concentrations of trace elements are limited. Estimates for white clover in association with grass, in mg kg^{-1}, are: Fe 50–70, Mn 25–35, B 25–35, zinc (Zn) 12–18, Cu 4–6 and Mo 0.05–0.15 (McNaught, 1970), Fe and Mo being constituents of the enzyme nitrogenase. Cu uptake by white clover is enhanced by VAM (Li et al., 1994).

Aluminium availability increases at low pH and so, in acid mineral soils, exchangeable Al concentration can be sufficiently high as to be toxic to white clover, limiting uptake of Ca and P and retarding root growth. Genotypes that have been selected for tolerance to high Al concentration take up Al more slowly than susceptible types. They have higher concentrations of K in plant DM (Crush and Caradus, 1992), and also Mg and P, with fertilizer P application reducing the toxic effect of Al and with tolerant genotypes having a higher yield response to P (Crush and Caradus, 1993). Increasing pH to above 5.5 by liming reduces the concentration of exchangeable Al in the soil.

Selenium (Se) toxicity can be a problem for white clover, Se replacing S in protein synthesis. Application of sulphate reduces toxicity by competing with Se for incorporation into proteins (Wu and Huang, 1992). Some genotypes are tolerant of high soil Se levels by taking the element up more slowly than susceptible types.

Manganese toxicity can be a problem to white clover in acid or water-logged soils, white clover being more sensitive to excess Mn than perennial ryegrass (Adams and Akhtar, 1994). Liming free-draining soils to pH above 5.5 will overcome Mn excess.

Heavy-metal toxicity may be a problem in some soils, especially those affected by industrial waste. White clover is more sensitive to cadmium (Cd) than perennial ryegrass (Yang et al., 1996a), while nickel (Ni) is less toxic to clover than Cd, but nevertheless reduces Cu, Fe and Mn uptake in white clover (Yang et al., 1996b).

As critical concentrations for a given nutrient can vary widely, they are not always strong indicators of deficiency or excess in a nutrient, because of the range of factors that influence these concentrations. Some methods have attempted to take the many factors that influence nutrient uptake into account in arriving at an ideal concentration for a specific nutrient for a given set of conditions. The diagnosis and recommendation integrated system (DRIS) is one of these approaches (Beaufils, 1973). Although originally developed to take account of all of the factors that have an influence on yield and to consider them simultaneously, its application is usually confined to consideration and interpretation of analysis of plant nutrient concentrations.

Table 2.2. Examples of DRIS norms for white clover in mixed swards derived from New Zealand and Spain (from Jones and Sinclair, 1991; Rodriguez Juliá, 1991).

Nutrients	New Zealand Lowland	New Zealand Highland	Northern Spain
N/P	13.18	12.48	9.60
N/K	1.67	1.52	1.64
S/N	0.06	0.06	0.05
Ca/N	0.33	0.43	0.27
K/P	7.72	8.12	6.25
S/P	0.72	0.70	0.47
Ca/P	4.48	5.30	2.64
K/S	10.91	11.76	12.50
Ca/K	0.51	0.63	0.45
Ca/S	6.03	7.66	5.84

The technique involves establishing relationships (usually ratios) between different nutrient concentrations in herbage typical of the highest-yielding stands in order to produce DRIS 'norms'. A range of ratios of the different nutrients in herbage, for which a diagnosis is required, are compared with the norm ratios mathematically; 'DRIS indices', calculated from the comparison between DRIS norms and actual ratios, identify the nutrients that are most likely to be limiting.

The DRIS approach has been applied to grass/white-clover swards in New Zealand and Spain using norms developed from different data sets, and examples are presented in Table 2.2 (Jones and Sinclair, 1991; Rodriguez Juliá, 1991). In the New Zealand study, P and S deficiencies, in particular, were more accurately diagnosed using DRIS than with critical concentrations. Although theoretically the same norms should be applicable, irrespective of location, the data in Table 2.2 show that the level of agreement varies, depending on the nutrients concerned.

Breeding

White clover is a natural tetraploid species, with 32 chromosomes, although it behaves as a diploid. It is an outbreeding species, each plant being genetically unique.

Although scientific breeding programmes for forage plants such as white clover only began during the early decades of the twentieth century, local ecotypes of widely adapted species, particularly white and red clovers, arose in the eighteenth and nineteenth centuries, as specialized farming practices developed. Under British conditions, the small-leaved, low-growing type called wild white clover was widespread, until farmers imported and sowed the larger-leaved but shorter-lived Dutch types. The best local ecotypes became popular and included small-leaved Kent and large-leaved Kersey. In the USA, large white clovers, both ladino and, later, non-ladino types, were introduced, together with intermediate types, which, being profusely floriferous, can be grown as reseeding winter annuals in south-east USA (Pederson, 1995).

In the UK, the development of forage-plant breeding, which started late in the nineteenth century, received a great impetus with the founding of the Welsh Plant Breeding Station, Aberystwyth, in 1919. Along with perennial ryegrass, white clover was among the first species to be improved by scientific selection and breeding. The old permanent pastures of the UK proved a source of valuable genotypes for the early breeding programmes. The first selections, referred to by 'S' numbers, were intended as temporary stopgaps, but the selections lasted many years and, for example, S 184, released in 1942, is still a recommended small-leaved cultivar in the UK, although seed is now often scarce. A number of cultivars have been released since then, notably during the past 25 years.

The British work strongly influenced cultivar development in New Zealand and, during the late 1920s, the range of white-clover ecotypes prevalent in the grasslands there were collected and, from these, the most productive, persistent type, New Zealand Certified White, was developed under grazing-management conditions and released in 1930. Improved selections were subsequently made and, in 1957, medium-leaved Grasslands Huia, which is well adapted to a range of pastoral farming conditions, was released (Williams, 1983). It has dominated the world market for many years on the basis of supply and price competitiveness, and still does, but with a reduced 35–40% of the global market volume of 8500–10,000 t year^{-1}; however, nine new improved cultivars have been released in the last 30 years and are increasingly making inroads into the market (Mather *et al.*, 1996).

In south-east Australia, the large-leaved ladino-type cultivar, Haifa, was developed, for adaptation to heat and low rainfall, from material collected in Israel and it has been of great benefit to dairy farming. A breeding programme for white clover has been in operation for some years, with the aim of improving yield, persistency and winter growth in the Australian environment (Lee *et al.*, 1993).

Conventionally, white-clover cultivars have been bred as synthetic varieties from open pollination of parent clones or lines, these parents having been selected initially on the basis of phenotype and then progeny-tested. The source of the parents can be from local collections, directly or after recurrent selection, or from introduced material, directly or after crossing with local material (Williams, 1903).

In the USA, the first white-clover type was a ladino ecotype originating from north Italy and this wrongly perpetuated the name ladino for all large types (Williams, 1987). Some examples of certified cultivars are the large types Regal, Canopy and Sacremento, intermediate Louisiana S-1 and small–intermediate Star (Pederson, 1995). Improvement of clover persistence by developing multiple pest resistance and enhancing agronomic traits is a major objective in breeding programmes (Pederson *et al.*, 1993).

White clover has been successfully hybridized with the *Trifolium* species *T. isthmocarpus, T. ambiguum, T. uniflorum* and *T. nigrescens* (Arcioni *et al.*, 1996). However, problems arise from the inability of hybrid embryos to develop *in vivo*, often requiring embryo–rescue techniques to produce the hybrid plants, and from the high proportion of infertile progeny resulting from hybridization, which limits the numbers available for use in breeding programmes. Nevertheless, geneticists in New Zealand have made a major breakthrough by developing fertile hybrids between white clover and Caucasian clover (*T. ambiguum*) with varying balances of parental chromosome sets, e.g. a hybrid Caucasian type, with the strong root system (based on rhizomes) and cold tolerance of Caucasian clover (Hussain and Williams, 1998).

Novel genetic-manipulation techniques offer further opportunities to improve both new and established cultivars, by adding particular features currently absent from their germplasm (Webb, 1996). Production of transgenic white clovers containing valuable genes, added to incorporate important features, such as disease and pest resistance, is well advanced (Burgess and Gatehouse, 1997). Current developments in plant-transformation technology in white clover allow such high-value genes to be added to élite commercial cultivars directly, using *Agrobacterium tumefaciens* as the vector, aided by selecting genotypes that have improved embryogenesis (Weissinger and Parrott, 1993). As an alternative, there are techniques that allow shoot tips of seedlings to be transformed directly (Voisey *et al.*, 1994), instead of the previous technique, which required transfer via an intermediate non-agronomic plant. This allows new plants from single cells to be regenerated within only 8 weeks, instead of taking over a year.

In recent years, the breeders' objectives have been to develop cultivars to cover the range of grazing and cutting systems used on modern farms, from continuous stocking with sheep to rotational grazing or cutting systems, and for specific environmental conditions. A high degree of success has already been achieved in these areas. Resistance to pests and diseases is an ongoing aim and, although it has been hampered by lack of knowledge about the incidence of certain pests and diseases or the severity of the damage they cause, the situation is being remedied by pest surveys (Lewis and Thomas, 1991; Clements and Murray, 1993) and disease surveys (Skipp and Lambert, 1984).

Obviously, the avoidance of bloat (tympanites) in stock grazing pastures rich in white clover would represent a major advance. Soluble proteins in white-clover herbage are the main agents of foaming in the rumen, which traps the gases from rumen fermentation and causes bloat. This does not occur in some other legume species, such as birdsfoot trefoil (*Lotus corniculatus*) or sulla (*Hedysarum coronarium*), which contain condensed tannins (CT). These CT occur in the flower petals of white clover, so the plant contains the genetic information for CT synthesis (Jones *et al.*, 1976). Exhaustive searches have failed to identify types of white clover that have CT present in their leaves, but it now seems increasingly possible to produce CT-containing white clover by genetic manipulation or else to develop transgenic hybrids with an optimal content of CT, and such research programmes are already under way (Webb and Rhodes, 1991). Apart from their bloat-prevention property, CT protect plant protein in the rumen, reducing their solubility and degradation by rumen bacteria and increasing the availability and absorption of amino acids (Waghorn et al., 1990).

In the UK, white-clover breeding objectives at Aberystwyth (Rhodes et al., 1989; Rhodes and Ortega, 1996) include:

- improved herbage yield and reliability through improved compatibility with grasses;
- good spring growth, combined with cold hardiness/winter survival;
- disease and pest resistance;
- high production and persistence under intensive grazing;
- improved seed yield;
- improved N_2 fixation and total sward yield;
- absence of animal-bloat-inducing traits.

In New Zealand, white-clover breeding programmes at AgResearch Grasslands, Palmerston North, aim to:

- improve yield and persistence under a wide range of grazing managements and livestock classes;
- improve growth at low P levels and obtain more efficient use of P at medium levels of fertilizer application;
- develop types resistant to nematodes, particularly root knot nematode (*Meloidogyne hapla*) and clover-cyst nematode (*Heterodera trifolii*);
- breed drought-tolerant cultivars, mainly by selecting deep-rooted forms;
- select for a range of stolon and root features and aid the plant's adaptiveness;
- improve early spring growth;
- select for better adaptation to temperature extremes, both hot and cold;
- select for better tolerance of high levels of mineral N application.

There are also breeding programmes, with differing degrees of intensity, by either government-sponsored organizations or private companies, e.g. in Australia, Japan, some states in the USA, some continental European countries and the countries of southern Latin America. The wide genetic variation identified in the species means that many specific aims are achievable by breeding, even under widely differing climatic and edaphic conditions, including tolerance to less-than-ideal environmental conditions. From evaluation of a world collection of 110 ecotypes and cultivars under sheep grazing, it is concluded that clover yield and percentage in the sward increased by 6% per decade during the past six decades, which is higher than rates of improvement for other forage crops (Caradus, 1994); genetic gains have been greatest for small-leaved and large-leaved non-ladino types and for cultivars originating in cool rather than in cold or warm climates (Caradus, 1994). There is scope for further genetic improvement, since clover can still be considered to be a wild species, compared with the level of domestication achieved with many crop species (Caradus, 1993).

Cultivars

In Europe, white-clover cultivars have been classified into three types (Van Bockstaele, 1985):

- prostrate, small-leaved wild white, with thin multibranched stolons;
- small-, medium small- and medium large-leaved cultivars from north-west Europe, but with short, less branched stolons and larger petioles than the wild white type;
- a large-leaved 'ladino' type, which has thick stolons and a robust root system, is a short-day plant and is more cold-sensitive than the other types.

In European recommended lists, white-clover cultivars are typically classified into small-, medium-, large- and very large-leaved types, according to leaflet size, although, increasingly, manipulation by breeders of factors relevant to grazing tolerance, e.g. more profuse nodal rooting, or the production of cultivars with leaf sizes between the arbitrarily defined classes above is blurring these distinct classes. In the USA, clovers are usually classified into small, intermediate or large types, as determined by a number of morphological features.

The evaluation of new white-clover cultivars is a continuing and essential process in many countries. Within the European Union, it is carried out primarily under the aegis of the statutory national list scheme. Assessment is based on seasonal and annual herbage production under cutting regimes, and quality is determined by digestibility and crude protein (CP). Other measured features include persistence, winter-hardiness and resistance to pests and diseases. This information, supplemented by the results of additional non-statutory trials and assessment under grazing conditions, where possible, is used to produce recommended lists.

There has been a synergistic relationship in the past between forage evaluators and breeders, which has had positive effects on the development of improved legume cultivars. None the less, in light of the range of breeding objectives and diversity of conditions and systems within which white-clover cultivars are used, there is a case for the assessment procedures for acceptance on to recommended lists being much more flexible than the present, rather rigid, cutting regimes. A similar argument can be put forward for other forage-legume (and grass) cultivars. Significant interactions between clover cultivars and grazing method have been clearly demonstrated (Evans and Williams, 1987; Swift *et al.*, 1992). In general, the small-leaved cultivars usually perform better under frequent defoliation, such as intensive continuous grazing by sheep, whereas the large-leaved cultivars are at their best under rotational grazing or periodic cutting.

The results of cultivar performance trials find application in the formulation of grass/white-clover seed mixtures, in relation to required method of utilization. If several objectives are specified for a sow-out, it is customary to make up a blend of white-clover leaf types, e.g. a small- and a medium-leaved cultivar for set-stocked sheep and a medium- and large-leaved cultivar for rotational cattle grazing and/or cutting for conservation. In dryland regions, a mixture of large-leaved erect white clover and small-leaved prostrate white clover is sometimes used, the large-leaved cultivar providing feed while the small-leaved cultivar remains below grazing height and continues to supply fixed N for the pasture. In addition, notable factors considered when compiling a seed mixture are agronomic performance, including seasonality of production, adaptation to the environment and pest and disease resistance.

A useful world check-list of 232 named white clovers, together with a summary of available information about their origin, breeding, descriptive features, agronomic potential and disease susceptibility, was produced in the 1980s and, for the sake of completeness, regional varieties or ecotypes, as well as bred cultivars, were included (Caradus, 1986). Another index of 120 varieties, germplasm and cultivars, which had appeared in English-language scientific literature in the 1980s, was published (Williams, 1987). The world check-list has since been updated to 319 known cultivars (Caradus and Woodfield, 1996).

Seed production

New Zealand is the world's major producer of white-clover seed, with an annual crop of 4500–5000 t from 15,000 ha and representing about 50% of the total world production (Table 2.3). By volume of sales (35–40%), Grasslands Huia is the major cultivar grown in the world, although new, improved cultivars, many of which are large-leaved as opposed to the currently most popular medium-leaved cultivars, are taking an ever-increasing share of the world market. Although skilled producers consistently average 400 kg ha^{-1} across all seed crops and 600 kg ha^{-1} or more in good growing seasons, seed yields in the 1990s have ranged from 100 to 1000 kg ha^{-1},

Table 2.3. Estimated annual production of white clover seed (tonnes) (from Mather *et al.*, 1996).

Source	Amount
New Zealand	4500–5000
USA	1500–2000
South America	1000–1500
Denmark	800–1000
Australia	600–700
Other countries	250–500
Total	8650–10,700

with a national average of 300 kg ha^{-1} (Mather *et al.*, 1996). Lower yields are associated with season, the growing of new cultivars, seed production being taken up by less experienced growers, and less suitable land being utilized because it is free of buried seed from previously grown cultivars (Clifford *et al.*, 1996). Contamination of the seed crop of a new cultivar by buried seed of the old one or by the production of off-types can be a significant problem in intensive seed-growing areas (Lancashire *et al.*, 1985).

The Official Quality Management Scheme requires, *inter alia*, that fields to be sown have been 5 years out of white clover before the new-cultivar crop is sown, annual cultivation is undertaken to eradicate any volunteer clovers, and there is a minimum spacing of 30 cm between sown rows to facilitate inter-row removal by herbicide application of volunteer clovers before the sown crop flowers. Assessment of buried seed loads, though not mandatory, is normally required by seed merchants as an indication of whether or not the field to be sown is sufficiently free from contamination (Clifford *et al.*, 1990); with 38 cultivars in production, such measures are clearly essential.

Denmark is the main white-clover seed producer in Europe, but output has declined in the last 15–20 years because of the wide variation in seed yields between individual crops and seasons – as little as 100 kg ha^{-1} in some areas. This has resulted in erratic financial returns to growers and, also, clover-seed growing is being replaced by more profitable arable crops, such as oilseed rape. In some of the north-west European countries where seed growing is practised, the strong likelihood of unfavourable weather during seed development and harvesting, which leads to reduced seed yields, is a major disadvantage (Evans *et al.*, 1986a; Connolly, 1990). There are other suitable seed-production areas, particularly in southern Europe. For example, in an experiment in southern Italy, average seed yields over five cultivars and two seed rates (2 and 4 kg ha^{-1}) were 351, 260 and 290 kg ha^{-1} for successive harvest years (Martiniello, 1992). However, apart from finding suitable soil and climatic conditions, willingness, time and experience are required for growers to develop skills in the seed-production process. Thus, in order to increase the availability of newly developed European cultivars, seed-production in Europe is supplemented by subcontracting growers in New Zealand. In the USA, white-clover seed yields average 420 kg ha^{-1}, with seed production concentrated in the states of California and Oregon, where the crops are usually grown under irrigation (Pederson, 1995).

In specialist growing areas, 2–4 kg ha^{-1} of white-clover seed are drilled evenly in rows 30–45 cm apart in late summer/early autumn. Sometimes, a few kilograms of perennial ryegrass are added, although seed yields are generally lower than from monocultures.

To ensure a clean crop, weed control may be necessary at different stages in the seed cropping cycle. Potentially damaging perennial weeds are best controlled at the presowing stage by crop rotation and/or herbicides. On the established crop, herbicide use appropriate to the weed spectrum or roguing of contaminant plants may be necessary. Apart from the clover-seed purity aspect, weed control is needed so that there is good clover growth *per se* and the crop-harvesting and drying processes are not hampered. Pesticide use against specific pests may also have to be considered.

The factors governing reproductive development and seed yields of white clover have been comprehensively reviewed (Thomas, 1987). Briefly, the components of seed yield are: number of flower heads (inflorescences) per unit area, number of florets per flower head, number of seeds per floret, weight of ripe seed and proportion of ripe seed harvested. Peak inflorescence emergence is in early summer, and spring defoliation by cutting or grazing to remove the leaf canopy during floral development prevents abortion of young inflorescences by shading; however, defoliation must be timed to avoid removing the first, and the highest seed-yielding, crop of flower heads (Hides *et al.*, 1984). Defoliation should be as early as possible in order to encourage uniform flowering. Delay in closing off adversely affects stolon density and seed-yield components, though to a lesser extent in the more stoloniferous small-leaved than in large-leaved cultivars (Marshall and Hides, 1990). Nevertheless, the inflorescences, which are produced in the axils of leaves arising from stolon nodes, emerge over an extended period, and so at any point in time only a proportion contain ripe harvestable seed (Hollington *et al.*, 1989). The use of the growth regulator paclobutrazol, to increase and concentrate inflorescence production and thereby seed yield, has been assessed over a range of conditions, but a review of its effects shows markedly inconsistent results – 0–350% increases in seed yield – with the conclusion that its potential remains unfulfilled (Hampton, 1996).

The number of florets per head is increased by long warm days, compared with cooler short days (Clifford, 1979), and the number of seeds set per floret is influenced by pollination effectiveness and ovule fertility. Honey- and bumble-bees are the main pollinators, but the use of hives of honey-bees is a more certain method. The weight of ripe seeds is a function of several factors, not least overall management to ensure good growing conditions, such as adequate soil fertility, defoliation management and irrigation, if required. The most suitable spring defoliation, whether cutting or grazing, varies among cultivars, with the generalization that they produce the greatest number of inflorescences and a higher proportion of ripe inflorescences with the defoliation method, similar to that for which they were selected (Hollington *et al.*, 1989; Marshall *et al.*, 1989). The early inflorescences, which usually contain the highest number of ripe seeds, tend to lodge within the canopy before the majority of inflorescences have reached maturity for harvesting; this effect is exacerbated by wet weather, which encourages excessive leaf growth, and so they rarely contribute much to the yield of harvested seed (Marshall and Hides, 1990).

The ideal harvesting date for maximum seed yield is when the number of ripe inflorescences is at a maximum (Evans *et al.*, 1986a) but this occurs only for a relatively short period, and delay in harvesting, through adverse weather conditions for example, will reduce seed yields, on account of the inflorescences lodging, seed shedding and seed sprouting. One method of ameliorating this effect would be the development of genotypes with strong, thick peduncles, which can keep inflorescences erect within the canopy (Marshall, 1995).

Direct combining or a process of cutting, swathing and threshing is an alternative method of seed harvesting. A crop desiccant may be applied before cutting to reduce prethreshing and drying times. The white-clover seed is subsequently dried to 100 g kg^{-1} moisture content or less for safe storage, drying being as rapid as possible so as to maintain germination capacity. Because of the small size and high bulk density of the seed, specialized machinery and skilled operators are needed.

Certified seed, which is sold to farmers by seed merchants, has to meet specific quality standards during and following the seed-multiplication process. Specific standards are defined internationally for germination capacity, including content of hard (dormant) seed, and also for seed purity, including dirt, chaff and broken seed, but sometimes even higher (voluntary) standards are available nationally. Taking the UK as an example, seed to be designated higher voluntary standard (HVS) require a 98% minimal analytical purity and 80% minimum germination, including no more than 40% of hard seed in the total pure seed. The maximum allowable content of other species is 1.5%, with no more than 0.5% of any one species (Ministry of Agriculture, Fisheries and Food, 1983). No dodder (*Cuscuta* spp.) seeds are permitted and only a limited number of dock (*Rumex* spp.) or blackgrass (*Alopecurus myosuroides*) seeds. A maximum percentage by weight (0.3%) is laid down for the presence of undesirable melilots (*Melilotus* spp.).

Agronomy and Management

Sowing rates and seed mixtures

Sowing rate of white clover is less critical than in most plant species, because of its capacity for vegetative spread by the procumbent stolons. Mixed grass/white-clover swards have been successfully established, using a wide range of seed rates of both white clover and companion grass. Having c. 1.5 million seeds kg^{-1} and a rapid rate of germination, even at low soil temperatures (Hampton *et al.*, 1987), white clover is quite easily established when sown at relatively low sowing rates.

A high clover : grass seed-rate ratio may be advantageous for initial clover establishment, but the effect does not persist much beyond the first or second harvest year (Cullen, 1964b). This short-lived effect has been confirmed in combinations of clover seed rates from 1 to 9 kg ha^{-1} with grass seed rates from 3 to 14 kg ha^{-1} (Laidlaw, 1978), and when using a range of companion grass seed rates with a constant clover seed rate (Cullen, 1964b; Frame and Boyd, 1986).

When sown under New Zealand conditions with standard rates of grasses, such as perennial ryegrass or cocksfoot (*Dactylis glomerata*), the average sowing rate of clover seed is 3–5 kg ha^{-1} (Charlton, 1992). For British conditions, 3–4 kg ha^{-1} is recommended (Frame and Newbould, 1986), although, because of cost, British seed companies tend to offer lower rates in seed mixtures, down to as little as 1–2 kg ha^{-1}, depending upon the purpose of the sward and envisaged fertilizer N use. Very low seed rates are commonly used in Australia when renewing pastures, mainly to reduce cost, because of the large paddock/field areas often involved.

While white clover and its frequently used companion grass, perennial ryegrass, are both easily established, because of their relatively high seed vigour (Charlton *et al.*, 1986), it is preferable to use the recommended higher seed rates for white clover if it is expected to contribute significantly to herbage feed supply in the early life of the sward. Where initial clover presence is poor, due to a low seed rate or poor establishment generally, it is possible to encourage the spread of clover by management geared to favour stolon proliferation. A seedling density of 150 clover plants m^{-2} at 3 months after sowing, giving a 30% ground cover by 1 year after sowing, is a proposed good establishment target (Haggar *et al.*, 1985).

Companion grasses

Different grass species and cultivars have contrasting competitive effects on white-clover performance in mixed swards. Thus, when seeking optimal performance of white clover, account should be taken of this, rather than judging mainly on the potential of the grass component, although it must always be borne in mind that the grass invariably supplies the greater proportion of the total herbage yield. It is well documented that some grass species, such as non-aggressive meadow fescue (*Festuca pratensis*), are highly compatible, while others, such as cocksfoot, are relatively incompatible. Others, such as timothy (*Phleum pratense*), are intermediate (Chestnutt and Lowe, 1970), although some modern cocksfoot cultivars are less aggressive and more compatible with modern white-clover cultivars than older cultivars of cocksfoot (Moloney, 1994).

Much experimental work has been conducted using simple grass/white-clover mixtures with only one grass species and cultivar, usually of perennial ryegrass. However, in the UK, for example, commercial seed mixtures often contain cultivars of several grass species or in simpler mixtures, a number of cultivars of the same grass species. Compatibility with a specific grass or mixture of grasses can also be confounded with intensity of defoliation, species of grazing animal and system of grazing.

Within perennial ryegrass, tetraploid cultivars permit better white-clover presence than diploid forms, under both frequent cutting, to simulate grazing (Frame and Boyd, 1986), and actual grazing (Fothergill and Davies, 1993). This is due to their growth habit, which is generally more open than that of diploids, thus allowing more space and better light penetration through the canopy. With both diploid and tetraploid ryegrasses, white clover presence decreases with increasing lateness of grass maturity, an effect attributable to increasing tillering capacity, while the effect of maturity group is greater than that of ploidy (Davies *et al.*, 1991; Gooding *et al.*, 1996). It has also been noted that, within perennial ryegrass, the least productive and least persistent cultivars permit the best performance of companion white clover (Camlin, 1981), but modern farming practice will not accept a policy of deliberately sowing inferior grass cultivars. In any case, such cultivars do not achieve national or recommended list status in countries that operate performance-evaluation systems.

The above-mentioned companion grasses are utilized in the cooler areas of the USA where white clover is grown, while in the warmer areas smooth brome grass (*Bromus inermis*), dallis grass (*Paspalum dilatatum*), Bermuda grass (*Cynodon dactylon*) or smooth-stalked meadow grass (*Poa pratensis*) are favoured companions in mixtures (Gibson and Cope, 1985).

In New Zealand, clover yield and persistence are reduced when grown with perennial ryegrass infected with the endophyte fungus, *Acremonium lolii*, as the competitiveness of the grass is increased, due to reduced pest damage and increased drought tolerance, conferred by the fungus (Prestidge *et al.*, 1992). Increased use of endophyte-infected grass could pose problems in maintaining satisfactory clover contents.

Grass species that develop particularly dense vegetative cover close to the ground, such as bent grasses (*Agrostis* spp.) or Yorkshire fog (*Holcus lanatus*), are incompatible companions, since white-clover growing points on the stolons are subject to considerable shading (Brougham *et al.*, 1978; Frame, 1990). The competitive root growth of Yorkshire fog, compared with other grasses, and its ability to regenerate by initiating new shoots and roots at its nodes and to form spreading clumps all militate against white clover (Watt, 1978).

In recent years in New Zealand, mixtures including three grass species with one or two clover cultivars and even a grazing herb, chicory (*Cichorium intybus*), have been successfully established in the drier regions and have given highly satisfactory performances in grazing systems for dairy and beef cattle, sheep, deer and horses (Milne and Fraser, 1990). Such mixtures include tall fescue (*Festuca arundinacea*), cocksfoot and phalaris or harding grass (*Phalaris aquatica*), all quite compatible with some cultivars of white clover (Moloney, 1990).

The seasonal distribution of grass production is also an influencing factor in the compatibility of grass and white clover, since growth rhythms of companion grass or grasses and clover may be complementary or coincident (Haynes, 1980). Breeders are developing white-clover and perennial-ryegrass cultivars from previously coexisting genotypes, in an attempt to achieve good mutual compatibility and performance in mixture, and this has been achieved in the early stages of sward life (Evans *et al.*, 1985; Rhodes, 1991).

Clover has a higher threshold temperature in spring for growth than grass and this puts it at a competitive disadvantage, the magnitude of which

Table 2.4. Effect of sowing depth on seed germination of white clover and perennial ryegrass (from ITEB/EDE de Bretagne, 1987).

Sowing depth (mm)	Germination (%) White clover	Perennial ryegrass
10	81	94
20	63	95
30	21	86
40	12	68

depends on the specific grass cultivar and the management applied. The disadvantage to clover is ameliorated by grazing the sward in autumn and winter (Laidlaw et al., 1992). A lower temperature threshold for white-clover growth and/or a greater growth response at low temperature, allied to better stolon survival under low winter temperatures, is a characteristic selected for in breeding programmes in the UK (Collins et al., 1991; Collins and Rhodes, 1995), which will prove advantageous in practice.

Sowing depth

With its small size and limited food reserves, white-clover seed should not be sown deeper than 10–15 mm, otherwise seedling emergence is reduced, delayed and weakened (Table 2.4; Cooper, 1977). Precision in sowing depth is only possible on a well-prepared, firm seed-bed with a fine tilth. A useful but oft-neglected practice is the use of a soil leveller during cultivation, to ensure a uniform soil surface. Adequate consolidation, both before and after drilling or broadcasting, is essential for good seed–soil contact, moisture conservation and soil cover of the seed, and can be ensured by the use of a Cambridge roller, with its ridged compression rings.

Ideally, the different components of a seed mixture should be sown at depths optimal for their seed sizes. This is not possible with commercial machinery, especially when sowing more complex, multispecies mixtures, but it is feasible for a simple grass/white-clover mixture. Experimentally, better white-clover establishment and early production have been demonstrated through drilling the larger grass seed at its optimal depth, followed by broadcasting the smaller white-clover seed (Herriott, 1970; Laidlaw and McBride, 1992).

Time of sowing

In temperate areas of Europe, where soil-moisture levels in summer remain adequate, the best establishment of white clover is obtained from spring sowing, since this takes advantage of good climatic conditions during summer and a long establishment period before winter. However, in some other countries where white clover is used widely, summer drought can desiccate seedlings and young plants and so autumn is the normal sowing time for the species (Charlton and Giddens, 1983).

Midsummer is often a risky sowing time in most temperate situations, because soil-moisture shortage is more likely then. Direct sowing in autumn offers additional flexibility of timing, but, relative to spring sowing, the optimum clover production normally expected from the first full harvest year is delayed until the following season; poor management, such as hard grazing after a harsh winter, could set back white-clover recovery (Laidlaw and McBride, 1992; Fraser and Kunelius, 1993).

Weed problems are usually less following late summer or autumn sowing, compared with spring. If pasture renewal follows an arable crop, date of crop harvesting is the main factor influencing time of sowing. 'Damping-off' of seedlings by *Phythium* and other fungal pathogens is another disadvantage of late autumn sowing, although chemical seed dressing against seedling diseases and pests is increasingly available. Seedling mortality may result from an inability to withstand the rigours of winter stress, such as low temperatures or snow cover, but clearly local climatic conditions will govern the severity of such hazards (Harris *et al.*, 1983).

Cover crop

Direct sowing offers the best chance of a good clover establishment, but the grass/white-clover mixture is often sown under a cover crop, such as a spring-sown cereal. This results in increased productivity per unit area in the establishment year and welcome returns from a cash crop of grain and straw. However, the competitive effect of the cereal undoubtedly inhibits white-clover establishment, and a late harvested cereal crop may result in etiolated and poorly developed clover plants, which lack cold-hardiness and resistance to disease or pest damage (Cullen, 1964a); again, a lodged cereal crop may cause damage and establishment failure of the clover. An alternative and potentially less damaging cover crop is a cereal/forage-pea (*Pisum arvense*) mixture for arable silage, which is harvested in midseason, before the cereal grain is ripe.

Oversowing

White clover can be introduced or reintroduced into existing natural or sown swards by some form of oversowing (Sheldrick *et al.*, 1987; Vidrih *et al.*, 1991). Grass/clover mixtures at half or two-third seed rates can also be added to renovate swards that have deteriorated botanically and productively. In oversowing techniques, which include surface seeding, direct drilling (no-till seeding) and partial rotavation, the seed is introduced into a more hostile environment than when using conventional ploughing and cultivation, with a resultant reduced chance of success and rate of improvement.

Surface sowing is the least efficient of pasture establishment (Charlton, 1978); yet initial establishment rates of over 75% of the original seed sown are achievable under satisfactory climate and management conditions, a major cause of seedling death in surface-sown white clover being desiccation of germinating seeds. The process of germination is moisture-dependent and irreversible, so many seedlings die off rapidly when exposed to drying winds. Warm moist conditions not only favour seed germination and seedling establishment but are also conducive to pests, such as slugs (Charlton and Sedcole, 1985). Surface temperatures of exposed areas within a pasture can rise rapidly to extreme levels, especially on slopes facing the sun, and seedlings in these niches are usually the first to suffer. Thus, timing of the operations should coincide with time of year when soil moisture is not likely to limit rapid establishment. Timing is also important in relation to existing sward density, since open sward conditions are required for successful surface sowing. For example, pastures dominated by dense, aggressive species, such as bent grasses, are at their lowest density in autumn after dry summer weather. Other oversowing techniques, such as slot or strip seeding, are also most successful on swards with a low-density vegetative cover.

Grazing, cutting, burning and partial chemical desiccation are means of thinning out dense swards; for example, cutting a silage or hay crop following a long rest interval is effective in reducing sward tiller density. Minimal cultivation, such as that achieved by discing or spike rotavation, is a technique that encourages good seed–soil contact by creating tilth, the spike rotavation being particularly useful where there is a mat of organic matter at the base of the sward. Heavy stock grazing is another means of introducing seed when oversowing, since seed is trampled into the soil and this assists establishment.

A number of direct drills of varying design have been developed and tested and some have proved particularly successful, e.g. the Hunter's Rotary Strip Seeder, which makes cultivated strips by small rotavators, independently mounted to follow ground contours (Pascal and Sheppard, 1985; Sheppard and Swift, 1995), or the Aitchison Seedmatic Drill, which has winged coulters, producing slots with an inverted T shape favourable to seed germination (Baker et al., 1993). Both drills can be fitted with an applicator to band spray a chemical, such as glyphosate, astride the slots, thereby desiccating the existing foliage and reducing competition to the introduced species.

In New Zealand, hill country has been improved by aerial sowing since the 1950s, using fixed wing aircraft or a helicopter, although, because of cost, the use of the technique has declined. Seed and fertilizer are better applied separately, since this improves seed distribution, although mixing both ingredients is cheaper (Charlton and Grant, 1977). Accurate flying and minimal cross-winds during application also improve the distribution pattern (Macfarlane et al., 1987).

A set of management guidelines that can lead to satisfactory establishment by oversowing are available (Naylor et al., 1983; Tiley and Frame, 1991). Briefly, these guidelines include:

- adequate bare space in which the introduced seed may germinate and flourish;
- ensuring soil–seed contact;
- satisfactory soil pH and nutrient status;

- controlling perennial weeds and pests, where these present a problem;
- sowing at times of year when soil moisture is not limiting;
- choosing techniques to suit soil/sward conditions;
- controlled grazing management to limit competition from existing sward and permit good light conditions for the developing seedlings.

Natural seeding of white clover in swards can be encouraged by rest periods from grazing of up to 6 weeks during clover flowering and seed setting. Also, by grazing the seeded clover with sheep or cattle, a proportion of the seed passes through the gut unaffected and is spread elsewhere by dung deposition (Suckling and Charlton, 1978; Young, 1989). In swards with a history of white-clover growth, buried clover seeds accumulate and these aid the regeneration of white clover in swards where it has become depleted – for example, by temporarily encouraging an open canopy through a rest from grazing, since a closed canopy militates against recruitment from shed seed (Chapman, 1987). However, clover seedling regeneration in established pasture is limited compared with vegetative propagation (Chapman and Anderson, 1987).

Fertilizer use

Soil-fertility requirements to ensure satisfactory white-clover establishment are well known, and official or commercial advisory recommendations are usually available for soils of differing fertility levels, as determined by soil analysis.

Lime is applied, if necessary, to raise and maintain the soil pH to satisfactory levels, although excessive quantities interfere with trace-element uptake, which may then affect white-clover growth and N_2 fixation adversely (Floate *et al.*, 1981). Minimum soil pH targets are 5.8 for mineral soils and 5.5 for organic soils.

Application of phosphate and potash fertilizers depends on how the sward is utilized. Swards which are mainly grazed require only small maintenance dressings annually, since there is recirculation of nutrients via the excreta of grazing animals, but, when the sward is cut for silage or hay, the nutrients removed in the crops require replenishment.

White-clover seedlings grown from direct seeding sometimes exhibit N deficiency until the root nodules are formed and N_2 fixation begins. Thus, the use of a moderate dressing of 'starter' N may be necessary, the quantity depending on the N status of the soil (Haystead and Marriott, 1979). Hill and upland soils often have a low N status, as have soils with a history of cereal cropping, whereas previously grazed grass swards under a high fertilizer-N regime normally have a high N status.

Application of fertilizer N has a major influence on the grass–white-clover relationship in an established mixed sward. Typically, increasing annual rates of fertilizer N, applied in repetitive dressings during the growing season, increase total herbage production but decrease the white-clover contribution

Table 2.5. Effect of repetitive fertilizer N application on herbage productivity from perennial ryegrass/white clover swards (mean of four white clover cultivars over 3 years) (from Frame and Boyd, 1987a).

Annual N application* (kg ha^{-1})	Total herbage DM (t ha^{-1})	White clover DM (t ha^{-1})	White clover (%)
0	7.8	4.1	53
120	8.7	2.4	28
240	10.0	1.1	11
360	11.7	0.5	4

*Distributed evenly over six cuts per year.

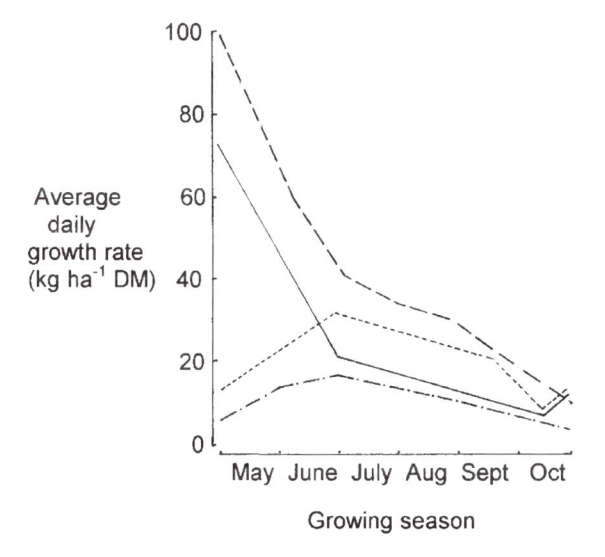

Fig. 2.5. Average daily growth rates of grass at 0 (——) and 200 (— —) and white clover at 0 (- - - - -) and 200 (— · — ·) kg N ha^{-1} annual fertilizer rates applied to grass/clover swards (from Frame and Newbould, 1986).

(Table 2.5, Fig. 2.5). The reason for the decline in white-clover growth due to N-fertilizer application to mixed swards is a subject of debate. The effect is unlikely to be due to direct effects of fertilizer N, as clover can grow at least as well in monoculture receiving fertilizer N, rather than relying solely on N_2 fixation (Davidson and Robson, 1985). Although some studies have concluded that white-clover laminae within the grass/clover canopy receive adequate irradiance to photosynthesize sufficiently to support the plant (Dennis and Woledge, 1985; Woledge, 1988) and the effect takes place too soon for shading to be implicated (Davies and Evans, 1990b), other studies have shown that clover in a mixed sward receiving N fertilizer captures less light per unit area of lamina than grass, especially with advancing growth of the grass (Sinoquet *et al.*, 1990). Growth-cabinet studies have demonstrated that clover has a lower radiation-use efficiency than grass (Faurie *et al.*, 1996). The cause of the adverse effect of N fertilizer on white clover in mixed swards is likely to be a combination of factors, including reduced photosynthesis, as discussed, and reduction in growing-point density (Laidlaw and Withers, 1989), due to reduction in assimilate allocation, inhibition in branching and, possibly, reduction in uptake of other nutrients, due to stimulated growth of grass. It would be interesting and possibly rewarding in the long term to assess the impact of slow-release or controlled slow-release N fertilizers on the grass–white-clover relationship.

Total herbage-response data vary considerably from year to year and from site to site. From the results of many cutting experiments using different N application rates from 0 to 400 kg ha^{-1} annually, the response from mixed swards cut under a simulated grazing regime varies but, on average, is 8–10 kg DM kg^{-1} N applied (Frame and Newbould 1986); this is approximately half the response from similarly treated grass swards or from mixed swards cut less frequently under a conservation regime (Holliday and Wilman, 1965; Frame and Newbould, 1986). In general, responses are lower in swards with a high white-clover content and negative responses may occur at low or moderate N rates, when the gain in grass production is insufficient to compensate for the loss in white-clover production (McEwen and Johnston, 1985). When a range of cutting frequencies is studied, the normal pattern of diminishing response to increasing N rates is less marked under frequent compared with infrequent cutting (Reid, 1978). In general, the decline in white-clover contribution due to increasing fertilizer-N application is intensified under grazing because of animal effects, particularly excretal N return, but there is a dearth of N-rate trials conducted under grazing.

One way of examining the N_2-fixing effect of white clover on total herbage yield is to determine its fertilizer-N equivalent, i.e. the quantity of fertilizer N that would be required annually on all-grass swards to produce the same total herbage production as from grass/white-clover swards without applied N. A mean of 172 kg N ha^{-1}, with a range of values between 124 and 278 kg N ha^{-1}, is cited from British experiments (Royal Society, 1983), with such a range being not unexpected, considering the different experimental conditions. This is illustrated in Fig. 2.6, which shows the generalized production response of grass and grass/clover swards to increasing fertilizer-N application.

The tactical application of fertilizer N in spring is a well-recognized technique for increasing early-season production from the grass component. However, there is an adverse effect on white-clover performance, but, at rates of 50–60 kg ha^{-1}, the effect is usually temporary (Laidlaw, 1980), although there has to be a substantial content of white clover in the sward to act as a 'buffer' against this adverse effect. The depressive effect of spring fertilizer N is marked in soils with a high water-holding capacity and on compact, 'cold' soils (Laissus, 1983). Subsequent management designed to favour white-

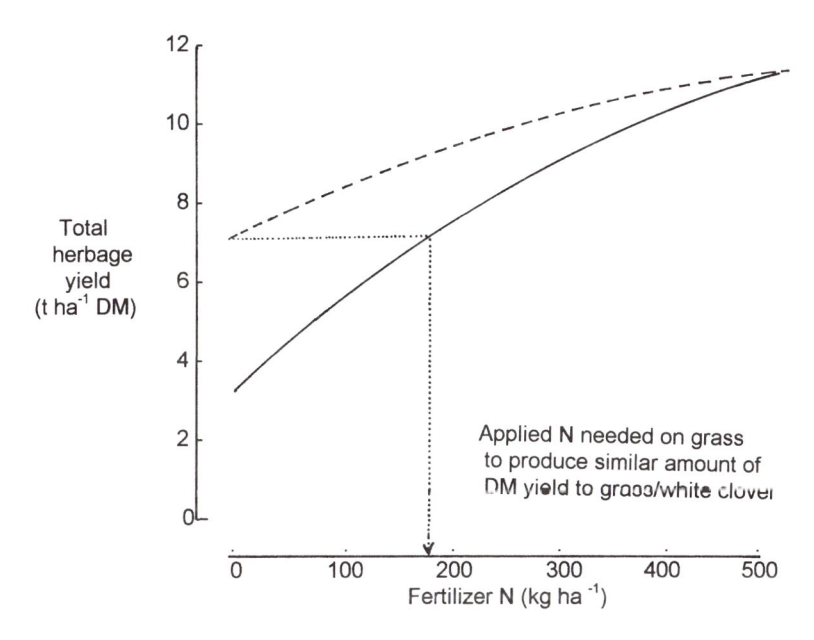

Fig. 2.6. Effect of increasing annual N fertilizer application on grass (——) and grass/white clover (– –) swards (from Frame and Newbould, 1984).

clover development also aids its recovery, e.g. resting a sward from continuous sheep grazing and taking a cut for conservation (Barthram and Grant, 1995).

Autumn is another period when fertilizer N may be used to increase late-season production, but declining temperatures and light intensities at this time – the reverse of the spring environment – result in a relatively lower grass response to the N. There is also a smaller decline in white-clover contribution than from spring N and, if N is applied in spring and autumn, clover contribution is depressed more when a high proportion of the N is applied in the spring (Table 2.6; Frame, 1996).

Table 2.6. Effect of spring and/or autumn fertilizer N application on white clover contents (%) in perennial ryegrass/white-clover swards (mean of 3 years) (from Frame and Boyd, 1987b).

Spring N (kg ha⁻¹)	Autumn N (kg ha⁻¹)			
	0	25	50	75
0	48	48	41	45
25	43	36	40	36
50	38	33	33	31
75	34	27	29	27

Animal slurries are increasingly used as a source of plant nutrients, particularly of N and K, in manurial programmes to complement purchased fertilizer, instead of being regarded as a waste product with disposal problems (Pain, 1991). Research results from the effect of slurry N on grass–white-clover swards are limited, but, in comparisons with fertilizer N at corresponding rates, cattle slurry and dilute urine resulted in greater contents of white clover (Drysdale, 1965; Nesheim *et al.*, 1990). The reasons for this have not been adequately explained but the K content in slurry partly offsets the clover-reducing effect of the N component (Chapman and Heath, 1987) and, while slurry damages leaf cells of clover more than those of grass, clover is capable of recovering rapidly from scorch damage (Wightman and Younie, 1994). Also, in some instances, the fertilizer-N value of slurry may have been overestimated and so the amount of fertilizer N applied in apparently comparable treatments will have a more deleterious effect on white clover than the slurry in these cases. Examination of the response of the grass component to the two forms of fertilizer would indicate if the N rates have been comparable.

The trace elements Fe, Mn, Zn, Cu, Mo, B and Cl are essential for clover growth. Cobalt (Co) is also needed for the survival and multiplication of the N_2-fixing rhizobia in white clover (and other legumes). Many soils have adequate contents of trace elements for white-clover growth, since the quantities required and herbage concentrations are very low. However, leached light-textured soils, other soils low in organic matter (OM) and leached peats have inherently low contents. Trace-element supply is strongly influenced by the application of lime, fertilizers and organic manures. An increase in soil pH from liming often lowers the availability of most trace elements. However, Mo availability is increased, but this, together with an enhancement in available S, renders ingested Cu less available for absorption during rumination by grazing stock (Suttle, 1983). Because of the interacting factors affecting trace element supply and availability, knowledge of soil type and characteristics and soil analysis, including OM content, should be allied to herbage analyses to solve problems that occur. The adverse effects of trace-element deficiencies are often more apparent on the health and performance of grazing animals than on sward growth (Grace, 1983; Suttle, 1983).

Over the years, fertilizers with the major nutrients, N, P and K, whether individually or in compound form, have become more refined and concentrated and so the previous 'bonus' of additional minor or trace elements has been reduced, thus making more consideration of the need for them necessary. It is inadvisable to apply a trace element to the sward in excess of that needed to rectify a deficiency, since toxicity in plants or stock may occur and an excess may be more difficult to treat than a deficiency. A specific element deficiency should be treated by fertilizer application to the sward and/or by supplying stock with mineral supplements, rather than the general application of 'cocktail' mixtures of trace elements to the sward.

Nitrogen loss from grass/white-clover swards

Nitrogen losses to the environment in grazing systems based on grass/white clover, particularly leaching losses, are usually less than those from intensive N-fertilized systems (Cuttle *et al.*, 1992). However, the role of white clover in the N cycle of a grazed sward is complex. High stock-carrying capacity can be maintained on grass/white-clover swards with very low clover content (less than 5%) and low N_2-fixation rates (20–30 kg N ha^{-1}) in a continuous stocking system (Parsons *et al.*, 1991c). Clearly, this is only possible if the fixed N cycles rapidly around the system and N loss is very small (Jarvis *et al.*, 1996). Gaseous losses of N, e.g. by denitrification, are lower in legume systems than in grass systems receiving fertilizer N which produce a similar amount of herbage (Watson *et al.*, 1992). The amount of N excreted per sheep on pure white-clover swards is as high as from individual sheep grazing grass swards receiving 420 kg N ha^{-1} and 45% more than from sheep on a well-balanced grass/clover sward (Orr *et al.*, 1995); however, the lower amount of herbage available for grazing in the clover monoculture than in the high-N grass sward keeps the N-excretion rate in the clover sward lower. A high rate of N excretion results in high rates of volatilization and leaching losses. The grass component in the mixed sward reduces the N concentration of the herbage in the diet consumed, the amount of N excreted is reduced and, since the grass is more able to utilize the excreted N better than white clover on its own, N leaching is minimized.

Weed control

Weed control should not be neglected at any stage in the management of grass/white-clover swards, but clover is probably most vulnerable to the adverse effects of weed competition during early establishment. Many weeds can effectively be controlled by cultural or mechanical measures, especially when dealt with at the correct stage in their life cycle. There are also likely to be advances in biological control in future. Nevertheless, several effective herbicides are available to control the most common weeds (Williams, 1984; Haggar *et al.*, 1990) and to widen the weed spectrum controlled in grassland generally, with mixed herbicide formulations being commonly used (Lockhart *et al.*, 1990).

Adequate cultivations during seed-bed preparation will ensure temporary weed-free conditions and tall-growing annuals can be controlled by topping or suppressed by grazing as the sward is establishing. Some weeds, such as chickweed (*Stellaria media*) may require the application of herbicide, but it is essential to select a 'clover-safe' type and to apply it when sown white-clover and grass seedlings are at a resistant stage of growth – normally one to three trifoliate leaves for white clover, and two to three leaves for grass, depending on the selected herbicide. Typical white-clover-safe herbicides are based on chemicals such as bentazone, benazolin, MCPB or 2,4-DB, and the choice depends on the spectrum of weeds present. It is important to follow codes of practice and official legislation when dealing with herbicides and other pesticides at all stages of handling, from storage through to spraying, and to follow manufacturer's label recommendations closely and accurately for their application.

The weeds to be controlled should be at the most susceptible stage and with adequate leaf area for reception of the herbicide. This set of conditions is not always easy to achieve, especially in undersown cereal crops. Retarded development or death of white clover may ensue if spraying is done at the correct growth stage for the cereal but too early for white clover, while application suited to white-clover growth stage but too late for the cereal can reduce grain yield. A novel approach proposed is to sow white clover in spring, use an appropriate herbicide to control weeds before and after clover emergence, and then direct-drill the companion grasses in late summer (Haggar *et al.*, 1985). If severe weed problems, especially from perennial weeds, are anticipated, another approach is to sow a short-term all-grass mixture, e.g. Italian ryegrass, which can tolerate the use of the necessary range of herbicides, before sowing a grass/white-clover mixture.

Invasion of grass/clover swards by perennial broad-leaved weeds is an ever-present hazard in established swards, and management to maintain a dense, vigorously growing sward is the best means of prevention. However, when a serious infestation occurs, e.g. by docks (*Rumex* spp.), it is necessary to apply effective herbicides, such as mecoprop-, dicamba- or fluroxypyr-based formulations. This will result in the loss of clover and so reintroduction by oversowing or else by complete sward renewal following total sward destruction, e.g. by glyphosate, will be necessary.

It is also possible to control isolated weed plants or small infestations by spot-spraying treatment. Alternatively, the rope-wick technique, which selectively smears weeds standing above the level of the sward with a vegetation-killing chemical, such as glyphosate, may be employed.

Pests

Pest damage to white clover may occur sporadically to frequently, and damage can range from extensive to intensive. Lists of the different pests that affect white clover are available (Burdon, 1983; Manglitz, 1985). The significance of pests in grassland is not always appreciated, because their activities are often covert and the damage insidious (Clements and Henderson, 1983), but attention is now focusing more strongly than before on their effects and how to counter them. Knowledge of pest biology and population dynamics assists in devising strategies, including integration of individual or independent control measures. Biological control measures are also receiving increased attention. In New Zealand, increases in pasture yield of 35% – mainly of white clover – and in N_2-fixation rate of 57% were obtained by high experimental rates of insecticides to control a range of insect and nematode pests, thus indicating their potential damaging effect (Watson *et al.*, 1985). Pesticides applied to control pests and diseases may not only increase white clover yields but also the incidence of beneficial VAM (McEwen *et al.*, 1989). Though technically possible, financial and environmental costs make the use of pesticides unlikely for the control of many of the pests, and developing resistant cultivars is recognized as the best way forward (Pottinger et al., 1993). A potentially effective strategy in breeding programmes is firstly to develop white clover populations resistant to individual pathogens and pests, and then incorporate this resistance into multiple pest-resistant populations (Pederson et al., 1993).

Slugs (*Deroceras reticulatum* and, to a lesser extent, *Arion* spp. in the UK and New Zealand and *Limax* spp. in the USA) are often implicated in damage to white clover, particularly on heavy-textured moist soils during establishment, when clover is most vulnerable to attack (Charlton, 1979). These nocturnally active molluscs attack clover seeds as soon as the seeds imbibe water to commence germination and hollow out the succulent interior of the seed, using their very efficient rasp-like radulae (teeth), and this can sometimes cause a complete failure of the sown clover. A finely cultivated seed-bed tilth, but subsequently well consolidated, not only aids white-clover establishment but restricts the activities of the slugs, partly by removing their cover, but also by checking the population, albeit temporarily.

Slug populations in pasture vary considerably during the year, and also depend on the type of pasture. They are usually at their population peak during spring in countries with mild, moist winter months, and can still be active when swards are sown down in early autumn (South, 1975). When pasture renewal with white clover-based mixtures is undertaken during or after warm and moist seasons, slug depredation of the establishing sward can be severe.

The use of an approved chemical protectant, such as methiocarb, against slugs may be necessary, particularly in wet seasons, and is essential when direct-drilling white clover into an existing grass sward, since the slugs find refuge in the slots during daytime and then feed on the seedling clovers in the slots at night. Certain legume species are more attractive to slugs that others, but white clover is one of the most appealing. Some clover cultivars known to attract slugs more than others, and therefore having a high susceptibility to damage (Ferguson et al., 1989), are usually identified as such on recommended lists. Chemical-composition differences probably play a part. The production of hydrocyanic acid following the injury of plant tissues containing a cyanogenic glucoside (cyanogenesis) is a trait of white clover, but glucoside content varies among cultivars and acyanogenic types of clover are preferred by slugs, snails, weevils and other herbivorous-feeding insects (Mowat and Shakeel, 1989; Kakes, 1990).

Other invertebrates, such as leatherjackets, the larvae of the crane fly (*Tipula* spp.), are also attracted to clover cultivars with the lowest cyanogenic potential. The leatherjacket larvae feed on and damage white-clover roots in the spring, particularly during cool moist conditions. Populations fluctuate markedly from year to year, but effective means are available to predict the onset of leatherjacket infestation and likely damage, and so control measures, e.g. by chlorpyrifos or other approved chemicals, can be implemented.

Large numbers of nematodes (eelworms) exist in the soil, mainly causing injury to plant roots, but sometimes, as with stem eelworm (*Ditylenchus dipsaci*), living within legume plant tissues and causing plant stunting, distortion and drying out (Cook and York, 1979). Slugs also act as a vector of stem eelworm from infected to healthy plants (Cook *et al.*, 1989). The clover-cyst nematode is widespread in soils used for agriculture in many countries, and it has been implicated in poor clover establishment or reduced clover production and persistence in established grass/clover swards. Clover-cyst nematode infection of white-clover seedlings is rapid, with heavy infestation being recorded within 3 weeks of sowing (Yeates *et al.*, 1975). In some cases, other nematodes, such as the root-knot nematode and free-living nematodes (*Pratylenchus* spp.), are associated with debilitated plant growth, due to root damage, either independently or in conjunction with the clover-cyst nematode. Root-knot nematodes cause substantial damage in southern USA. Breeding for resistance is a major objective in a number of breeding programmes in different countries (Mercer *et al.*, 1990).

Weevils (*Sitona* spp.) are a major pest of white clover (Clements, 1994). Notching of clover leaf margins is a typical symptom of damage by adult weevils and, while established clover plants may tolerate attack, establishing seedlings may be severely damaged, particularly since the larvae feed on the roots and root nodules. Control measures are rarely adopted, but there are effective pesticides, such as chlorpyrifos, which control the larvae (Mowat and Shakeel, 1988). Other weevil pests include clover-leaf (*Hypera punctata*) and alfalfa weevils (*Hypera postica*). The lucerne flea (*Sminthurus viridis*), a serious pest in warm climates, could be a potential problem for white clover in temperate regions in fine, sunny summers.

In many parts of New Zealand, high infestations of the grass grub (*Costelytra zealandica*) feed on plant roots, causing loss of clover plants and reducing herbage production and ground cover. Biological methods of control, whether by predators, parasitoids or pathogens, are being evolved for this and other pests (Chapman, 1990). Alternative grasses, such as cocksfoot, tall fescue and phalaris (*Phalaris aquatica*), are more tolerant of grass grub than the traditional perennial ryegrass and are used where the pest is severe. *Phalaris*, in particular, is toxic to grass grub and, sown in mixture with susceptible ryegrass, enhances the performance and persistence of the renewed pasture (Moloney, 1990). Porina (*Wiseana cervinata*), a lepidopterous root-feeding pest, also causes serious damage to hill-land pastures, denuding them of white clover (Barlow *et al.*, 1986). In such pastures, greater lotus (*L. uliginosus*), being tolerant of the pest, is included in seed mixtures as an insurance legume against complete clover loss during severe attacks and to reduce populations of porina, since the lotus roots are toxic to the pest (Sutherland, 1975).

In southern Australia, the red-headed cockchafer (*Adoryphorus couloni*), has similar effects on white-clover-based pastures. The red-legged earth mite (*Halotydeus destructor*) is another notable pest there, and screening of clover varieties and ecotypes for resistance is part of a national breeding programme for white-clover improvement. In the USA, a number of weevils (*Hypera* spp. and *Tychius picirostris*) feed on clover leaves, flower-heads or developing seed (Pederson, 1995).

Diseases

The study of diseases and their effects on clover performance has been limited, relative to other damaging factors. Nevertheless, disease is a clinical, and sometimes subclinical, cause of impaired performance and frequently there is no economic method of chemical control. Sometimes, several fungal diseases may occur simultaneously, and disease complexes with other pathogens and pests may occur (Nelson and Campbell, 1993). The main diseases of white clover have been reviewed (Leath, 1985, 1989). In parts of Europe and the USA, clover rot (*Sclerotinia trifoliorum*) is a serious disease that results in reduced herbage production and plant persistence. Infection by the fungus, which may be present in the soil and/or seed, occurs mainly in autumn. It spreads by wind-borne spores, produced from sclerotia formed in the plant tissues. The effects become obvious in the spring and include plant wilting, death and rotting of stolons and roots. Chemical control by propiconazole has proved effective (McGimpsey and Mercer, 1984), but choosing a resistant cultivar and, in the long term, breeding for further resistance are more feasible options.

In what is referred to as a root-rot complex, caused by the interactive effects of several pathogenic fungi, there is progressive necrotic breakdown of clover roots and stolons. *Fusarium* spp. are notably implicated, although mainly as secondary pathogens. Various forms of stress – environmental, managerial or pest attack – are believed to be the initial predisposing causes of root deterioration by a range of soil-inhabiting fungi, and so amelioration of the stress factors, where possible, is the main way of combating the disease.

Several foliar and stem fungal diseases attack and damage clover in a number of different ways. Attacks are often most severe in humid or wet climatic conditions. Apart from visible plant damage, the physiological performance of the host plant is adversely affected (Skipp and Lambert, 1984). Typical symptoms include lack of plant vigour, stunting, leaf yellowing, leaf fall and stem damage or death. There is a lack of quantification of the adverse effects, since much of the information about the diseases has come from surveys using incidence and severity criteria, rather than from experimental treatments, with and without fungicide application, on disease outbreaks (Clements, 1994). Clover varieties vary in susceptibility to fungal diseases, but the information available is not yet comprehensive (Lewis and Asteraki, 1987). Breeding programmes for resistance against certain foliar diseases have enjoyed some success (Gibson and Cope, 1985), but, because of the complex nature of some of the diseases, progress has been slow.

Both black blotch (syn. sooty blotch) (*Cymodothea trifolii*) and leaf spot (*Pseudopeziza trifolii*) result in leaf loss and reduced herbage production. Nutritional quality is diminished and the diseased leaves have increased content of oestrogenic coumestans, which, if ingested in large quantities, reduce reproductive performance in breeding sheep (Wong et al., 1971). Pepper spot (syn. burn) (*Leptosphaerulina trifolii*) causes leaf spotting, shrivelling of tissue and decreased production. Leaf diseases that are less common (although localized flare-ups may occur) include downy mildew (*Peronospora trifoliorum*), powdery mildew (*Erysiphe trifolii*) and rusts (*Uromyces* spp.). Other less common stem diseases are spring black stem (*Ascochyta imperfecta*) and summer black stem (*Cercospora zebrina*).

A number of viruses infest white clover and multiple infection by several viruses is a common occurrence (Barnett and Diachun, 1985; McLaughlin *et al.*, 1992). Aphids are the primary vectors of virus transmission and alternative host plants include other herbage and crop legumes, but the netted slug (*D. reticulatum*) can also transmit white-clover mosaic virus from infected to healthy plants (Cook *et al.*, 1989). Damage to the plants ranges from subclinical to severe, affecting herbage seed production and plant persistency to varying degrees. For example, 30% less production can result from combined infection of clover mosaic and clover yellow-vein viruses, compared with healthy white clover (Scott, 1982). A review of the adverse effects of five viruses (alfalfa mosaic, bean yellow mosaic, clover yellow vein, peanut stunt and white clover mosaic) noted maximum reductions of 35–59% for DM yield, 22–72% for nodule number and 37–38% for nitrogenase activity (Forster *et al.*, 1997); a maximum seed-yield loss of 90% was recorded due to the alfalfa mosaic virus.

Phyllody, witches' broom, and clover red leaf (English stolbus) are three white-clover diseases of the 'virus yellows' type caused by mycoplasmas (*Mycoplasma* spp.) rather than viruses, the main agents of transmission being leafhoppers (e.g. *Macrosteles* spp.). Phyllody infection results in poor rhizobial nodulation, lack of growth vigour, susceptibility to winter cold and failure of seed crops to set seed, while the other two mycoplasmas cause morphological abnormalities in the plants (O'Rourke, 1976; Leath, 1985). Plant breeding, with an emphasis on genetic means of interspecific or intergeneric hybridization and genetic manipulation, offers the best means of achieving resistance to virus and mycoplasma infections, although their low incidence and a lack of knowledge of their biology mean resistance is not yet a priority aim of breeders.

Herbage production

A potential DM production of between 18.5 and 22.5 t ha^{-1} from perennial-ryegrass/white-clover swards in the UK has been estimated, based on a 50 : 50 mixture of grass and white clover, with each performing to half its theoretical monoculture potential, and with optimum overall management (Frame and Newbould, 1984). In comparison, estimates of the potential DM production from monoculture grass swards in Western Europe lie within the range 27–30 t ha^{-1} (Alberda, 1971; Cooper and Breese, 1971). Calculations of the potential harvestable DM production from monoculture white-clover swards in the UK suggest around 16 t ha^{-1}, but, in experiments, 12 t ha^{-1} has been achieved under irrigation and 8–10 t ha^{-1} from unirrigated swards (Frame and Newbould, 1984).

For perennial-ryegrass/white-clover swards in New Zealand, an annual DM production ceiling of 22–28 t ha^{-1}, set by genotype and climate, is quoted, compared with production from white-clover monocultures of 10 t ha^{-1} (Smetham, 1973) and 10.6 t ha^{-1} (Suckling, 1960). Other calculations predicted annual herbage DM production of 24.7 t ha^{-1} from irrigated ryegrass/white clover (Brougham, 1959).

Under a cutting regime to simulate grazing, the highest recorded annual DM yield achieved in the UK from a perennial-ryegrass/white-clover sward receiving no N fertilizer was 15.5 t ha^{-1} (Frame and Newbould, 1984). However, best levels achieved from cutting experiments, whether in simulation of grazing, by frequent cutting, or of silage, by infrequent cutting, are usually in the range 7–11 t ha^{-1}. If clover pests or diseases are present, production levels can be substantially lower. Similar yield levels (6.2–11.6 t ha^{-1}) were recorded in Atlantic Canada (Fraser and Kunelius, 1995).

An annual DM production range of 6.7–14.9 t ha^{-1} over nine sites throughout New Zealand has been recorded, with a mean production of 8.4 t ha^{-1} total herbage and a white-clover content averaging 37% (Hoglund et al., 1979). Annual DM yields from a 23-year-long cutting trial under dryland conditions ranged from 4 to 10 t ha^{-1}, with clover contents rarely above 5%, when unirrigated and 8–13 t ha^{-1}, with clover contents mainly between the 20 and 30% levels, when irrigated (Rickard and McBride, 1986). From evaluations under grazing, using small paddocks, the highest production levels achieved at research stations over a range of environments were 12 t ha^{-1} for hill sites, 16.2 t ha^{-1} for unirrigated dryland and 22.8 t ha^{-1} for more favourable lowland sites, these estimates being a third higher than the quantities of herbage utilized in farming-system studies (Brougham, 1977). More general estimates indicate 10–20 t ha^{-1} on fertile land, the actual level depending on rainfall and temperature, while on unimproved or reverted hill land, dominated by bent grass (*Agrostis capillaris*) and with little white-clover content, the yield range was 3–5 t ha^{-1}, in comparison with 5–10 t ha^{-1} on poorly developed lowland farms or well-developed hill farms, with varying proportions of good and poor grass species (Coop, 1986).

It is well documented that close mowing (25–75 mm above ground level) of grass/white-clover (and grass-only) swards increases herbage production, compared with less severe or lax defoliation (75–150 mm), provided that adequate recovery periods are allowed between defoliations (Frame and Boyd, 1987a). Detailed studies show that a reduction in cutting height (to as low as 20 mm) reduces the length, width and weight of clover leaflets, petiole length, stolon diameter, the length of stolon internodes and the height of stolons above ground level, but increases the length of stolon m^{-2}, the number of stolon nodes m^{-2} and the proportion of stolon nodes that produced branches (Wilman and Acuña, 1993); the latter factors outweigh the former, leading to an inverse relationship between cutting height and the proportion of clover in the sward (Acuña and Wilman, 1993).

Cutting studies show that total herbage production from mixed swards generally increases as the interval between defoliations is lengthened, but the proportion of white clover in the total herbage may decrease or be little affected (Frame and Newbould, 1986). Lengthening the interval between defoliations reduces the density of grass tillers, since tiller development from axillary basal buds is temporarily suppressed, improving light conditions at the base of the sward after cutting, and white-clover petiole length, leaf weight, stolon length and stolon diameter increase (Wilman and Asiegbu, 1982).

There is a dearth of production data from grass/white-clover swards in farm situations, but the variability in production is undoubtedly greater than from experimental swards, where more precise management control is exercised. A survey of lowland permanent grassland in England and Wales showed that clover content was very low (Forbes *et al.*, 1980) and so total herbage production would also be low, unless adequate rates of fertilizer N were applied to the mainly grass swards. For example, white clover averaged 2% white-clover ground cover on permanent grassland on dairy farms, with only 10% of the farms having enough white clover – interpreted as over 5% ground cover – to make an impact on herbage production, compared with 25% on beef-cattle fattening farms and 30% on suckler-cow farms. Surveys of upland permanent grassland also reported low clover presence (McAdam, 1983; Hopkins *et al.*, 1988). Even in New Zealand, with its reliance on white clover for the N economy of pastures, its contribution to pasture yield could be greatly improved, since estimates showed 10–15% on flat and rolling land and about 5% on hill land (Lancashire, 1990).

It is noteworthy that clover monocultures are not very responsive to application of fertilizer N and, for example, yields of 8.6 t ha^{-1} (nil N) and 9.2 t ha^{-1} (mean of five annual N application rates from 150 to 750 kg ha^{-1}), i.e. an increase of 7%, have been recorded (Reid, 1983). As the respiratory costs of N_2 fixation are considerably higher than those of NO_3^- uptake and reduction within the plant, it would be expected that N-fertilized white-clover monocultures would yield 10–12% higher than corresponding monocultures relying on N_2 fixation. The lower difference than expected may be due to the harvested yield being only part of the total biomass, which would include stubble and roots.

Grazing

Grass/clover swards may be utilized successfully by a range of grazing systems, from some form of intermittent (rotational) to continuous (set) stocking or a blend of both in the same season. In Atlantic Canada, white clover is proving a suitable component of swards in which the herbage is allowed to stockpile from late summer for use in late autumn, thus extending the grazing season (Fraser *et al.*, 1993; Kunelius and Narasimhalu, 1993). Experience has shown that white clover is a flexible enough plant to thrive under most systems of defoliation. However, the major influences within the grazing process include not only defoliation, but also trampling and excretal return. The effects of these factors are interdependent and, at any given time, one may be overriding. In addition, each factor may have beneficial or detrimental effects on white clover and on the other sward components.

Frequency, intensity and timing of grazing are all factors that can induce botanical change in swards. Grazing hard with sheep for short periods in winter or early spring increases the white-clover content of swards, compared with lenient grazing, due to improved growing-point density and hence white-clover growth and development (Laidlaw and Stewart, 1987; Laidlaw et al., 1992). This is brought about by a higher irradiance reaching the base of the sward and, in addition, a higher R :F R ratio of the light spectrum (Thompson and Harper, 1988).

Overgrazing in late winter/early spring is detrimental to white-clover persistence, since too much photosynthetic tissue is removed and, if low temperatures prevail, winter kill of stolons may occur. Conversely, white clover is basically at a disadvantage compared with grass if the sward is left ungrazed over the winter/early-spring period, since the white-clover petioles are shorter than the grass and the clover laminae are lower in the sward canopy, because of reduction in maximum petiole length, brought about by low temperatures (Marriott et al., 1988; Woledge et al., 1990). In the summer, the bulk of the horizontally placed white-clover foliage is present in the upper layers of the sward canopy and able to intercept light efficiently. At the same time, white-clover leaves are highly accessible to grazing animals, and the proportion of white clover consumed from the mixed sward is higher than the total herbage on offer (Curll et al., 1985). This is particularly so with sheep, which graze more fastidiously than cattle and selectively graze the white clover (Ridout and Robson, 1991). However, there is a running debate about the constancy of this preference, since sheep may also prefer the opposite species to the one grazed before when offered a choice, and the dynamics of mixed-sward growth may lead to sheep actively selecting whichever species is most actively growing at a position in time (Newman et al., 1992). In a grazing-preference study, sheep on the first day chose a diet contrasting with the clover content of the diet fed prior to the opportunity to select but thereafter chose a diet tending to the clover content of the previous diet, although all sheep selected a diet with a high clover content, irrespective of original diet (Parsons et al., 1994).

Prolonged periods of undergrazing in summer, due to lower stocking rates than the sward is capable of sustaining, militate against white-clover performance, mainly because of increased shading, which decreases the density of white-clover growing points at ground level and reduces stolon branching (Dennis and Woledge, 1987). Although, when grazed, usually a whole lamina (three leaflets) of a leaf is removed – unlike grass, in which a leaf is usually only partially removed – clover has the ability to invest proportionately more photosynthate into production of leaf area than grass (Parsons et al., 1991a,b). Laminae of small-leaved genotypes tend to be lower in the canopy than those of larger types in continuously stocked swards grazed to the same surface height. Consequently, more leaves per stolon remain ungrazed in the smaller-leaved types, which are the most suitable for sheep grazing (Davies and Jones, 1987).

Grazing may result in the removal of stolon apices, in addition to leaves, especially in large-leaved types, in which adventitious roots are less effective in anchoring the large stolons to the soil surface. While removal of secondary stolon apices stimulates branching in undefoliated stolons, removal of stolon apices in combination with defoliation results in a net reduction in apices (Brink, 1996).

Compared with cutting, the clover contribution in mixed swards is reduced by grazing, and more so by sheep than cattle grazing (Garwood et al., 1982; Evans et al., 1992). Relative to other methods of defoliation, white-clover plants under sheep grazing have fewer, thinner stolons, shorter stolon internodes and petioles and smaller leaves, effects which an increasing intensity of defoliation exacerbates. Following a respite from sheep grazing, the sizes of the above-ground parts of white clover recover to dimensions similar to those of white clover that was cattle-grazed or cut, because of clover's inherent phenotypic plasticity (Briseño de la Hoz and Wilman, 1981). Differences in grazing behaviour between sheep and cattle can thus be exploited to manipulate clover content in swards. Interestingly, goats prefer coarse or stemmy grasses and weeds in a sward and so offer a means of enhancing clover content (Grant et al., 1984). Grass/white-clover swards goat-grazed in the first half of the season resulted in higher clover contents than when sheep-grazed, and this subsequently resulted in higher herbage intakes and liveweight gains in weaned lambs later in the season (del Pozo et al., 1996).

There is evidence that continuous sheep stocking militates against white-clover persistence and production, compared with the rotational-grazing systems (Widdup and Turner, 1983; Curll et al., 1985), but white-clover performance benefits by switching from high to low stocking rates (Curll, 1982) or inserting a rest interval and subsequent silage cut (Curll and Wilkins, 1985). The adverse effect of uninterrupted continuous stocking by sheep on white-clover persistence is in contrast to continuously stocked cattle-grazed swards, in which year-to-year variation is usually relatively small, unless the initial content of clover is high (Gibb et al., 1989).

Cattle-grazed swards continuously stocked to give a range of sward surface heights (SSH) do not differ markedly in clover content (Gibb and Baker, 1989) and, even when imposed over 4 years, there was only a slight trend for clover content to decline at all heights over that period (Laidlaw et al., 1995b). As swards grazed with cattle to different SSH have a high proportion of their foliage at a common surface height, the differences in mean height between the swards are due mainly to different proportions of the area being rejected, those areas that are grazed to a common height buffering the mean SSH effects (Teuber and Laidlaw, 1995). In contrast, swards continuously stocked with sheep may suffer a progressive year-to-year decline in clover content (Orr *et al.*, 1990), explained possibly by selective grazing of clover by sheep, especially in swards that are set at a high target SSH (Milne *et al.*, 1982). Also, sheep-grazed swards do not have obvious rejected areas and so clover may not have an opportunity to recover from repeated defoliation, whereas in cattle-grazed swards the rejected areas are gradually brought back into the grazed area, provided stocking rate is well controlled.

Generally, continuously stocked swards do not have high clover contents in their herbage mass, i.e. in the herbage sampled to ground level, but the content of clover in the diet has been measured for sheep (Curll, 1982; Clark and Harris, 1985) or calculated for cattle (Steen and Laidlaw, 1995) and found to be consistently higher than the content in the herbage mass; for example, at sward contents of about 10%, the content in the diet could be double that value or more. The elevated content of clover in the diet is due to the disproportionate amount of clover in the grazed horizon, relative to the total sward, and so the content of clover in the diet can be high, even if there is no active selection for clover.

Continuous stocking in spring increases the number of white-clover growing points, compared with rotational grazing, and the growth potential of these points may be realized better later in the season by a switch to rotational grazing, with its built-in rest periods (Hay and Baxter, 1984). Nevertheless, sustained satisfactory performance from white clover is achievable from continuously sheep-stocked swards, provided suitable small-leaved clover cultivars are blended with compatible grasses and grazing pressure is controlled to maintain specific SSH (Davies *et al.*, 1991; Gooding *et al.*, 1996). It may be concluded that flexibility of defoliation, with the aim of favouring the needs of the clover, is a major key to sustaining its satisfactory performance in mixed swards.

Trampling damages sward plants physically, particularly on finely textured, poorly structured clay and silt soils prone to poaching in wet conditions. Under severe poaching, white-clover plant parts may be destroyed, displaced or buried, although some of the buried stolon fragments may subsequently generate new clover plants (Hay, 1983) and clover has a capacity to recover from trampling damage (Vertès *et al.*, 1988). Trampling may also affect all sward constituents indirectly, through its adverse effects on the soil/root environment, soil compaction, reduced water infiltration and restricted uptake of soil nutrients by plant roots. White clover is more susceptible to trampling damage than grasses such as perennial or hybrid ryegrass, although more resistant than timothy or cocksfoot (Edmond, 1964). The joint effect of severe defoliation and trampling under high stocking with sheep has a greater adverse effect on clover content in the sward than trampling *per se* (Curll and Wilkins, 1983).

Excretal return has a strong influence on the composition of a grazed sward, mainly through nutrient cycling of N, P, K and S. In particular, N is involved, since between 75 and 90% of the N ingested is excreted. About three-quarters of the excretal N is in the urine, and N deposition in urine patches is equivalent to 300–600 kg N ha^{-1}. This N is readily available and, although a substantial proportion is lost by volatilization and leaching, the remaining N gives the grass a competitive advantage over the clover in the patches.

Urine application to a mixed sward in spring or summer shows a marked decline in the population of white-clover growing points, stolon survival and the rate of production of new branch stolons; however, these adverse effects do not occur following urine application in autumn, because both grass and white clover grow slowly then, neither seizing a competitive advantage (Marriott *et al.*, 1987, 1991).

In dry weather, both grass and white clover foliage may suffer from urine burning or scorch in patches, but white clover is more sensitive. White clover may also be adversely affected by the free ammonia from volatilization (Curll, 1982) and from the temporary increased pH of the urine-affected soil (Thomas *et al.*, 1986).

In comparisons of cutting and sheep-grazing treatments, clover contribution is markedly reduced under grazing, partly because of the boost to grass competitiveness, since the N effect overrides the benefit to clover of the urinary and faecal P and K (Frame, 1976). The excretal-return pattern from sheep, though patchy, is more evenly distributed than from cattle grazing, and may also play a part in causing more white-clover depression under sheep, relative to cattle grazing.

Choice of cultivar is important for maintaining white clover in a grazed sward. Under set stocking conditions, small-leaved white clover trades off leaf size against leaf number and stolon production, the latter being important in persistence, whereas large-leaved types are less able to adjust to the demands of regular and frequent defoliation by sheep (Table 2.7). Therefore, small-leaved types display a degree of plasticity which, although producing lower DM yields than large-leaved types in rotationally grazed swards, allows them

Table 2.7. Mean performance of two contrasting leaf-sized white clovers under rotational grazing and set stocking with sheep (mean of 5 years) (from Brock and Hay, 1996).

	Rotational grazing		Set stocking	
	Small-leaved cv. Tahora	Large-leaved cv. Kopu	Small-leaved cv. Tahora	Large-leaved cv. Kopu
Area of leaf (cm^2)	1.27	3.05	0.51	1.00
Leaves (m^{-2})	7320	3390	19,270	3610
Growing points (m^{-2})	3040	1400	11,000	2140
Leaf dry weight (g m^{-2})	30.9	30.9	46.1	11.6
LAI	0.84	1.00	0.92	0.35
Stolon dry weight (g m^{-2})	34.5	24.2	69.4	17.8
Leaf : stolon ratio	0.97	1.33	0.68	0.67

Conservation

Grass/white-clover swards are capable of sustained use for conservation, e.g. silage cropping, giving c. 80% of the yields from grass swards receiving 350 kg N ha^{-1} annually, while still maintaining annual clover contents of 20–40% (Roberts et al., 1989; Bax and Thomas, 1992); first-cut yields were similar for both types of sward, averaging 4.5 t DM ha^{-1}. A conservation cut can also be integrated into a mainly grazed situation with benefit to clover, and late summer, when clover is growing most vigorously, is a more advantageous time to introduce a silage cut than early summer (Gooding et al., 1996). Experience from cutting trials also suggests that clover-rich swards can be exploited for cut-forage (zero-grazing) systems. Flexibility of use for grazing and/or cutting is a key factor in the acceptance of grass/white-clover swards in farming practice hitherto orientated towards systems based on N-fertilized grass. Clover's shorter growing season than grass, due to its temperature requirements, is not a disadvantage when used for mainly conservation purposes.

A number of changes within the sward occur as the time interval between defoliations is increased from a frequent to an infrequent regime. Total herbage production from cut mixed swards increases, the production of white clover being boosted particularly in early season, but the annual proportion of white clover in the total herbage may decrease or be little affected (Frame and Newbould, 1986). It is recognized that close cutting of swards confers a production advantage, particularly with respect to the structure of the grass component, with its high proportion of growth in lower layers of the canopy. In the case of the clover component, an acknowledged benefit is the improved light conditions at the base of the sward, leading to increased stolon branching and number of clover growing points per unit area and promoting the development of photosynthetically active leaves (Dennis and Woledge, 1987). In practice, it is often necessary to forego the production advantage from cutting swards closely for conservation and concomitantly reduce the benefit to clover, since sward damage by scalping and undesirable soil contamination of the harvested herbage occur from close cutting if the field surface is irregular.

Timing of the conservation cut influences the scope of the benefit to white clover. Progressively later rest intervals during the season result in progressively lower grass densities, which coincide with peak clover-stolon development (Laidlaw and Vertès, 1993). The early-season flush of grass tillers causes losses in stolon weight and hence the breaking up of complex branching systems into smaller and simpler units (Brock et al., 1988), with shading of clover and cooler temperatures for growth also playing a part (Barthram and Grant, 1995). The benefit to clover performance from later- rather than early-season rest periods, prior to cutting for conservation, is greater for early- than for late-maturing ryegrasses and for tetraploid than for diploid ryegrasses, while, within clover leaf-size categories, the benefit increases with increasing leaf size (Gooding et al., 1996).

Taking silage or hay crops removes considerable amounts of N from the available soil supply and, for example, a first silage cut of 4 t DM ha^{-1} at 300 g N kg^{-1} DM removes 120 kg N ha^{-1}, while two crops remove c. 200 kg N ha^{-1}. Such removals of N benefit white-clover growth and development, since grass competition is reduced (Frame and Paterson, 1987).

White-clover monocultures have been assessed for field-scale silage making, although not over the long term. Annual DM yields of 6–8 t ha^{-1} were obtained from young established monocultures, and, using good silage-making techniques, e.g. wilting the crop, short chopping and use of an effective additive, high-quality silage for milk production was produced (Castle et al., 1983). These DM yields were below the highest levels cited from plot experiments – 10.3 t DM ha^{-1} from unirrigated monocultures and 11.8 t ha^{-1} from irrigated monocultures in the UK and 10.6 t ha^{-1} from unirrigated swards in New Zealand (Frame and Newbould, 1986); however, account has to be taken of the fact that most of these yields included invasive grasses, though to a greater or lesser degree in some cases, since carbetamide was applied in late winter in order to reduce grass ingress.

Herbage Quality

The chemical composition of white-clover herbage, presented in Tables 2.1 and 2.8, is usually higher in contents of pectin, N, Ca, Mg and most of the minor or trace elements, lower in DM, cellulose, hemicellulose, water-soluble carbohydrates, lignin, Na and Mn, and similar in lipids, P, K, S and Zn, compared with grasses. Stage of growth has a strong influence on the chemical composition of forage legumes generally, particularly in upright-growing plants, such as lucerne and red clover, where the leaf : stem ratio decreases with advancing maturity. However, in white clover there is a continual generation of new leaves and petioles from the stolon network, concurrent with advancing maturity of previously generated foliage, and this minimizes compositional changes. Apart from stage of growth, the soil pH and nutrient status, as influenced by fertilizer application, and the season of year, all these

Table 2.8. Chemical composition of white clover and perennial ryegrass harvested at similar digestibilities throughout the growing season (from Thomson et al., 1985).

	White clover	Perennial ryegrass
Nitrogen	44	28
Total cell content	216	427
Cellulose	173	240
Hemicellulose	8	161
Lignin	38	27
Pectin	40	8

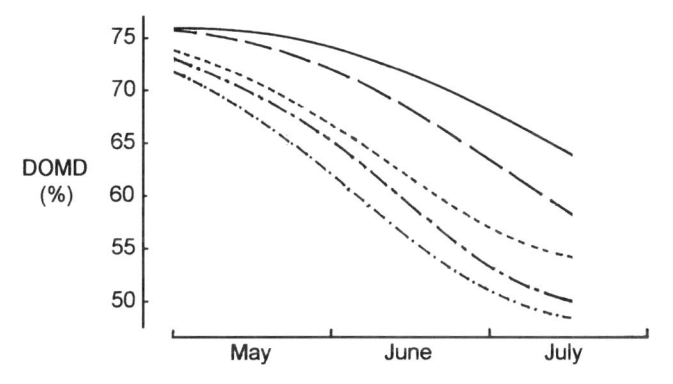

Fig. 2.7. Effect of increasing maturity on the digestibility of primary growths of white clover (——) in relation to other forage legumes, i.e. late red clover (— —), early red clover (------), sainfoin (— - —) and lucerne (–·–··–). Digestibility is expressed as per cent digestible organic matter in the dry matter, i.e. DOMD syn. D value (from Harkess, 1970).

factors influence the chemical composition of white clover and, of course, other sward components.

In the assessment of the quality of herbage on offer to dairy cows continuously stocked over the grazing season, perennial-ryegrass/white-clover swards had consistently higher OM digestibility (OMD) and P, K and Ca concentrations but generally lower N and Mg contents than the herbage from ryegrass swards receiving 350 kg N ha^{-1} annually (Frame et al., 1992); the superiority in OMD ranged from 1.7 to 5.8 percentage units, with the lower values occurring in summer. The digestibility of white clover is higher than that of other temperate forage legumes, and in primary growths the decline in digestibility with increasing maturity is less marked (Fig. 2.7).

When white-clover and grass herbage are compared at equivalent degrees of maturity, white clover has higher cell contents, especially of the more digestible nitrogenous and other soluble compounds, and lower cell-wall contents, particularly of the less digestible hemicellulose fraction (Table 2.8). However, the digestibilities of clover flower-heads and peduncles (flower-head stalks) are lower than those of grass leaves and stems – and also of clover leaves and their petioles – which may explain why mixed grass/clover herbage may not be superior to grass herbage in grazed swards in summer (Wilman and Altimimi, 1984; Søegaard, 1994).

Since white clover is invariably grown in association with grasses, the question arises as to the target optimal content required to make an impact on animal nutrition. The precise target has not been identified, but several researchers have suggested c. 30% of the total herbage DM (Curll, 1982; Stewart, 1984); however, even with attempted manipulation of management, the white-clover content is variable over the season, rarely attaining 30% in spring but sometimes well above this value later in the season. The need for such high contents is being increasingly queried, and good animal performance has been obtained from grazed swards with much lower contents (Orr *et al.*, 1990; Fothergill and Davies, 1993). Even in New Zealand, where the N economy of pastures relies almost entirely on white-clover N_2 fixation, white-clover contents vary markedly in farm swards, depending upon environmental and management conditions; however, 20–30% or more is common in grazing experiments at research centres (Brougham, 1977; Lancashire, 1990). High contents of clover are typical of cutting experiments, whether on a multitreatment small-plot scale (Frame and Paterson, 1987) or on a field scale (Roberts *et al.*, 1989).

Antiquality factors

In countries where white clover and other legumes, such as lucerne and red clover, play an important role in grazed swards, bloat (tympanites) is a hazard. Similarly, in Western Europe, for example, where there is increasing interest in utilizing legume-based swards, bloat is perceived by farmers as a major risk if such pastures are proposed or adopted. The complex of plant and animal factors inducing it are not yet fully understood, in spite of considerable research in the past, and yet, ironically, there is little or no current research. It is encouraging that bloat has not been a serious problem in Europe in several system trials using grass/white-clover swards and beef cattle or dairy cows (Stewart and Haycock, 1984; Bax and Schils, 1993) or in development projects on commercial dairy farms (Bax and Browne, 1995; Frankow-Lindberg *et al.*, 1996).

Bloat is a significant risk in cattle and, to a lesser extent, in sheep. In New Zealand grassland farming, dairy-cow mortality can exceed 0.8% in some years (Carruthers *et al.*, 1987). The main system of grazing there is rotational, and cattle may go from grazed-down swards to lush clover-rich swards. A technique used to minimize any bloat problems in north-west France is to extend the rotation to 4–6 weeks, in order to offer more mature herbage, rather than 3–4 weeks, although the shorter rotation is used at times of rapid spring growth (Pflimlin, 1993). Set stocking is another beneficial approach, since there is less variation in the amount of herbage available than in rotational grazing, provided grazing pressure and SSH are controlled.

Several other preventive and control measures are available, some of which are more effective than others. Provision of a limited quantity of hay or straw usually reduces the risk, by supplying additional fibre in the diet. A number of specific prophylactic measures have been listed (Essig, 1985; Carruthers *et al.*, 1987), but one of the most effective is a rumen antifoaming agent, such as the non-ionic surfactant, poloxalene. It can be incorporated into the drinking-water, block licks or a concentrate supplement fed in the parlour during milking, if appropriate to the system. Lipids, including animal fats and vegetable or mineral oils, sprayed on to pastures or used to drench cattle routinely, have also proved effective.

The cyanogenic potential of some white-clover cultivars can be a cause of some concern, since hydrocyanic acid (HCN) is metabolized within the grazing animal to inorganic thiocyanate, which is goitrogenic. The concern varies among countries and, in Switzerland, notably, the selection criteria for clover-cultivar entry to the national recommended list includes content of cyanogenic glycosides as HCN, in addition to production, competitiveness and disease resistance. If the mean annual HCN content of a cultivar is above 370 mg HCN kg^{-1} OM, the concentration in the standard cv. Milkanova, the cultivar is rejected (Lehmann et al, 1991). There is a positive association between increasing leaf size and cyanogenesis levels for most of the New Zealand white-clover cultivars (Caradus et al., 1996). North American cultivars have notably lower HCN potential (HCNp) than most European cultivars, although there are considerable differences among cultivars bred within countries (Wheeler and Vickery, 1989). Cultivars bred in the UK are generally found in the group with so-called high content (Mowat and Shakeel, 1989; Lehmann et al., 1991). Over a range of 15 white-clover cultivars, HCNp was not highly correlated with herbage acceptability to grazing sheep, leading to the conclusion that plant breeding should aim for cultivars with moderate to low HCNp to retain pest protection but minimize goitrogenic potential (Hill et al., 1995).

The HCNp of white clover is increased by moisture stress, low light intensity, cool grazing conditions and low soil-P supply (Vickery et al., 1987). Differing response of cultivars to these factors and differences in experimental technique probably explain the lack of agreement in the relative ranking of cultivars in lists compiled according to their HCNp. The age and type of the plant tissues sampled also have a bearing, since HCNp decreases as the plant tissues mature and is higher in the leaves than in other plant parts.

Improving range lands

In many countries of the world, there are regions or areas of range lands with unimproved indigenous or seminatural pasture which are extensively managed. Selected parts are suitable for pasture improvement, using a combination of inputs and techniques, such as liming, fencing, and the introduction of bred grasses and legumes, following different degrees of soil/sward preparation. White clover, with its N_2-fixing ability and high intake/feeding-value characteristics, is the cornerstone of many of the most significant improvement packages, e.g. in Scotland (Frame et al., 1985; Frame and Tiley, 1993) or in New Zealand (White, 1990).

The 6 Mha of grassland in the UK classified as 'rough grazing' and equivalent in many ways to open range land will serve as an example. These rough grazings are chiefly associated with the lower slopes of mountainous areas. Hill grazings mainly at 250–1000 m above sea level (a.s.l.), are used for sheep farming, with only 10% or so of the grazings fenced and improved or improvable by some form of cultivation and seeding. Upland farms, mainly at 100–250 m a.s.l., generally have suckler beef herds as the main enterprise and upwards of 50% of the land may be improved or improvable. Upgrading of sheep nutrition is required at critical times in the annual cycle of animal nutritional requirements, namely, before and during mating in autumn and during lactation in spring (Eadie, 1978). Improved pasture, chiefly based on grass/white-clover swards, is utilized at these times and rough grazing at other less critical times, i.e. a 'two-pasture' system.

The soils range from poorly drained acid peat, with a pH of 3.5–4.0 and vegetation type of deer's hair sedge/bog cotton grass/heather (*Trichophorum/Eriophorum/Calluna* spp.), to freely drained brown earths, with a pH 3.5–6.0 and a species-rich bent/fescue (*Agrostis/Festuca* spp.) sward, in which indigenous white clover is usually present, at least in the less acid soils (Hill Farming Research Organisation, 1979). Major constraints to white-clover (and grass) production include acid soils, impeded drainage and deficiencies in available soil N and P. Associated with the northerly latitudes, weather constraints to pasture growth include low temperatures, winter frost, snow cover and low evapotranspiration (Taylor, 1976).

In general, assessment of white-clover and grass cultivars in hill and upland conditions under sheep grazing has been limited, but investigations show that the small-leaved prostrate stoloniferous cultivars persist best and are the most productive. The development of white clovers capable of growing at low temperatures will be a major benefit (Collins and Rhodes, 1995). Perennial ryegrass, in combination with timothy, is a favoured companion grass for sowing with white clover, but small quantities of cocksfoot are added in drier environments and of red fescue (*Festuca rubra*) in harsh, high-altitude conditions. The use of winter-hardy tetraploid perennial ryegrasses is increasing, since their open growth habit permits good white-clover growth and spread. In upland trials in western Scotland, grasses such as cocksfoot, bent grass (*Agrostis castellana*) and Yorkshire fog were less compatible with white clover than perennial ryegrass or red fescue (Tiley *et al.*, 1986).

Substantial quantities of N are bound up organically in hill and upland soils, but its availability is limited. Hence, it is often advisable to add a 'starter' dressing of fertilizer N (60–90 kg ha^{-1}) to encourage the establishment of both components of the sown-grass/white-clover mixtures (Haystead and Marriott, 1979).

Rhizobial inoculation of white-clover seed prior to sowing, although not normally needed in the lowlands, is recommended in the hills and uplands, especially deep peats and wet peaty podzol soils, since some contain no rhizobia and others may contain rhizobia that are mainly ineffecive at forming functional root nodules (Newbould *et al.*, 1982). A method of spray inoculation of emerged seedlings has been devised should the white-clover plants show N deficiency due to lack of developed root nodules but its success is not wholly predictable (Young and Mytton, 1983; Mytton and Hughes, 1984). A major problem of inoculation is the high degree of competitiveness exhibited by the relatively ineffective indigenous rhizobia to introduced strains. Microbial genetics has a role to play in transferring efficient N$_2$-fixing genes to indigenous competitive strains.

Liming, to reduce soil acidity, and phosphate-fertilizer application, to overcome soil-P deficiency, are key essentials. An important consequence of liming is the stimulation of OM mineralization, resulting in increases in the inorganic pools of N and P. On soils deficient in Mg, or where there is a history of hypomagnesaemia (grass tetany or staggers) in stock, magnesian limestone is preferable to ground limestone (calcium carbonate). Ground mineral phosphate (GMP) is efficient on acid soils, but less effective and slower-acting than water-soluble types, such as superphosphate, when the pH is more than 6.0 or where lime has been recently applied (Archer, 1978). Potash fertilizer is required particularly on organic soils, which have a poor retention capacity for K.

Herbage production in spring is at a premium for ewe lambing and lactation, and early application of fertilizer N at 40–60 kg ha^{-1} increases the quantity available. Provided there is a satisfactory white-clover presence, as represented by a well-distributed stolon network, white clover recovers from the enhanced competition from the grass engendered by applied fertilizer N, and the normally heavy grazing intensity at that time also reduces grass competition (Fothergill and Davies, 1993). Fertilizer N is also generally used, typically at applications of 60–90 kg ha^{-1}, on sown upland swards for conservation, for which there is less emphasis on maintaining good white-clover contents. Because of its convenience, big-bale silage making, often contracted by the farmer, is increasingly being adopted on upland farms, in spite of the difficulty in wilting, because of weather, and therefore attaining the desirable high DM content (300–400 g kg^{-1} DM) in the herbage for effective ensiling.

Intake and performance

Over a range of forage types – fresh, dried, hay or silage – and using sheep, young cattle or lactating dairy cows, there is a consistently higher voluntary intake of DM, by *c.* 20%, of white clover compared with grass (Thomson, 1984). Physical, chemical and plant anatomical features all contribute to the intake superiority of white clover. For example, during grazing, sheep spend about 30% less time in prehending and masticating white-clover DM, compared with the equivalent DM intake of perennial ryegrass (Penning *et al.*, 1991). Also, weight per bite is usually heavier when sheep graze white clover than when grazing grass, due to greater bulk density of the grazed horizon in clover swards and area per bite (Edwards *et al.*, 1995). However, if the metabolic demand of ewes is low, e.g. non-lactating, the lower requirement for handling time and the larger bite size on clover swards than on perennial ryegrass may be balanced by fewer bites being taken, compared with milking ewes (Penning *et al.*, 1995). Heifers spend longer grazing and ruminating when grazing grass than when grazing clover, and so, again, any benefit of larger bite size and reduced handling time with clover is countered by a shorter grazing time (Orr *et al.*, 1996).

The rate of particle degradation in the rumen is faster with white clover than with ryegrass (Moseley and Jones, 1984; Ulyatt *et al.*, 1986), and there is enhanced ruminal digestion with the legume (Beever and Thorp, 1996). However, white-clover petioles may take longer than expected to pass through the rumen, as they are generally not chewed longitudinally and so digestion may be slow (Mtengeti et al., 1995), and, therefore, the rapid passage of white clover out of the rumen may be mainly due to leaf laminae. Ruminal fluids of cattle fed perennial ryegrass or white clover contained viable populations of microorganisms, in the ratio of 1 : 5 : 27 for grass : mature clover : immature clover (Theodorou et al., 1984). Further evidence is emerging from studies of the anatomy of species plant parts, including the differing arrangement and proportions of all types and their differing physical and chemical characteristics (Wilson, 1993).

In addition to a faster rate of intake for white clover than for grass at comparable digestibility levels, ingested nutrients in white clover may be utilized more efficiently (Beever et al., 1985) and more efficient use made of metabolizable energy (ME) for animal production (Rattray and Joyce, 1974). However, the greater efficiency in ME use seems to be apparent only when ME intakes are high (Cammell et al., 1986). The availability of total non-ammonium N supplied to the small intestine is usually higher when fed white clover than when fed grass, probably due to increased net synthesis of microbial protein, as a consequence of higher content of protein in the white clover (Beever and Thorp, 1996).

In summarizing trials on the comparative feeding value of white-clover and grass forages, sheep grazing white clover grew 65% faster than sheep offered grass, while sheep grazing mixed swards with variable contents of white clover grew 25% faster (Thomson, 1979); growing cattle gave about 30% higher and lactating cows 15% higher production from clover swards. In New Zealand, the liveweight gain of young sheep grazing pure swards of white clover was 86% superior to that on perennial ryegrass, and was higher than that on all the compared plant species, which included lucerne (*Medicago sativa*) and greater lotus (*Lotus pedunculatus*) at 70 and 43% superiority to grass, respectively (Ulyatt, 1981). Following the introduction and implementation of the 'two-pasture' system, mentioned above, from an experimental scale to a commercial farm scale, improved animal performance from hill and upland farms in terms of ewe stocking rates, lambing percentages and weights of weaned lambs has resulted (Armstrong and McCreath, 1985; Johnson and Merrell, 1994).

Considering a more specific example, a comparison of white-clover versus perennial-ryegrass monocultures grazed by dairy cows showed a direct response in milk yield of 300 l from white clover between weeks 4 and 18 of lactation (Fig. 2.8), together with a residual response of 630 l post-week 18, when all cows were offered grass silage *ad libitum* and each received 500 kg DM of concentrates (Thomson, 1984; Thomson et al., 1985); with whole lactation yields of 5660 l for clover and 4730 l for grass, the response to white clover of 930 l represented a 20% increase.

Taking the common situation of utilizing mixed swards, these have pro-

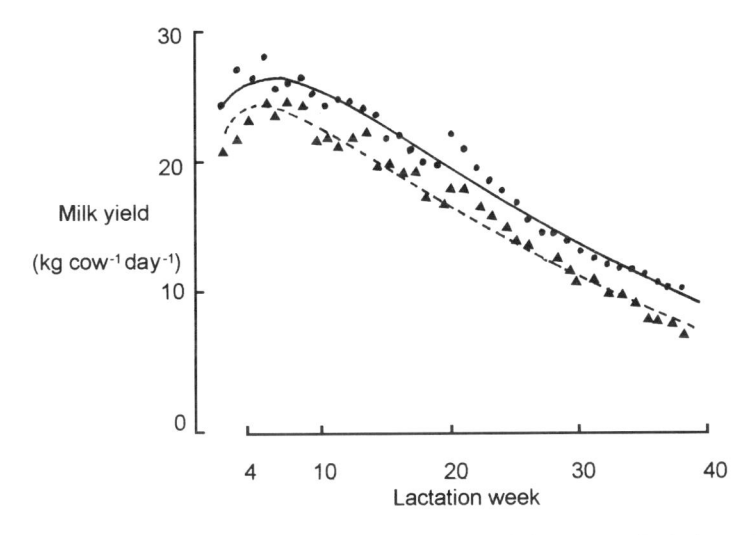

Fig. 2.8. Milk yield of cows grazing *ad libitum* perennial ryegrass (--▲--) or white clover (--●--) between weeks 4 and 18 of lactation and then fed a common silage and concentrate diet until week 40 (from Thomson, 1984).

vided the backbone of grassland farming in some regions, notably southern Australasia. However, there is scope for improved sward and animal productivity, not least by increasing clover contribution to swards and intake by stock (Chapman *et al.*, 1996). In other regions where grass/white-clover swards are being reappraised against N-fertilized grass swards, the mixed swards have proved superior for lamb production, notably during the post-weaning period, and output ha^{-1} is equivalent to that from moderately N-fertilized (150–200 kg ha^{-1} annually) grass swards (Vipond and Swift, 1992; Table 2.9), with greater individual animal performance compensating for a lower carrying

Table 2.9. Daily liveweight gain of lambs (g day^{-1}) and annual lamb production (kg ha^{-1}) from perennial ryegrass/white clover (nil N) and perennial ryegrass (200 kg N ha^{-1} year^{-1}) at two sites (from Davies and Munro, 1988).

	Lowland site		Upland site	
Period	Grass/clover	Grass	Grass/clover	Grass
Lamb liveweight gain (g day^{-1})				
Preweaning	231	211	201	186
Postweaning	140	81	112	86
Lamb production (kg ha^{-1})				
Preweaning	503	536	493	520
Postweaning	417	251	329	304
Total lamb production (kg ha^{-1})	920	787	822	824

capacity. Results have been less consistent with growing beef cattle, and daily liveweight gains may only be slightly greater on the mixed swards, but, nevertheless, carrying capacity and beef production ha^{-1} are 75–80% of grass swards receiving 180–400 kg ha^{-1} annually (Clark, 1988).

In a field-scale grazing trial in summer with spring-calved dairy cows, individual milk yields from grass/clover swards were superior to yields from N-fertilized grass swards, although milk production ha^{-1} was less, because of lower stocking rates (Ryan, 1989). In a comparison of whole-farm, whole-year systems with spring-calved cows, based on either grass/white-clover swards or grass swards receiving 350 kg N ha^{-1} year^{-1}, the clover system has proved reliable and viable in the long term, indicating its suitability for extensive low-input systems (Bax and Schils, 1993). Individual animal performance did not differ between the swards and the lack of response to the mixed swards, anticipated from the results of small-scale grazing or indoor-feeding trials, was attributed to the autumn-calving pattern; consequently the nutritional advantage of the grass/clover herbage (Frame *et al.*, 1992) was not manifest, as it might have been with spring-calving cows in early lactation. Notably, the herbage from the grazed grass/white-clover swards had lower CP contents than that from the N-fertilized grass swards in early season, because of lower clover contents, as did the herbage at the first silage cut, which provided the bulk of the winter feed.

An average of 25–30% clover DM in the total herbage DM over the season has often been propounded, although the evidence for such precision is lacking. Such levels may not be possible in spring, unless clover cultivars capable of fast growth in competition with the normally earlier-growing grass are available; some progress in this direction has been made (Rhodes and Ortega, 1996). In most of the published sward comparisons, perennial ryegrass has been both the favoured companion grass for white clover and the N-fertilized grass against which clover swards have been measured. This begs the question of the potential of other major grass species, whether sown singly or in a mixture of grasses sown with white clover, a New Zealand example being the success in drier areas of a mixture of tall fescue, cocksfoot and phalaris (*P. aquatica*) with white clover for a range of animal enterprises (Moloney, 1990).

Alternative Uses

Organic farming is an established and increasingly developing part of the agricultural scene in many countries. Crop and animal production systems have evolved in accordance with specific sets of standards, which vary from country to country, although progress on unifying the standards is being made in Western Europe. The standards apply to soil management, rotations, fertilizer and organic-manure application, together with the control of weeds, pests and diseases. White clover has an important role to perform in stockless systems of arable cropping, organic or otherwise, particularly in relation to sustaining or building up soil fertility, whether as green manure or as a legume-rich phase within a crop rotation (Barney, 1987; Ten Holte and Van Keulen, 1989). It also has a role when undersown in suitable spaced arable crops, such as maize, in protecting the soil from erosion and minimizing soil damage from harvesting operations (Lampkin, 1990). Pure-sown or in mixture with grass, white clover acts as a protective ground cover or soil-stabilization plant, where its stoloniferous growth habit is advantageous, and as a weed-suppressing, living mulch in orchards, gardens and vineyards (Parente and Frame, 1993).

There is increased interest in the use of white clover as an understorey to supply the N requirements of a cereal crop, with the crop benefiting from both current N_2 fixation and previously fixed N. Provided there are survival and growth of the white clover, the process is repeatable with a series of cereal crops. A degree of white-clover suppression is needed so that the clover-fixed N becomes available to the cereal through mineralization, following death and decay of root nodules and plant tissues. White clover makes a particularly suitable understorey, because its stolon network is capable of recovery and ramification when given the right light conditions. Suppression by a herbicide, which also aids crop drilling and establishment, is one method, while the shading by the cereal crop, until its grain is harvested, is another. Using this rationale, spring barley, spring oats and winter barley, harvested as arable silage or for cereal grain, have given yields similar to those grown by conventional ploughing and seed-bed cultivations (Jones, 1992; Jones and Clements, 1993). Any white-clover material harvested with the arable silage will improve the protein content of the silage. The continued presence of vegetation on the land minimizes N leaching, since cereal and white-clover roots take up any mineralized N.

In an organic system of beef production, utilizing grass/white-clover swards, most of the parameters measured, such as daily and annual liveweight gains, were only slightly lower than those from a grass sward receiving 270 kg ha^{-1} fertilizer N annually (Younie, 1989). However, in spite of the return of K-rich animal slurry following silage feeding of the stock in winter, there was a net depletion of available soil K in the organic system, a consequence that would eventually bias the maintenance of a satisfactory white-clover content in the sward.

White clover is also a valuable constituent in swards for conventional or organic free-range systems of poultry or pig husbandry (Lampkin, 1990), for specialist deer farming (Stevens et al., 1992) and for horse breeding or husbandry enterprises. The leaves are utilizable in human nutrition as a salad or vegetable and 'wine' can be made from the flower-heads, but more commonly the flowers are a valuable source for substantial amounts of clover honey for honey-bees. Nor should the enhancement of landscape beauty by white clover be overlooked, when the plants are flowering.

References

Acuña P.G.H. and Wilman, D. (1993) Effects of cutting height on the productivity and composition of perennial ryegrass–white clover swards. *Journal of Agricultural Science, Cambridge* 121, 29–37.

Adams, W.A. and Akhtar, N. (1994) The possible consequences for herbage growth of waterlogging compacted pasture soils. *Plant and Soil* 162, 1–17.

Alberda, T. (1971) Potential production of grassland. In: Wareing, P.F. and Cooper, J.P. (eds) *Potential Crop Production*. Heinemann, London, pp. 159–171.

Andrew, C.S. (1976) Effect of calcium, pH and nitrogen on the growth and chemical composition of some tropical and temperate pasture legumes. l. Nodulation and growth. *Australian Journal of Agricultural Research* 27, 611–623.

Andrew, C.S. and Norris, D.O. (1961) Comparative responses to calcium of five tropical and and four temperate pasture legume species. *Australian Journal of Agricultural Research* 12, 40–55.

Aparicio-Tejo, P.M., Sanchez-Diaz, M.F. and Pena, J.I. (1980) Nitrogen fixation, stomatal response and transpiration in *Medicago sativa, Trifolium repens* and *T. subterraneum* under water stress and recovery. *Physiologia Plantarum* 48, 1–4.

Archer, F.C. (1978) Comparison of different forms of phosphate fertilizers. Part 2. Grassland. *Journal of Soil Science* 29, 277–285.

Arcioni, S., Damiani, F., Mariani, A. and Pupilli, F. (1996) Somatic hybridisation and embryo rescue for the introduction of wild germplasm. In: McKersie, B.D. and Brown, D.C.W. (eds) *Biotechnology and Improvement of Forage Legumes*. CAB International, Wallingford, pp. 61–90.

Armstrong, R.H. and McCreath, J.B. (1985) *Hill Sheep Development Programme*. Scottish Agricultural Colleges/Hill Farming Research Organisation Report, 90 pp.

Baker, C.J., Choudhary, M.A. and Saxton, K.E. (1993) Inverted 'T' drill openers for pasture establishment by conservation tillage. In: Baker, M.J. (ed.) *Proceeedings of the XVII International Grassland Congress, New Zealand and Australia*, Vol. 1. New Zealand Grassland Association *et al.*, Palmerston North, pp. 331–334.

Barlow, N.D., French, R.A. and Pearson, J.F. (1986) Population ecology of *Wiseana cervinata*, a pasture pest in New Zealand. *Journal of Applied Ecology* 23, 415–431.

Barnett, O.W. and Diachun, S. (1985) Virus diseases of clovers. In: Taylor, N.L. (ed.) *Clover Science and Technology*. ASA/CSSA/SSSA, Madison, Wisconsin, pp. 235–268.

Barney, P.A. (1987) The use of *Trifolium repens, Trifolium subterraneum* and *Medicago lupulina* as overwintering leguminous green manures. *Biological Agriculture and Horticulture* 4, 225–234.

Barthram, G.T. and Grant, S.A. (1995) Interactions between variety and the timing of conservation cuts on species balance in *Lolium perenne–Trifolium repens* swards. *Grass and Forage Science* 50, 98–105.

Bax, J.A. and Browne, I. (1995) *The Use of Clover on Dairy Farms*. Research Summary, Milk Development Council, London, 21pp.

Bax, J.A. and Schils, R.L.M. (1993) Animal responses to white clover. *FAO/REUR Technical Series* 29, 7–16.

Bax, J. and Thomas, C. (1992) Developments in legume use for milk production. In: Hopkins, A. (ed.) *Grass on the Move*. Occasional Symposium No. 26, British Grassland Society, Reading, pp. 40–53.

Beaufils, E.R. (1973) Diagnosis and recommendation integrated system (DRIS): a general scheme for experimentation and calibration based on principles developed from research in plant nutrition. *Soil Science Bulletin*, no. 1. University of Natal, Pietermaritzburg, 132 pp.

Beever, D.E. and Thorp, C. (1996) Advances in the understanding of factors influencing the nutritive value of legumes. In: Younie, D. (ed.) *Legumes in Sustainable Farming Systems*. Occasional Symposium No. 30, British Grassland Society Reading, pp. 194–207..

Beever, D.E., Thomson, D.J., Ulyatt, M.J., Cammell, S.B. and Spooner, M.C. (1985) The digestion of fresh perennial ryegrass (*Lolium perenne* L. cv. Melle) and white clover (*Trifolium repens* L. cv. Blanca) by growing cattle fed indoors. *British Journal of Nutrition* 54, 763–775.

Belaygne, C., Wery, J., Cowan, A.A. and Tardieu, F. (1996) Contribution of leaf expansion, rate of leaf appearance and stolon branching to growth of plant leaf area under water deficit in white clover. *Crop Science* 36, 1240–1246.

Bircham, J.S. and Gillingham, A.S. (1986) A soil water balance model for sloping land. *New Zealand Journal of Agricultural Research* 29, 315–323.

Boller, B.C. and Nösberger, J. (1983) Effects of temperature and photoperiod on stolon characteristics, dry matter partitioning, and nonstructural carbohydrate concentration of two white clover ecotypes. *Crop Science* 23, 1057–1062.

Boller, B.C. and Nösberger, J. (1985) Photosynthesis of white clover leaves as influenced by canopy position, leaf age and temperature. *Annals of Botany* 56, 19–27.

Brink, G.E. (1996) White clover response to stolon apex removal. In: Phillips, T. (ed.) *Proceedings of the 14th Trifolium Conference, Lexington, Kentucky*. Lexington, Kentucky, p. 31.

Briseño de la Hoz, V.M. and Wilman, D. (1981) Effects of cattle grazing, sheep grazing, cutting and sward height on a grass/white clover sward. *Journal of Agricultural Science, Cambridge* 97, 699–706.

Brock, J.L. and Hay, M.J.M. (1996) A review of the role of grazing management on the growth and performance of white clover cultivars in lowland New Zealand pastures. In: Woodfield, D.R. (ed.) *White Clover: New Zealand's Competitive Edge*. Grassland Research and Practice Series No. 6, New Zealand Grassland Association, Palmerston North, pp. 65–70.

Brock, J.L. and Hay, R.J.M. (1993) An ecological approach to forage management. In: Boulter, M.J. (ed.) *Proceedings of the XVII International Grassland Congress, New Zealand and Australia*, Vol. 1. New Zealand Grassland Association et al., Palmerston North, pp. 837–842.

Brock, J.L. and Kim, M.C. (1994) Influence of the stolon/soil surface interface and plant morphology on the survival of white clover during severe drought. *Proceedings of the New Zealand Grassland Association* 56, 187–191.

Brock, J.L., Hay, M.J.M., Thomas, V.J. and Sedcole, J.R. (1988) Morphology of white clover plants under intensive sheep grazing. *Journal of Agricultural Science, Cambridge* 111, 273–283.

Brougham, R.W. (1959) The effects of season and weather on the growth rate of a ryegrass and clover pasture. *New Zealand Journal of Agricultural Research* 2, 283–296.

Brougham, R.W. (1977) Maximising animal production from temperate grassland. In: Gilsenan, B. (ed.) *Proceedings of an International Meeting on Animal Production from Temperate Pastures, Dublin, Ireland*. Irish Grassland and Animal Production Association/The Agricultural Institute, Dublin, pp. 140–146.

Brougham, R.W., Ball, P.R. and Williams, W.M. (1978) The ecology and management of white clover based pastures. In: Wilson, J.R. (ed.) *Plant Relations in Pastures*. CSIRO, Canberra, pp. 309–324.

Burch, G.J. and Johns, G.G. (1978) Root absorption of water and physiological responses to water deficits by *Festuca arundinacea*. Schreb. and *Trifolium repens* L. *Australian Journal of Plant Physiology* 5, 859–871.

Burdon, J.J. (1983) Biological flora of the British Isles: *Trifolium repens*. *Journal of Ecology* 71, 307–330.

Burgess, E.P.J. and Gatehouse, A.M.R. (1997) Engineering for insect pest resistance. In: McKersie, B.D. and Brown, D.C.W. (eds) *Biotechnology and the Improvement of Forage Legumes*. CAB International, Wallingford, pp. 229–258.

Camlin, M.S. (1981) Competitive effects between ten cultivars of perennial ryegrass and three cultivars of white clover grown in association. *Grass and Forage Science* 36, 169–178.

Cammell, S.B., Thomson, D.J., Beever, D.E., Haines, M.J., Dhanoa, M.S. and Spooner, M.C. (1986) The efficiency of energy utilization in growing cattle consuming fresh perennial ryegrass (*Lolium perenne* cv. Melle) and white clover (*Trifolium repens* cv. Blanca). *British Journal of Nutrition* 55, 669–680.

Caradus, J.R. (1977) Structural variation of white clover root systems. *New Zealand Journal of Agricultural Research* 20, 213–219.

Caradus, J.R. (1986) World checklist of white clover varieties. *New Zealand Journal of Experimental Agriculture* 14, 119–164.

Caradus, J.R. (1990) The structure and function of white clover root systems. *Advances in Agronomy* 43, 1–46.

Caradus, J.R. (1993) Progress in white clover agronomic performance through breeding. In: Boulter, M.J. (ed.) *Proceedings of the XVII International Grassland Congress, New Zealand and Australia*, Vol. 1. New Zealand Grassland Association *et al.*, Palmerston North, pp. 396–397.

Caradus, J.R. (1994) Genetic improvement in white clover representing six decades of plant breeding. *Crop Science* 34, 1205–1213.

Caradus, J.R. and Chapman, D.F. (1996) Selection for and heritability of stolon characteristics in two cultivars of white clover. *Crop Science* 36, 900–904.

Caradus, J.R. and Crush, J.R. (1996) Effect of recurrent selection for aluminium tolerance on shoot morphology and chemical content of white clover. *Journal of Plant Nutrition* 19, 1485–1492.

Caradus, J.R. and Woodfield, D.R. (1986) Evaluation of root type in white clover genotypes and populations. In: Williams, T.A. and Wratt, G.S. (eds) *DSIR Plant Breeding Symposium, Lincoln*. Special Publication No 5, Agronomy Society of New Zealand, Christchurch, pp. 322–325.

Caradus, J.R. and Woodfield, D.R. (1996) World checklist of white clover varieties. *New Zealand Journal of Agricultural Research* 40, 115–206.

Caradus, J.R., Hay, R.J.M. and Woodfield, D.R. (1996) The positioning of white clover cultivars in New Zealand. In: Woodfield, D.R. (ed.) *White Clover: New Zealand's Competitive Edge*. Grassland Research and Practice Series No. 6,. New Zealand Grassland Association, Palmerston North, pp. 45–49.

Carran, R.A. (1991) Calcium magnesium imbalance – a cause of negative yield response to liming. *Plant and Soil* 134, 107–114.

Carruthers, V.R., O'Connor, M.B., Feyter, C., Upsell, M.P. and Ledgard, S.F. (1987) Results from the Ruakura Bloat survey. In: *Proceedings of the Ruakura Farmers Conference 1987*. Ministry of Agriculture and Fisheries, Wellington, pp. 44–46.

Castle, M.E., Reid, D. and Watson, J. (1983) Silage and milk production: studies with diets containing white clover silage. *Grass and Forage Science* 38, 193–200.

Chapman, D.F. (1987) Natural re-seeding and *Trifolium repens* demography in grazed hill pastures. II. Seedling appearance and survival. *Journal of Applied Ecology* 24, 1037–1043.

Chapman, D.F. and Anderson, C.B. (1987) Natural re-seeding and *Trifolium repens* demography in grazed hill pastures. I. Flowerhead appearance and fate, and seed dynamics. *Journal of Applied Ecology* 24, 1025–1035.

Chapman, D.F., Robson, M.J. and Snaydon, R.W. (1992) The carbon economy of clonal plants of *Trifolium repens* L. *Journal of Experimental Botany* 43, 427–434.

Chapman, D.F., Parsons, A.J. and Schwinning, S. (1996) Management of clover in grazed pastures: expectations, limitations and opportunities. In: Woodfield, D.R. (ed.) *White Clover: New Zealand's Competitive Edge*. Grassland Research and Practice Series No. 6, New Zealand Grassland Association, Palmerston North, pp. 55–64.

Chapman, R. and Heath, S.B. (1987) The effect of cattle slurry on clover in grass/clover swards. In: Meer, H.G. Van der, Dijk, T.A. Van and Ennik, G.C. (eds) *Animal Manure on Grassland and Forage Crops: Fertilizer or Waste?* Martinus Nijhoff, Dordrecht, pp. 337–340.

Chapman, R.B. (1990) Insect pests. In: Langer, R.H.M. (ed.) *Pastures: Their Ecology and Management*. Oxford University Press, Auckland, pp. 448–467.

Charlton, J.F.L. (1978) Establishment of pasture legumes in North Island hill country. II. Seedling establishment and plant survival. *New Zealand Journal of Experimental Agriculture* 5, 385–390.

Charlton, J.F.L. (1979) Effects of slugs during establishment of oversown legumes in box experiments. In: Crosby, T.K. and Pottinger, R.P. (eds) *Proceedings of the 2nd Australian Conference on Grassland Invertebrate Ecology*. Government Printer, Wellington, pp. 253–255.

Charlton, J.F.L. (1992) Some basic concepts of pasture seed mixtures for New Zealand farms. *Proceedings of the New Zealand Grassland Association* 53, 37–40.

Charlton, J.F.L. and Giddens, N.G. (1983) Establishment of hill country white clover selections from oversowing. *Proceedings of the New Zealand Grassland Association* 44, 149–155.

Charlton, J.F.L. and Grant, D.A. (1977) Distribution of legume seed by aircraft in unploughable hill country. *New Zealand Journal of Experimental Agriculture* 5, 85–89.

Charlton, J.F.L. and Sedcole, J.R. (1985) Survival of a white clover cultivar in New Zealand hill pasture. In: *Proceedings of the XV International Grassland Congress, Kyoto, Japan*. The Science Council of Japan and the Japanese Society of Grassland Science, Tochigi-ken, pp. 678–679

Charlton, J.F.L., Hampton, J.G. and Scott, D.J. (1986) Temperature effects on germination of New Zealand herbage grasses. *Proceedings of the New Zealand Grassland Association* 47, 165–172.

Chaudri, A.M., McGrath, S.P. and Giller, K.E. (1992) Survival of the indigenous population of *Rhizobium leguminosarum* biovar. *trifolii* in soil spiked with Cd, Zn, Cu and Ni salts. *Soil Biology and Biochemistry* 24, 625–632.

Chestnutt, D.M.B. and Lowe, J. (1970) Agronomy of white clover/grass swards: a review. In: Lowe, J. (ed.) *White Clover* Occasional Symposium No. 6, British Grassland Society, Hurley, pp. 191–213.

Chu, A.G.P. and Robertson, A.G. (1974) The effects of shading and defoliation on nodulation and nitrogen fixation by white clover. *Plant and Soil* 41, 509–519.

Clark, D.A. and Harris, P.S. (1985) Composition of the diet of sheep grazing swards of differing white clover content and spatial distribution. *New Zealand Journal of Agricultural Research* 28, 233–240.

Clark, H. (1988) Beef and sheep output from grass/white clover swards. In: *Proceedings of an RASE/ADAS Conference*, NAC, Stoneleigh, pp. 12–16.

Clements, R.O. (1994) *A Review of Damage Caused by Pests and Diseases to White Clover.* MAFF Commissioned Review, IGER, North Wyke, 47 pp.

Clements, R.O. and Henderson, J.F. (1983) An assessment of insidious pest damage to 26 varieties of seven species of herbage legumes. *Crop Protection* 2, 491–495.

Clements, R.O. and Murray, P.J. (1993) Sitona damage to clover in the UK. In: Prestidge, R.A. (ed.) *Proceedings of the 6th Australasian Grassland Invertebrate Ecology Conference, Hamilton, New Zealand.* AgResearch, Hamilton, pp. 260–264.

Clifford, P.T.P. (1979) Effect of closing date on potential seed yields from 'Grasslands Huia' and 'Grasslands Pitau' white clovers. *New Zealand Journal of Experimental Agriculture* 7, 303–306.

Clifford, P.T.P., Baird, I.J., Grbavac, N. and Sparks, G.A. (1990) White clover soil seed loads: effect on requirements and resultant success of cultivar-change crops. *Proceedings of the New Zealand Grassland Association* 52, 95–98.

Clifford, P.T.P., Sparks, G.A. and Woodfield, D.R. (1996) The intensifying requirements for white clover cultivar change. In: Woodfield, D.R. (ed.) *White Clover: New Zealand's Competitive Edge.* Grassland Research and Practice Series No. 6, New Zealand Grassland Association, Palmerston North, pp. 19–24.

Collins, R.P. and Rhodes, I. (1995) Stolon characteristics related to winter survival in white clover. *Journal of Agricultural Science, Cambridge* 124, 11–16.

Collins, R.P., Glendining, M.J. and Rhodes, I. (1991) The relationships between stolon characteristics, winter survival and annual yields in white clover (*Trifolium repens* L.). *Grass and Forage Science* 46, 51–61.

Collins, R.P., Connolly, J., Fothergill, M., Frankow-Lindberg, B.E., Guckert, A., Guinchard, M.P., Lüscher, A., Nösberger, J., Rhodes, I., Robin, C., Stäheli, B. and Stoffel, S. (1996) Variation in the overwintering of white clover cultivars in cool wet areas of Europe. In: Parente, G., Frame, J. and Orsi, S. (eds) *Grassland and Land Use Systems. Proceedings of the 16th General Meeting of the European Grassland Federation, Grado, Italy,* published as *Grassland Science in Europe,* Vol. 1. ERSA, Gorizia, pp. 201–204.

Connolly, V. (1990) Seed yield and yield components in ten white clover cultivars. *Irish Journal of Agricultural Research* 29, 41–48.

Cook, R. and York, P.A. (1979) Nematodes of herbage legumes. In: *Welsh Plant Breeding Station Annual Report for 1978.* WPBS, Aberystwyth, pp. 177–207.

Cook, R., Thomas, B.J. and Mizen, K.A. (1989) Dissemination of white clover mosaic virus and stem nematode, *Ditylenchus dipsaci,* by the slug *Deroceras reticulum.* In: Henderson, I. (ed.) *Slugs and Snails in World Agriculture.* Monograph 41, British Crop Protection Council, Croydon, pp. 107–112.

Coop, I.E. (1986) Pasture and crop production. In: McCutcheon, S.N., McDonald, M.F. and Wickham, G.A. (eds) *Sheep Production,* Vol. II, *Feeding, Growth and Health.* New Zealand Institute of Agricultural Science, Wellington, pp. 110–136.

Cooper, C.S. (1977) Growth of the legume seed. *Advances in Agronomy* 29, 119–139.

Cooper, J.E., Wood, M. and Holding, A.J. (1983) The influence of soil acidity factors on rhizobia. In: Jones, D.G. and Davies, D.R. (eds) *Temperate Legumes.* Pitman, London, pp. 319–335.

Cooper, J.P. and Breese, E.L. (1971) Plant breeding: forage grasses and legumes. In: Wareing, P.F. and Cooper, J.P. (eds) *Potential Crop Production*. Heinemann, London, pp. 295–318.
Copeland, R. and Pate, J.S. (1970) Nitrogen metabolism of nodulated white clover in the presence and absence of nitrate nitrogen. In: Lowe, J. (ed.) *White Clover*. Occasional Symposium No. 6, British Grassland Society, Hurley, pp. 71–77.
Cowling, D.W. (1982) Biological nitrogen fixation and grassland production in the United Kingdom. *Philosophical Transactions, Royal Society of London*, B 296, 397–404.
Crush, J.R. (1976) Endomycorrhizas and legume growth in some soils of the Mackenzie Basin. *New Zealand Journal of Agricultural Research* 19, 473–476.
Crush, J.R. (1987) Nitrogen fixation. In: Baker, M.J. and Williams, W.M. (eds) *White Clover*. CAB International, Wallingford, pp. 185–201.
Crush, J.R. (1993) Hydrogen evolution from root nodules of *Trifolium repens* and *Medicago sativa* plants grown under elevated atmospheric CO_2. *New Zealand Journal of Agricultural Research* 36, 177–183.
Crush, J.R. and Caradus, J.R. (1992) Response to soil aluminium of 2 white clover (*Trifolium repens* L.) genotypes. *Plant and Soil* 146, 39–43.
Crush, J.R. and Caradus, J.R. (1993) Effect of different aluminium and phosphorus levels on aluminium tolerant and aluminium susceptible genotypes of white clover (*Trifolium repens* L.). *New Zealand Journal of Agricultural Research* 36, 99–107.
Crush J.R. and Caradus, J.R. (1995) Cyanogenesis potential and iodine concentration in white clover (*Trifolium repens* L.) cultivars. *New Zealand Journal of Agricultural Research* 38, 309–316.
Crush, J.R., Campbell, D.B. and Caradus, J.R. (1993) Effect of temperature on nitrogen fixation rates in seven white clover cultivars. In: Baker, M.J. (ed.) *Proceedings of the XVII International Grassland Congress, New Zealand and Australia*, Vol. 1. New Zealand Grassland Association et al., Palmerston North, pp. 119–121.
Cullen, N.A. (1964a) The effect of nurse crops on the establishment of pasture. II. Cereal nurse crops. *New Zealand Journal of Agricultural Science* 7, 52–59.
Cullen, N.A. (1964b) Species competition in establishing swards: suppression effects of ryegrass on establishment and production of associated grasses and clovers. *New Zealand Journal of Agricultural Science* 7, 678–693.
Curll, M.L. (1982) The grass and clover content of pastures grazed by sheep. *Herbage Abstracts* 52, 403–411.
Curll, M.L. and Wilkins, R.J. (1983) The comparative effects of defoliation, treading and excreta on a *Lolium perenne*–*Trifolium repens* pasture grazed by sheep. *Journal of Agricultural Science, Cambridge* 100, 451–460.
Curll, M.L. and Wilkins, R.J. (1985) The effect of cutting for conservation on a grazed perennial ryegrass–white clover pasture. *Grass and Forage Science*, 40, 19–30.
Curll, M.L., Wilkins, R.J., Snaydon, R.W. and Shanmugalingham, V.S. (1985) The effects of stocking rate and fertilizer on a perennial ryegrass–white clover sward. 1. Sward and sheep performance. *Grass and Forage Science* 40, 129–140.
Cuttle, S.P., Hallard, M., Daniel, G. and Scurlock, R.V. (1992) Nitrate leaching from sheep grazed grass/clover and fertilized grass pastures. *Journal of Agricultural Science, Cambridge* 119, 335–342.
Davidson, I.A. and Robson, M.J. (1984) The effect of temperature and nitrogen supply on the physiology of grass/clover swards. In: Thomson, D.J. (ed.) *Forage Legumes*. Occasional Symposium No.16, British Grassland Society, Hurley, pp. 56–60.
Davidson, I.A. and Robson, M.J. (1985) Effects of nitrogen supply on the grass and

clover components of simulated mixed swards grown under favourable environmental conditions. 2. Nitrogen fixation and nitrate uptake. *Annals of Botany* 55, 697–703.

Davies, A. (1992) White clover. *Biologist* 39, 129–133.

Davies, A. and Evans, M.E. (1982) The pattern of growth in swards of two contrasting varieties of white clover in winter and spring. *Grass and Forage Science* 37, 199–207.

Davies, A. and Evans, M.E. (1990a) Effects of spring defoliation and fertilizer nitrogen on the growth of white clover in ryegrass/clover swards. *Grass and Forage Science* 45, 345–356.

Davies, A. and Evans, M.E. (1990b) Axillary bud development in white clover in relation to defoliation and shading treatments. *Annals of Botany* 66, 349–357.

Davies, A. and Jones, D.R. (1987) Tissue fluxes in white clover varieties grown in swards continuously grazed by sheep. In: Pollott, G.E. (ed.) *Efficient Sheep Production from Grass.* Occasional Symposium No. 21, British Grassland Society, pp. 185-187.

Davies, D.A. and Munro, J.M.M. (1988) Assessment of grass–clover pastures for lowland and upland lamb production. In: *Proceedings of the 12th General Meeting of the European Grassland Federation, Dublin, Ireland.* Irish Grassland Association, Belclare, pp. 164–167.

Davies, D.A., Fothergill, M. and Jones, D. (1991) Assessment of contrasting perennial ryegrasses, with and without white clover, under continuous sheep stocking in the uplands. 3. Herbage production, quality and intake. *Grass and Forage Science* 46, 39–50.

del Pozo, M., Wright, I.A., Whyte, T.K. and Colgrove, P.M. (1996) Effects of grazing by sheep or goats on sward composition in ryegrass/white clover pasture and on subsequent performance of weaned lambs. *Grass and Forage Science* 51, 142–154.

Dennis, W.D. and Woledge, J. (1982) Photosynthesis by white clover leaves in mixed clover–ryegrass swards. *Annals of Botany (London), (New Series)* 49, 627–635.

Dennis, W.D. and Woledge, J. (1983) The effect of shade during leaf expansion on photosynthesis by white clover leaves. *Annals of Botany (London), (New Series)*, 5l, 111–118.

Dennis, W.D. and Woledge, J. (1985) The effect of nitrogenous fertilizer on the photosynthesis and growth of a white clover/perennial ryegrass sward. *Annals of Botany* 55, 171–178.

Dennis, W.D. and Woledge, J. (1987) The effect of nitrogen in spring on shoot number and leaf area in mixtures. *Grass and Forage Science* 42, 265–269.

Drysdale, A.D. (1965) Liquid manure as a grassland fertilizer. III. The effect of liquid manure on the yield and botanical composition of pasture, and its interaction with nitrogen, phosphate and potash fertilizers. *Journal of Agricultural Science, Cambridge* 654, 333–340.

Dunlop, J. and Hart, A.L. (1987) Mineral nutrition. In: Baker, M.J. and Chapman, M.W. (eds) *White Clover.* CAB International, Wallingford, pp. 153–184.

Eadie, J. (1978) Increasing output in hill farming. *Journal of the Royal Agricultural Society of England* 139, 103–114.

Eagles, C.F. and Othman, O.B. (1981) Growth at low temperatures and cold hardiness in white clover. In: Wright, E.C. (ed.) *Plant Physiology and Herbage Production.* Occasional Symposium No. 13, British Grassland Society, Hurley, pp. 109–13.

Eagles, C.F. and Othman, O.B. (1988a) Variation in growth of overwintered stolons of contrasting white clover populations in response to temperature, photoperiod and spring environment. *Annals of Applied Biology* 112, 563–574.

Eagles, C.F. and Othman, O.B. (1988b) Seasonal variation of morphological characters and growth in contrasting white clover populations. *Annals of Applied Biology* 112, 575–583.

Edmond, D.B. (1964) Some effects of sheep treading on the growth of ten pasture species. *New Zealand Journal of Agricultural Research* 7, 1–16.

Edwards, G.R., Parsons, A.J., Penning, P.D. and Newman, J.A. (1995) Relationship between vegetation state and bite dimensions of sheep grazing contrasting plant species and its implications for intake rate and diet selection. *Grass and Forage Science* 50, 378–388.

Engin, M. and Sprent, J.I. (1973) Effects of water stress on growth and nitrogen-fixing ability of *Trifolium repens*. *New Phytologist* 72, 117–126.

Essig, H.W. (1985) Quality and antiquality components. In: Taylor, N.L. (ed.) *Clover Science and Technology*. ASA/CSSA/SSSA, Madison, Wisconsin, pp. 309–324.

Evans, D.R. and Williams, T.A. (1987) The effect of cutting and grazing managements on dry matter yield of white clover varieties (*Trifolium repens*) when grown with S23 perennial ryegrass. *Grass and Forage Science* 42, 153–159.

Evans, D.R., Hill, J., Williams, T.A. and Rhodes, I. (1985) Effects of coexistence on the performance of white clover/perennial ryegrass mixtures. *Oecologia (Berlin)* 66, 536–539.

Evans, D.R., Williams, T.A. and Davies, W.E. (1986a) Potential seed yield of white clover varieties. *Grass and Forage Science* 41, 221–227.

Evans, D.R., Thomas, T.A., Williams, T.A.and Davies, E.W. (1986b) Effect of fertilizers on the yield and chemical composition of pure sown white clover and on soil nutrient status. *Grass and Forage Science* 41, 295–302.

Evans, D.R., Williams, T.A. and Evans, S.A. (1992) Evaluation of white clover varietes under grazing and their role in farm systems. *Grass and Forage Science* 47, 342–352.

Evans, P.S. (1977) Comparative root morphology of some pasture grasses and clovers. *New Zealand Journal of Agricultural Research* 20, 331–335.

Evans, P.S. (1978) Plant root distribution and water use patterns of some pasture and crop species. *New Zealand Journal of Agricultural Research* 20, 331–335.

Evers, G.W. (1989) Intermediate white clover – A model for clover persistence on the Gulf Coast of the USA. In: *Proceedings of the XVI International Grassland Congress, Nice, France*, Vol. 1. Association Française pour la Production Fourragère, Versailles, pp. 381–382.

Faurie, O., Soussana, J.F. and Sinoquet, H. (1996) Radiation interception, partitioning and use in grass/clover mixtures. *Annals of Botany* 77, 35–45.

Ferguson, C.M., Lewis, G.C., Hanks, C.B., Parsons, D.M.J. and Asteraki, E.J. (1989) Incidence and severity of damage by slugs and snails to leaves of twelve white clover cultivars. Tests of Agrochemicals and Cultivars No. 10. *(Annals of Applied Biology* 114, (Suppl.), 138–139.)

Floate, M.J.S., Rangeley, A. and Bolton, G.R. (1981) An investigation of problems of sward improvement on deep peat with special reference to potassium responses and interactions with lime and phosphorus. *Grass and Forage Science* 36, 81–90.

Forbes, T.J., Dibb, C., Green, J.O., Hopkins, A. and Peel, S. (1980) *Factors Affecting the Productivity of Permanent Grassland: A National Farm Study*. GRI/ADAS, Maidenhead.

Forster, R.L.S., Beck, A.L. and Lough, T.J. (1997) Engineering for resistance to virus diseases. In: McKersie, B.W. and Brown, D.C.W. (eds) *Biotechnology and the Improvement of Forage Legumes*. CAB International, Wallingford, pp. 291–315.

Fothergill, M. and Davies, D.A. (1993) White clover contribution to continuously stocked sheep pastures in association with contrasting perennial ryegrasses. *Grass and Forage Science* 48, 369–379.

Fothergill, M., Davies, D.A., Morgan, C.T. and Jones, J.R. (1996) White clover crashes. In: Younie, D. (ed.) *Legumes in Sustainable Farming Systems*. Occasional Symposium No. 30, British Grassland Society, Reading, pp. 172–176.

Foulds, W. (1978) Response to soil moisture supply in three leguminous species. *New Phytologist* 80, 535–545.

Frame, J. (1976) A comparison of herbage production under cutting and grazing (including comments on deleterious factors such as treading). In: Hodgson, J. and Jackson, D.K. (eds) *Pasture Utilization by the Grazing Animal*. Occasional Symposium No. 8, British Grassland Society, Hurley, pp. 39–49.

Frame, J. (1990) Herbage productivity of a range of grass species in association with white clover. *Grass and Forage Science* 45, 57–64.

Frame, J. (1996) The effect of fertilizer nitrogen application in spring and/or autumn on the production from a perennial ryegrass/white clover sward. *FAO/REU Technical Series* 42, 88–91.

Frame, J. and Boyd, A.G. (1984) Response of white clover to climatic factors In: Riley, H. and Skjelvåg, A.O. *Proceedings of the 10th General Meeting of the European Grassland Federation, Ås, Norway*. The Norwegian State Agricultural Research Stations, Ås, pp. 171–175.

Frame, J. and Boyd, A.G. (1986) Effect of cultivar and seed rate of perennial ryegrass and strategic fertilizer nitrogen on the productivity of grass/white clover swards. *Grass and Forage Science* 41, 359–366.

Frame, J. and Boyd, A.G (1987a) The effect of fertilizer nitrogen rate, white clover variety and closeness of cutting on herbage productivity from perennial ryegrass/white clover swards. *Grass and Forage Science* 42, 85–96.

Frame, J. and Boyd, A.G. (1987b) The effect of strategic use of fertilizer nitrogen in spring and/or autumn on the productivity of a perennial ryegrass/white clover sward. *Grass and Forage Science* 42, 429–438.

Frame, J. and Newbould, P. (1984) Herbage production from grass/white clover swards. In: Thomson, D.J. (ed.) *Forage Legumes*. Occasional Symposium No. 16, British Grassland Society, Hurley, pp. 15–35.

Frame, J. and Newbould, P. (1986) Agronomy of white clover. *Advances in Agronomy* 40, 1–88.

Frame, J. and Paterson, D.J. (1987) The effect of strategic nitrogen application and defoliation systems on the productivity of a perennial ryegrass/white clover sward. *Grass and Forage Science* 42, 271–280.

Frame, J. and Tiley, G.E.D. (1993) The role of white clover (*Trifolium repens* L.) in hill and upland livestock systems in Scotland. In: Gaston, A., Kernick, M. and Le Houérou, H.-L. *Proceedings of the Fourth International Rangeland Congress, Montpellier, France*, CIBAD/SCIST, Montpellier, pp. 368–371.

Frame, J., Newbould, P. and Munro, J.M.M. (1985) Herbage production from the hills and uplands. In: Charles, A.H. and Haggar, R.J. (eds) *Changes in Sward Composition and Productivity*. Occasional Symposium No.10, British Society of Animal Production, Edinburgh, pp. 9–37.

Frame, J., Bax, J. and Bryden, G. (1992) Herbage quality of perennial ryegrass/white clover and N-fertilized ryegrass swards in intensively managed dairy systems. In:*Proceedings of the 14th General Meeting of the European Grassland Federation, Lahti, Finland.* European Grassland Federation, pp. 180–183.

Frankow-Lindberg, B.E. (1987) Effects of cultivar, nitrogen application and cutting interval on above-ground growth of white clover grown with meadow fescue at two contrasting temperatures. *Swedish Journal of Agricultural Research* 17, 31–40.

Frankow-Lindberg, B.E., Danielsson, D.A. and Moore, C. (1996) The uptake of white clover technology in farming practice. *FAO/REU Technical Series* 42, 37–43.

Fraser, J. and Kunelius, H.T. (1993) Influence of seeding on the yield of white clover/orchard grass mixtures in Atlantic Canada. *Journal of Agricultural Science, Cambridge* 120, 197–203.

Fraser, J. and Kunelius, H.T. (1995) Herbage yield and composition of white clover/grass associations in Atlantic Canada. *Journal of Agricultural Science, Cambridge* 125, 371–377.

Fraser, J., Sutherland, K. and Martin, R.C. (1993) Effects of autumn harvest date on the performance of white clover/grass mixtures in Nova Scotia. *Journal of Agricultural Science, Cambridge* 121, 315–321.

Free, J.B. (1993) *Insect Pollination of Crops.* Academic Press, London.

Garwood, E.A., Tyson, K.C. and Roberts, D. (1982) The production and persistency of perennial ryegrass/white clover and high-N perennial ryegrass swards under grazing and cutting. *Grass and Forage Science* 37, 174–176.

Gibb, M.J. and Baker, R.D. (1989) Effect of changing grazing severity on the composition of perennial ryegrass/white clover swards stocked with beef cattle. *Grass and Forage Science* 44, 329–334.

Gibb, M.J., Baker, R.D. and Sayer, A.M.E. (1989) The impact of grazing severity on perennial ryegrass/white clover swards stocked continuously with beef cattle. *Grass and Forage Science* 44, 315–328.

Gibson, P.B. and Cope, W.A. (1985) White clover. In: Taylor, N.L. (ed.) *Clover Science and Technology.* ASA/CSSA/SSSA, Madison, Wisconsin, pp. 471–490.

Gooding, R.F. (1993) Cutting and Grazing Systems for Grass/White Clover *Trifolium repens* L. Associations. Ph.D. thesis, University of Glasgow.

Gooding, R.F., Frame, J. and Thomas, C. (1996) Effects of sward type and rest periods from sheep grazing on white clover presence in perennial ryegrass/white clover associations. *Grass and Forage Science* 51, 180–189.

Goodman, P.J. (1988) Nitrogen fixation, transfer and turnover in upland and lowland grass-clover swards using ^{15}N isotope dilution. *Plant and Soil* 112, 247–254.

Gordon, A.J., Kessler, W. and Minchin, F.R. (1990) Defoliation-induced stress in nodules of white clover. 1. Changes in physiological parameters and protein synthesis. *Journal of Experimental Botany* 41, 1245–1253.

Grace, N.D. (1983) (ed.) *The Mineral Requirements of Grazing Ruminants.* Occasional Publication No. 9, New Zealand Society of Animal Production, Hamilton.

Grant, S.A. and Barthram, G.T. (1991) The effects of contrasting cutting regimes on the components of grass and clover growth in microswards. *Grass and Forage Science* 46, 1–13.

Grant, S.A. and Marriott, C.A. (1989) Some factors causing temporal and spatial variation in white clover performance in grazed swards. In: *Proceedings of the XVI*

International Grassland Congress, Nice, France, Vol. II. Association Française pour la Production Fourragère, Versailles, pp. 1041–1042.

Grant, S.A., Bolton, G.R. and Russel, A.J.F. (1984) The utilization of sown and indigenous plant species by sheep and goats grazing hill pastures. *Grass and Forage Science* 39, 361–370.

Greenwood, R.M. (1961) Pasture establishment on a podsolised soil in Northland. III. Studies on the rhizobial populations and the effects of inoculation. *New Zealand Journal of Agricultural Research* 4, 375–389.

Haggar, R.J., Standell, C.J., and Birnie, J.E. (1985) Occurrence, impact and control of weeds in early sown leys. In: Brockman, J.S. (ed.) *Weeds, Pests and Diseases of Grassland and Herbage of Legumes.* Occasional Symposium No.18, British Grassland Society, Hurley, pp. 11–18.

Haggar, R.J., Soper, D. and Cormack, W.F. (1990) Weed control in agricultural grassland. In: Hance, R.J. and Holly, K. (eds) *Weed Control Handbook: Principles,* 8th edn. Blackwell Scientific Publications, Oxford, pp. 387–405.

Hale, C.N. (1977) Some factors affecting the survival of *Rhizobium trifolii* on white clover. *Proceedings of the New Zealand Grassland Association* 38, 182–186.

Halliday, J. and Pate, J. (1976) The acetylene reduction assay as a means of studying nitrogen fixation in white clover under sward and laboratory conditions. *Journal of the British Grassland Society* 31, 29–35.

Hampton, J.G. (1996) Paclobutrazol and white clover seed production: a non-fulfilled potential. In: Woodfield, D.R. (ed.) *White Clover: New Zealand's Competitive Edge.* Grassland Research and Practice Series No. 6, New Zealand Grassland Association, Palmerston North, pp. 35–39.

Hampton, J.G., Charlton, J.F.L., Bell, D.D. and Scott, D.J. (1987) Temperature effects on the germination of New Zealand herbage legumes. *Proceedings of the New Zealand Grassland Association* 48, 177–183.

Harkess, R.D. (1970) Fundamentals of grassland management, Part 9: the herbage legumes. *Scottish Agriculture* 49, 202–216.

Harper, J.L. (1978) Plant relations in pastures. In: Wilson, J.R. (ed.) *Plant Relations in Pastures.* CSIRO, Melbourne, pp. 3–14.

Harris, W., Rhodes, I. and Mee, S.S. (1983) Observations on environmental and genotypic influences on the overwintering of white clover. *Journal of Applied Ecology* 20, 609–624.

Hartwig, U. and Nösberger, J. (1994) What triggers the regulation of nitrogenase activity in forage legume nodules after defoliation? *Plant and Soil* 161, 109–114.

Hartwig, U., Boller, B.C. and Nösberger, J. (1987) Oxygen supply limits nitrogenase activity of clover nodules after defoliation. *Annals of Botany* 59, 285–291.

Hay, M.J.M. (1983) Seasonal variation in the distribution of white clover (*Trifolium repens* L.) stolons among three horizontal strata in two grazed swards. *New Zealand Journal of Agricultural Research* 26, 29–34.

Hay, M.J.M. (1985) Seasonal variation in the vertical distribution of white clover (*Trifolium repens* L.) in two contrasting pastures. *Proceedings of the New Zealand Grassland Association* 46, 195–198.

Hay, M.J.M. and Newton P.C.D. (1996) Effect of severity of defoliation on the viability of reproductive and vegetative axillary buds of *Trifolium repens* L. *Annals of Botany* 78, 117–123.

Hay, M.J.M. and Sackville Hamilton, R. (1996) Influence of xylem vascular architecture on the translocation of phosphorus from nodal roots in a genotype of *Trifolium repens* during undisturbed growth. *New Phytologist* 132, 575–582.

Hay, M.J.M., Chapman, D.F., Hay, R.J.M., Pennell, C.G.L., Woods, P.W. and Fletcher, R.H. (1987) Seasonal variation in the vertical distribution of white clover in grazed swards. *New Zealand Journal of Agricultural Research* 30, 1–8.

Hay, M.J.M., Chu, A.C.P., Knighton, M.V. and Wewala, S. (1989) Variation with season and node position in carbohydrate content of white clover stolons. In: *Proceedings of the XVI International Grassland Congress, Nice, France*, Vol. II. Association Française pour la Production Fourragère, Versailles, pp. 1059–1060.

Hay, R.J.M. and Baxter, G.S. (1984) Spring management of pasture to increase summer white clover growth. *Proceedings of the Lincoln College Farming Conference* 34, 132–137.

Haynes, R.J. (1980) Competitive aspects of the grass–legume association. *Advances in Agronomy* 33, 227–261.

Haystead, A. and Marriott, C.A. (1979) Effects of rate and times of application of starter dressings of nitrogen fertilizer to surface-sown perennial ryegrass–white clover on hill peat. *Grass and Forage Science* 34, 241–247.

Haystead, A., King, J. and Lamb, W.I.C. (1979) Photosynthesis, respiration and nitrogen fixation in white clover on hill peat. *Grass and Forage Science* 34, 241–247.

Haystead, A., Malajczuk, N. and Grove, T.S. (1988) Underground transfer of nitrogen between pasture plants infected with vesicular–arbuscular mycorrhizal fungi. *New Phytologist* 108, 417–423.

Herriott, J.B.D. (1970) The influence of seeding method on sward productivity. In: Lowe, J. (ed.) *White Clover.* Occasional Symposium No. 10, British Grassland Society, Hurley, pp. 239–245.

Hides, D., Lewis, J. and Marshall, A. (1984) Prospects for white clover seed production in the UK. In: Thomson, D.J. (ed.) *Forage Legumes.* Occasional Symposium No. 16, British Grassland Society, Hurley, pp. 36–39.

Hill, M.J., Hockney, M.J., Mulcahy, C.A. and Rapp, G. (1995) The effect of hydrogen cyanide potential (HCNp) and sward morphology on the relative acceptability to sheep of white clover and Caucasian clover herbage. *Grass and Forage Science* 50, 1–9.

Hill Farming Research Organisation (1979) Science and hill farming. In: *Silver Jubilee Report 1954–1979.* Hill Farming Research Organisation, Penicuik, pp. 9–21.

Hoglund, J.H., Crush, J.R., Brock, J.L., Ball, R. and Carran, R.A. (1979) Nitrogen fixation in pasture. XII. General discussion. *New Zealand Journal of Experimental Agriculture* 7, 45–51.

Holding, A.J. and King, J. (1963) The effectiveness of indigenous populations of *Rhizobium trifolii* in relation to soil factors. *Plant and Soil* 18, 191–198.

Holliday, R. and Wilman, D. (1965) The effect of fertilizer nitrogen and frequency of defoliation on yield of grassland herbage. *Journal of the British Grassland Society* 20, 32–40.

Hollington, P.A., Marshall, A.H. and Hides, D.H. (1989) Effect of seed crop management on potential seed yield of contrasting white clover varieties. II. Seed yield components and potential seed yield. *Grass and Forage Science* 44, 189–193.

Hopkins, A., Wainwright, J., Murray, P.J., Bowling, P.J. and Webb, M. (1988) 1986 survey of upland grassland in England and Wales: changes in age structure and botanical composition since 1970–72 in relation to grassland management and physical features. *Grass and Forage Science* 43, 185–198.

Hopkins, A., Davies, D.A. and Doyle, C. (1994) *Clovers and Other Grazed Legumes in UK Pasture Land.* IGER Technical Review No. 1, Institute of Grassland and Environmental Research, Aberystwyth, 61 pp.

Hussain, S.W. and Williams, W.M. (1998) Development of a fertile genetic bridge between *Trifolium ambiguum* M. Bic. and *T. repens* L. *Theoretical and Applied Genetics* (in press).

Hutchinson, K.J., King, K.L. and Wilkinson, D.R. (1995) Effects of rainfall, moisture stress, and stocking rate on the persistence of white clover over 30 years. *Australian Journal of Experimental Agriculture* 35, 1039–1047.

ITEB and EDE de Bretagne (1987) Des pâtures riches en trèfle blanc. Pourquoi? Institute Technique d'Elevage Bovin and Etablissements d'Utilité Agricole d'Elevage, September issue.

Jackman, R.H. and Mouat, M.C.H. (1972) Competition between grass and clover for phosphate. II. Effect of root activity, efficiency of response to phosphate, and soil moisture. *New Zealand Journal of Agricultural Research* 15, 667–675.

Jarvis, S.C., Wilkins, R.J. and Pain, B.F. (1996) Opportunities for reducing the environmental impact of dairy farming managements: a systems approach. *Grass and Forage Science* 51, 21–31.

Johnson, J. and Merrell, B.G. (1994) Practical pasture management in hill and upland systems. In: Lawrence, T.L.J., Parker, D.S. and Rowlinson, P. (eds) *Livestock Production and Land Use in Hills and Uplands. Occasional Publication No. 18*, British Society of Animal Production, Edinburgh, pp. 31–41.

Jones, D.R. and Davies, A. (1988) The effects of simulated continuous grazing on development and senescence of white clover. *Grass and Forage Science* 43, 421–425.

Jones, L. (1992) Preliminary trials using a white clover (*Trifolium repens* L.) understorey to supply the nitrogen requirements of a cereal crop. *Grass and Forage Science* 47, 366–374.

Jones, L. and Clements, R.O. (1993) Development of a low-input system for growing wheat (*Triticum vulgare*) in a permanent understorey of white clover. *Annals of Applied Biology* 123, 109–119.

Jones, R.M. (1980) Survival of seedlings and primary taproots of white clover in subtropical pastures in south-east Queensland. *Tropical Grasslands* 14, 19–21.

Jones, R.M. (1982) White clover (*Trifolium repens*) in subtropical south-east Queensland. 1. Some effects of site, season and management practices on the population dynamics of white clover. *Tropical Grasslands* 16, 118–127.

Jones, M.B. and Sinclair, A.G. (1991) Application of DRIS to white clover based pastures. *Communications in Soil and Plant Analysis* 22, 1895–1918.

Jones, W.T., Broadhurst, R.B. and Lyttleton, J.W. (1976) The condensed tannins of pasture legume species. *Phytochemistry* 15, 1407–1409.

Jongen, M., Fay, P. and Jones, M.B. (1996) Effects of elevated carbon dioxide and arbuscular mycorrhizal infection on *Trifolium repens*. *New Phytologist* 132, 413–423.

Kakes, P. (1990) Properties and functions of the cyanogenic system in higher plants. *Euphytica* 48, 25–43.

Kelling, K.A. and Matocha, J.E. (1990) Plant analysis as an aid in fertilizer forage crops. In: Kelling, K.A., Matocha, J.E. and Westerman, R.L. (eds) *Soil Testing and Plant Analysis*, 3rd edn. Book Series No. 3, Soil Science Society of America, Madison, Wisconsin, pp. 603–643.

Kemball, W.D. and Marshall, C. (1994) The significance of nodal rooting in *Trifolium repens* L. P-32 distribution and local growth responses. *New Phytologist* 127, 83–91.

Kemball, W.D. and Marshall, C. (1995) Clonal integration between parent branch stolons in white clover – a developmental study. *New Phytologist* 129, 513–521.

Kemball, W.D., Sackville Hamilton, N.R. and Charnock, R.B. (1996) Population dynamics of white clover stolons in clover rich patches. In: Younie, D. (ed.) *Legumes in Sustainable Farming Systems*. Occasional Symposium No. 30, British Grassland Society Reading, pp. 183–184.

Kendall, W.A. and Stringer, W.C. (1985) Physiological aspects of clover. In: Taylor, N.L. (ed.) *Clover Science and Technology*. ASA/CSSA/SSSA, Madison, Wisconsin, pp.111–160.

Kessler, W. and Nösberger, J. (1994) Factors limiting white clover growth in grass/clover systems. In: Mannetje, L. 't and Frame, J. (eds) *Grassland and Society. Proceedings of the 15th General Meeting of the European Grassland Federation, Wageningen, The Netherlands*. Wageningen Pers, Wageningen, pp. 525–538.

King, J., Lamb, W.I.C. and McGregor, M.T. (1978) Effect of partial and complete defoliation on regrowth of white clover plants. *Journal of the British Grassland Society* 33, 49–55.

Kleter, H.J. (1968) Influence of weather and nitrogen fertilization on white clover percentage of permanent grassland. *Netherlands Journal of Agricultural Science* 16, 43–52.

Kunelius, H.T. and Narasimhalu, P.R. (1993) Effect of autumn harvest date on herbage yield and composition of grasses and white clover. *Field Crops Research* 31, 341–349.

Laidlaw, A.S. (1978) Control of white clover content in swards by varying sowing rates of perennial ryegrass and white clover seeds. *Record of Agricultural Research in Northern Ireland* 26, 21–27.

Laidlaw, A.S. (1980) The effects of nitrogen fertilizer applied in spring in swards of ryegrass sown with four cultivars of white clover. *Grass and Forage Science* 35, 295–299.

Laidlaw, A.S. and McBride, J. (1992) The effect of time of sowing and sowing method on production of white clover in mixed swards. *Grass and Forage Science* 47, 203–210.

Laidlaw, A.S. and Stewart, T.A. (1987) Clover development in the sixth to ninth year of a grass/clover sward as affected by out-of-season management and spring fertilizer nitrogen application. *Research and Development in Agriculture* 4, 155–160..

Laidlaw, A.S. and Vertès, F. (1993) White clover under grazing conditions: morphology and growth. *FAO/REUR Technical Series* 29, 24–29.

Laidlaw, A.S. and Withers, J.A. (1989) The effect of accumulated herbage mass in winter and spring on white clover development. In: *Proceedings of the XVI International Grassland Congress, Nice, France*, Vol. II, Association Française pour la Production Fourragère, Versailles, pp. 1045–1046.

Laidlaw, A.S., Teuber, N.G. and Withers, J.A. (1992) Out-of-season management of grass/clover swards to manipulate clover content. *Grass and Forage Science* 47, 220–229.

Laidlaw, A.S., Patterson, J.D. and Withers, J.A. (1995a) Canopy structure and white clover development. *FAO/REU Technical Series* 42, 69–71.

Laidlaw, A.S., Withers, J.A. and Toal, L.G. (1995b) The effect of surface heights of swards continuously stocked with cattle on herbage production and clover content over four years. *Grass and Forage Science* 50, 48–54.

Laidlaw, A.S., Christie, P. and Lee, H.W. (1996) Effect of white clover cultivar on apparent transfer of nitrogen from clover to grass and estimation of relative turnover rates of nitrogen in roots. *Plant and Soil* 179, 243–253.

Laissus, R. (1983) How to use nitrogen fertilisers on a grass–white clover sward. In: Corrall, J. (ed.) *Efficient Grassland Farming*. Occasional Symposium No. 14, British Grassland Society, Hurley, pp. 223–226.

Lampkin, N. (1990) *Organic Farming*. Farming Press Books, Ipswich.

Lancashire, J.A. (1990) Special address: 150 years of grassland development in New Zealand. *Proceedings of the New Zealand Grassland Association* 52, 9–15.

Lancashire, J.A., Rolston, M.P. and Scott, D.J. (1985) Contamination of white clover seed crops by buried seeds. In: Hare, M.D. and Brock, J.L. (eds) *Producing Herbage Seeds*. Grassland Research and Practice Series No. 2, New Zealand Grassland Association, Palmerston North, pp. 61–65.

Leath, K.T. (1985) General diseases. In: Taylor, N.L. (ed.) *Clover Science and Technology*. ASA/CSSA/SSSA, Madison, Wisconsin, pp. 205–233.

Leath, K.T. (1989) Diseases and forage stand persistence in the United States. In: Marten, G.C., Matches, A.G., Barnes, R.F., Brougham, R.W., Clements, R.J. and Sheath, G.W. (eds) *Persistence of Forage Legumes*. American Society of Agronomy, Madison, Wisconsin, pp. 465–479.

Ledgard, S.F. (1991) Transfer of fixed nitrogen from white clover to associated grasses in swards grazed by dairy cows estimated using ^{15}N methods. *Plant and Soil* 131, 215–223.

Lee, C.K., Reed, K.F.M., Cunningham, P.J. and Rowe, J.G. (1993) Breeding white clover for increased persistence and winter growth. In: Baker, M.J. (ed.) *Proceedings of the XVII International Grassland Congress, New Zealand and Australia*, Vol. 1. New Zealand Grassland Association *et al.*, Palmerston North, pp. 413–415

Lehmann, J., Meister, E., Gutzwiller, A., Jans, F., Charles, J.P. and Blum, J. (1991) Peut-on utiliser des variétés de trèfle blanc (*Trifolium repens* L.) à forte teneur en acide cyanhydrique? [Should one use white clover *Trifolium repens* L. varieties rich in hydrogen cyanide?]. *Revue Suisse d'Agriculture* 23, 107–112.

Lewis, G.C. and Asteraki, E.J. (1987) Incidence and severity of damage by pests and diseases to leaves of twelve cultivars of white clover. Tests of Agrochemicals and Cultivars No. 8. *Annals of Applied Biology* 110 (Suppl.), 140–141.

Lewis, G.C. and Thomas, B.J. (1991) Incidence and severity of pest and disease damage to white clover foliage at 16 sites in England and Wales. *Annals of Applied Biology* 118, 1–8.

Li, X.L., Marschner, H. and George, E. (1994) Acquisition of phosphorus and copper by VA mycorrhizal hyphae and root-to-shoot transport in white clover. *Plant and Soil* 136, 49–57.

Lockhart, J.A.R., Samuel, A. and Greaves, M.P. (1990) The evolution of weed control in British agriculture. In: Hance, R.J. and Holly, K. (eds) *Weed Control Handbook: Principles*, 8th edn. Blackwell Scientific Publications, Oxford, pp. 43–74.

Lotscher, M. and Hay, M.J.M. (1996) Distribution of phosphorus and calcium from nodal roots of *Trifolium repens*: the relative importance of transport via xylem or phloem. *New Phytologist* 133, 445–452.

Lotscher, M. and Nösberger, J. (1996) Influence of position and number of nodal roots on outgrowth of axillary buds and development of branches in *Trifolium repens* L. *Annals of Botany* 78, 459–465.

Macfarlane, M.J. and Sheath, G.W. (1984) Clovers – what types for dry hill country? *Proceedings of the New Zealand Grassland Association* 45, 140–150.

Macfarlane, M.J., Korte, C.J. and Gillingham, A.G. (1987) The effect of increasing

winds on the distribution of oversown seed and fertiliser. *Proceedings of the New Zealand Grassland Association* 48, 131–136.
McAdam, J.H. (1983) *Characteristics of Grassland on Hill Farms in N. Ireland – Physical Features, Botanical Composition and Productivity*. Report, Queens University of Belfast/Department of Agriculture for Northern Ireland, Belfast.
McEwen, J. and Johnston, A.E. (1985) Factors affecting the production and composition of mixed grass/clover swards containing modern high yielding clovers. In: *Proceedings of the 18th Colloquium, Edinburgh*. International Potash Institute, Berne, pp. 47–61.
McEwen, J., Day, W., Henderson, I.F., Johnston, A.E., Plumb, R.T. and Poulton, P.R. (1989) Effects of irrigation, N fertilizer, cutting frequency and pesticides on ryegrass, ryegrass–clover mixtures, clover and lucerne grown on heavy and light land. *Journal of Agricultural Science, Cambridge* 112, 227–247.
McGimpsey, H.C. and Mercer, P.C. (1984) Effects of fungicides, including propiconazole and benomyl on clover rot. Tests of Agrochemicals and Cultivars No. 5. *Annals of Applied Biology* 104 (Suppl.), 52–53.
McLaughlin, M.R., Pederson, G.A., Evans, R.R. and Ivy, R.L. (1992) Virus diseases and stand decline in a white clover pasture. *Plant Disease* 76, 158–162.
McNaught, K.J. (1970) Diagnosis of mineral deficiencies in grass-legume pastures by plant analysis. In: Norman, M.J.T. *Proceedings of the XI International Grassland Congress, Surfers' Paradise, Australia*. University of Queensland Press, St Lucia, pp. 334–338.
Mächler, F. and Nösberger, J. (1977) Effect of light intensity and temperature on apparent photosynthesis of altitudinal ecotypes of *Trifolium repens* L. *Oecologia* 31, 73–78.
Manglitz, G.R. (1985) Insects and related pests. In: Taylor, N.L. (ed.) *Clover Science and Technology*. ASA/CSSA/SSSA, Madison, Wisconsin, pp. 269–294.
Marriott, C.A. and Haystead, A. (1993) Nitrogen fixation and transfer. In: Davies, A., Baker, R.D., Grant, S.A. and Laidlaw, A.S. (eds) *Sward Measurement Handbook*, 2nd Edn. British Grassland Society, Reading, pp. 245–264
Marriott, C.A., Smith, M.A. and Baird, M.A. (1987) The effect of sheep urine on clover performance in a grazed upland sward. *Journal of Agricultural Science, Cambridge* 109, 177–185.
Marriott, C.A., Thomas, R.J., Smith, M.A., Logan, K.A.B., Baird, M.A. and Ironside, A.D. (1988) The effect of temperature and nitrogen interactions on growth and nitrogen assimilation of white clover. *Plant and Soil* 111, 43–51.
Marriott, C.A., Smith, M.A. and Brunton, M.A. (1991) Effects of urine in white clover. *FAO/REUR Technical Series* 19, 103–108.
Marshall, A.H. (1995) Peduncle characteristics, inflorescence survival and reproductive growth of white clover (*Trifolium repens* L.). *Grass and Forage Science* 50, 324–330.
Marshall, A.H. and Hides, D.H. (1990) White clover seed production from mixed swards: effect of sheep grazing on stolon density and on seed yield components of two contrasting white clover varieties. *Grass and Forage Science* 45, 35–42.
Marshall, A.H., Hollington, P.A. and Hides, D.H. (1989) Effect of seed crop management on the potential seed yield of contrasting white clover varieties. I. Inflorescence production. *Grass and Forage Science* 44, 181–188.
Marten, G.C., Matches, A.G., Barnes, R.F., Brougham, R.W., Clements, R.J. and Sheath, G.W. (eds) (1989) *Persistence of Forage Legumes*. ASA/CSSA/SSSA, Madison, Wisconsin, pp. 569–572
Martiniello, P. (1992) White clover adaptability for seed yield and seed yield components in a Mediterranean environment. *Herba* 5, 60–64.

Mather, R.D.J., Melhuish, D.T. and Herlihy, M. (1996) Trends in the global marketing of white clover cultivars. In: *White Clover: New Zealand's Competitive Edge*. Grassland Research and Practice Series No. 6, New Zealand Grassland Association, Palmerston North, pp. 7–14.

Mercer, C.F., Van den Bosch, J., Grant, J.L. and Black, I.K. (1990) Breeding *Trifolium repens* for resistance to *Meloidogyne naasi*. I. Characterization by host preference. *Journal of Nematology* 5, 41–43.

Milne, G. and Fraser, T.J. (1990) Establishment of 1600 hectares in dryland species around Oamaru and Timaru. *Proceedings of the New Zealand Grassland Association* 52, 133–138.

Milne, J.A., Hodgson, J., Thompson, R., Souter, W.G. and Barthram, G.T. (1982) The diet ingested by sheep grazing swards differing in white clover and perennial ryegrass content. *Grass and Forage Science* 37, 209–218.

Ministry of Agriculture, Fisheries and Food (1983) *Certification of Seed of Grasses and Herbage Legumes*. MAFF, London, 74 pp.

Moloney, S.C. (1990) Performance of tall fescue, cocksfoot and phalaris based pastures compared with perennial ryegrass, in on-farm trials. *Proceedings of the New Zealand Grassland Association* 53, 41–46.

Moloney, S.C. (1994) Selection, management and use of cocksfoot cultivars in North Island pastoral farming. *Proceedings of the New Zealand Grassland Association* 55, 119–125.

Moseley, G. and Jones, J.R. (1984) The physical digestion of perennial ryegrass (*Lolium perenne*) and white clover (*Trifolium repens*) in the foregut of sheep. *British Journal of Nutrition* 52, 381–390.

Mowat, D.J. and Shakeel, M.A. (1988) The effect of pesticide application on the establishment of white clover in a newly-sown ryegrass–white clover sward. *Grass and Forage Science* 43, 371–375.

Mowat, D.J. and Shakeel, M.A. (1989) The effect of different cultivars of clover on numbers of, and leaf damage by, some invertebrate species. *Grass and Forage Science* 44, 11–18.

Mtengeti, E.J., Wilman, D. and Moseley, G. (1995) Physical structure of white clover, rape, spurrey and perennial ryegrass in relation to rate of intake by sheep, chewing activity and particle breakdown. *Journal of Agricultural Research, Cambridge* 125, 43–50.

Mytton, L.R. and Hughes, D.M. (1984) Inoculation of white clover with different strains of *Rhizobium trifolii* on a mineral hill soil. *Journal of Agricultural Science, Cambridge* 102, 455–459.

Naylor, R.E.L., Marshall, A.H. and Matthews, S. (1983) Seed establishment in directly drilled sowings. *Herbage Abstracts* 53, 73–91.

Nelson, S.C. and Campbell, C.L. (1993) Comparative spacial analysis of foliar epidemics on white clover caused by viruses, fungi and a bacterium. *Phytopathology* 83, 288–301.

Nesheim, L., Boller, B.C., Lehmann, J. and Walther, U. (1990) The effect of nitrogen in cattle slurry and mineral fertilizers on nitrogen fixation by white clover. *Grass and Forage Science*, 45, 91–97.

Newbould, P. (1982) Biological nitrogen fixation in upland and marginal areas of the UK. *Transactions of the Philosophical Society of London*, B 296, 405–417.

Newbould, P., Holding, A.J., Davies, G.J., Rangeley, A., Copeman, G.J.F., Davies, A., Frame, J., Haystead, A., Herriott, J.B.D., Holmes, J.C., Lowe, J.F., Parker, J.W.G.,

Waterson, H.A., Wildig, J., Wray, J.P. and Younie, D. (1982) The effect of *Rhizobium* inoculation on white clover in improved hill soils in the United Kingdom. *Journal of Agricultural Science, Cambridge* 99, 591–610.

Newman, J.A., Parsons, A.J. and Harvey, A. (1992) Not all sheep prefer clover: diet selection revisited. *Journal of Agricultural Science, Cambridge* 119, 275–283.

Newton, P.C.D. and Hay, M.J.M. (1994) Patterns of nodal rooting in *Trifolium repens* L. and correlations with stages in the development of axillary buds. *Grass and Forage Science* 49, 270–276.

Newton, P.C.D. and Hay, M.J.M. (1996) Clonal growth of white clover: factors influencing the viability of axillary buds and the outgrowth of a viable bud to form a branch. *Annals of Botany* 78, 111–115.

Newton, P.C.D., Hay, M.J.M., Thomas, V.J., Glasgow, E.M. and Dick, H.B. (1990) Patterns of axillary bud activity in white clover. *Proceedings of the New Zealand Grassland Association* 52, 247–249.

Newton, P.C.D., Clark, H., Bell, C.C. and Glasgow, E.M. (1996) Interaction of soil moisture and elevated CO_2 on the above ground growth rate, root length density and gas exchange of turves from temperate pasture. *Journal of Experimental Botany* 47, 771–779.

Norris, I.B. (1989) *Trifolium repens*. In: Halevy, A.H. (ed.) *Handbook of Flowering.*, Vol. IV. CRC Press, Boca Raton, Florida, pp. 630–636.

Nösberger, J., Lüscher, A., Hebeisen, T., Zanetti, S. and Fischer, B. (1995) The effect of elevated CO_2 on growth of perennial ryegrass and white clover. In: Pollott, G.E. (ed.) *Grassland into the 21st Century*. Occasional Symposium No. 29, British Grassland Society, Reading, pp. 243–244.

Nutman, P.S. (1962) The relationship between root hair infection by *Rhizobium* and nodulation in *Trifolium* and *Vicia*. *Proceedings of the Royal Society of London* 156, 122–137.

O'Rourke, C.J. (1976) *Diseases of Grasses and Forage Legumes in Ireland*. Agricultural Institute, Dublin, pp. 79–82.

Orr, R.J., Parsons, A.J., Penning, P.D. and Treacher, T.T. (1990) Sward composition, animal performance and the potential production of grass/white clover swards continuously stocked with sheep. *Grass and Forage Science* 45, 325–336.

Orr, R.J., Penning, P.D., Parsons, A.J. and Champion, R.A. (1995) Herbage intake and N excretion by sheep grazing monocultures or a mixture of grass and white clover. *Grass and Forage Science* 50, 31–40.

Orr, R.J., Rutter, S.M., Penning, P.D., Yarrow, N.H. and Champion, R.A. (1996) Grazing behaviour and herbage intake rate by Friesian dairy heifers grazing ryegrass or white clover. In: Younie, D. (ed.) *Legumes in Sustainable Farming Systems*. Occasional Symposium No. 30, British Grassland Society, Reading, pp. 221–224.

Pain, B. (1991) Improving the utilization of slurry and farm effluents. In: Mayne, C.S. (ed.) *Management Issues for the Grassland Farmer in the 1990s*. Occasional Symposium No. 25, British Grassland Society, Reading, pp. 121–133.

Parente, G. and Frame, J. (1993) Alternative uses of white clover. *FAO/REUR Technical Series* 29, 30–36.

Parsons, A.J., Harvey, A. and Woledge, J. (1991a) Plant–animal interactions in a continuously grazed mixture. I. Differences in the physiology of leaf expansion and the fate of leaves of grass and clover. *Journal of Applied Ecology* 28, 619–634.

Parsons, A.J., Harvey, A. and Johnson, I.R. (1991b) Plant–animal interactions in a continuously grazed mixture. II. The role of differences in the physiology of plant growth and of selective grazing on the performance and stability of species in a mixture. *Journal of Applied Ecology* 28, 635–658.

Parsons, A.J., Penning, P.D., Lockyer, D.R. and Ryden, J.C. (1991c) Uptake, cycling and fate of nitrogen in grass–clover swards continuously grazed by sheep. *Journal of Agricultural Science, Cambridge* 116, 47–61.

Parsons, A.J., Newman, J.A., Penning, P.D., Harvey, A. and Orr, R.J. (1994) Diet preference of sheep – effects of recent diet, physiological state and species abundance. *Journal of Animal Behaviour Science* 63, 465–478.

Pascal, J.A. and Sheppard, B.W. (1985) The development of a strip-seeder for sward improvement. *Research and Development in Agriculture* 2, 125–134.

Patterson, J.D., Laidlaw, A.S. and McBride, J. (1995) The influence of autumn management and companion grass on the development of white clover over winter in mixed swards. *Grass and Forage Science* 50, 345–352.

Pederson, G.A. (1995) White clover and other perennial clovers. In: Barnes, R.F., Miller, D.A. and Nelson, C.J. (eds) *Forages*, 5th edn, Vol. I. *An Introduction to Grassland Agriculture*. Iowa State University Press, Ames, Iowa, pp. 227–236.

Pederson, G.A., Windham, G.L., McLaughlin, M.R., Pratt, R.G. and Brink, G.E. (1993) Breeding for multiple pest resistance as a strategy to improve white clover persistence. In: Baker, M.J. (ed.) *Proceedings of the XVII International Grassland Congress, New Zealand and Australia*, Vol. 1. New Zealand Grassland Association *et al.*, Palmerston North, pp. 926–927.

Penning, P.D., Rook, A.J. and Orr, R.J. (1991) Patterns of ingestive behaviour of sheep continuously stocked on monocultures of ryegrass or white clover. *Applied Animal Behaviour Science* 31, 237–250.

Penning, P.D., Parsons, A.J., Orr, R.J., Harvey, A. and Champion, R.A. (1995) Intake and behaviour responses by sheep in differing physiological states, when grazing monocultures of grass and clover. *Applied Animal Behaviour Science* 45, 63–78.

Pflimlin, A. (1993) Conduite et utilisation des associations graminée–trèfle blanc. *Fourrages* 135, 407–428.

Pottinger, R.P., Barbetti, M.J. and Ridsdill-Smith, T.J. (1993) Invertebrate pests, plant pathogens and beneficial organisms of improved temperate pasture. In: Baker, M.J. (ed.) *Proceedings of the XVII International Grassland Congress, New Zealand and Australia*, Vol. I. New Zealand Grassland Association *et al.*, Palmerston North, pp. 909–918.

Powell, C.L. (1976) Mycorrhizal fungi stimulate clover growth in New Zealand hill country soils. *Nature* 264, 436–438..

Prestidge, R.A., Thorn, E.R., Marshall, S.L., Taylor, M.J., Willoughby, B. and Wildermoth, D.D. (1992) Influence of *Acremonium lolii*-infected perennial ryegrass on germination, survival and growth of white clover. *New Zealand Journal of Agricultural Research* 35, 225–234.

Rangeley, A. and Newbould, P. (1985) Growth response to lime and fertilizers and critical concentrations in herbage of white clover in Scottish hill soils. *Grass and Forage Science* 40, 265–277.

Rattray, P.V. and Joyce, J.P. (1974) Nutritive value of white clover and perennial ryegrass for young sheep. IV. Utilization of dietary energy. *New Zealand Journal of Agricultural Research* 17, 401–406.

Reid, D. (1978) The effects of frequency of defoliation on the yield response of a perennial ryegrass sward to a wide range of nitrogen applications. *Journal of Agricultural Science, Cambridge* 79, 291–301.

Reid, D. (1983) The combined use of fertilizer nitrogen and white clover as nitrogen sources for herbage growth. *Journal of Agricultural Science, Cambridge*, 100, 613–623.

Rhodes, I. (1991) Progress in white clover breeding. *FAO/REUR Technical Series* 19, 1–9.

Rhodes, I. and Ortega, F. (1996) Progress in forage legume breeding. In: Younie, D. (ed.) *Legumes in Sustainable Farming Systems*. Occasional Symposium No. 30, British Grassland Society, Reading, pp. 62–71.

Rhodes, I., Collins, R.P., Evans, D.R. and Glendining, M.J. (1989) Breeding reliable white clover for low input pastures. In: *Proceedings of the XVI International Grassland Congress, Nice, France*, Vol. I. Association Française pour la Production Fourragère. Versailles, pp. 315–316.

Rhodes, I., Collins, R.P. and Evans, D.R. (1994) Breeding white clover for tolerance to low temperature and grazing stress. *Euphytica* 77, 239–242.

Rickard, D.S. and McBride, S.D. (1986) *Irrigated and Non-irrigated Pasture Production at Winchmore*. Technical Report 21, Winchmore Irrigation Research Station, Winchmore.

Ridout, M.S. and Robson, M.J. (1991) Diet composition of sheep grazing grass/white clover swards: a re-evaluation. *New Zealand Journal of Agricultural Research* 34, 89–93.

Roberts, D.J., Frame, J. and Leaver, J.D. (1989) A comparison of a grass/white clover sward with a grass sward plus nitrogen under a three-cut regime. *Research and Development in Agriculture* 6, 147–150.

Robin, C., Hay, M.J.M., Newton, P.C.D. and Greer, D.H. (1994) Effect of light quality (red:far red ratio) at the apical bud of the main stolon on morphogenesis of *Trifolium repens* L. *Annals of Botany* 74, 119–123.

Rodriguez Juliá, M. (1991) Desarollo y evaluación del sistema integrado de diagnostico y recomendación (DRIS) para la fertilización de las praderas permanentes. PhD Thesis, University of the Basque State, 255 pp.

Rogers, J.B. (1993) Investigation of the role of mycorrhizas in the transfer of nitrogen from white clover to grass. PhD Thesis, Queen's University of Belfast, 216 pp.

Rogers, M.E. and Noble, C.L. (1992) Variation in growth and ion accumulation between two selected populations of *Trifolium repens* L. differing in salt tolerance. *Plant and Soil* 146, 131–136.

Rogers, M.E., Noble, C.L., Nicolas, M.E. and Halloran, G.M. (1993) Variation in yield potential and salt tolerance of selected cultivars and natural populations of *Trifolium repens* L. *Australian Journal of Agricultural Research* 44, 785–798.

Rogers, M.E., Noble, C.L., Nicolas, M.E. and Halloran, G.M. (1994) Leaf, stolon and root growth of white clover (*Trifolium repens* L.) in response to irrigation with saline water. *Irrigation Science* 15, 183–194.

Royal Society (1983) *The Nitrogen Cycle of the United Kingdom*. Report of a Royal Society Study Group. Royal Society, London.

Ryan, M. (1989) Development of a legume-based dairy system. *Developments in Plant and Soil Sciences* 37, 159–167.

Ryle, G.J.A., Powell, C.E. and Gordon, A.J. (1979) The respiratory costs of nitrogen fixation in soybean, cowpea and white clover. II. Comparisons of the cost of nitrogen fixation and the utilization of combined nitrogen. *Journal of Experimental Botany* 30, 145–153.

Ryle, G.J.A., Powell, C.E., Timbrell, M.K. and Gordon, A.J. (1989) Effect of temperature and nitrogenase activity in white clover. *Journal of Experimental Botany* 40, 733–739.

Ryle, G.J.A., Woledge, J., Tewson, V. and Powell, C.E. (1992) Influence of elevated CO_2 and temperature on the photosynthesis and respiration of white clover dependent on N_2 fixation. *Annals of Botany* 70, 213–220.

Sanchez-Diaz, M. and Sanchez-Marin, M. (1974) Resistance of leaf tissue to desiccation and stomatal closure in lucerne (*Medicago sativa* L.) and white clover (*Trifolium repens* L.) in relation to water stress. *Annales de Edafología y Agrobiología* 33, 743–754.

Schenk, U., Jager, H.J. and Weigel, H.J. (1996) Nitrogen supply determines responses of yield and biomass partitioning of perennial ryegrass to elevated atmospheric carbon dioxide concentrations. *Journal of Plant Nutrition* 19, 1423–1440.

Schwinning, S. and Parsons, A.J. (1996) Interactions between grasses and legumes: understanding variability in species composition. In: Younie, D. (ed.) *Legumes in Sustainable Farming Systems*. Occasional Symposium No. 30, British Grassland Society, Reading, pp. 153–163.

Scott, D.J. and Hampton, J.G. (1985) Aspects of seed quality. In: Hare, M.D. and Brock, J.L. (eds) *Producing Herbage Seeds*. Grassland Research and Practice Series No. 2, New Zealand Grassland Association, Palmerston North, pp. 43–52.

Scott, R.S. (1976) The phosphate nutrition of white clover. *Proceedings of the New Zealand Grassland Association* 38, 151–159.

Scott, S.W. (1982) Tests for resistance to white clover mosaic virus in red and white clover. *Annals of Applied Biology* 100, 393–398.

Sheldrick, R.D., Lavender, R.H. and Parkinson, A.E. (1987) The effect of subsequent management on the success of introducing white clover to an existing sward. *Grass and Forage Science* 42, 359–371.

Sheppard, B.W. and Swift, G. (1995) *Successful Strip Seeding*. Scottish Centre of Agricultural Engineering Research Summary No. 1 (2nd revision), Edinburgh, 14 pp.

Sinclair, A.G., Smith, L.C., Morrison, J.D. and Dodds, K.G. (1996a) Effects and interactions of phosphorus and sulphur on a mown white clover/ryegrass sward. 1. Herbage dry matter production and balanced nutrition. *New Zealand Journal of Agricultural Research* 39, 421–433.

Sinclair, A.G., Morrison, J.D., Smith, L.C. and Dodds, K.G. (1996b) Effects and interactions of phosphorus and sulphur in a mown white clover/ryegrass sward. 2. Concentrations and ratios of phosphorus, sulphur and nitrogen in clover herbage in relation to balanced plant nutrition. *New Zealand Journal of Agricultural Research* 39, 435–445.

Sinoquet, H., Moulia, B., Gastal, F., Bonhomme, R. and Varlet-Grancher, C. (1990) Modelling the radiative balance of the components of a binary mixed canopy: application to a white clover/tall fescue mixture. *Acta Oecologia* 11, 469–486.

Skipp, R.A. and Lambert, M.G. (1984) Damage to white clover foliage in grazed pasture caused by fungi and other organisms. *New Zealand Journal of Agricultural Research* 27, 313–320.

Smetham, M.L. (1973) Pasture legume species and strains. In: Langer, R.H.M. (ed.) *Pastures and Pasture Plants*. A.H. and A.W. Reed, Wellington, pp. 85–127.

Smith, S.R. and Giller, K.E. (1992) Effective *Rhizobium leguminosarum* biovar. *trifolii* present in five soils contaminated with heavy metals from long-term application of sewage sludge or metal mine spoil. *Soil Biology and Biochemistry* 24, 781–788.

Snaydon, R.W. and Bradshaw, A.D. (1969) Differences between natural populations of *Trifolium repens* L. in response to mineral nutrients. II. Calcium, magnesium and potassium. *Journal of Applied Ecology* 6, 185–202.

Søegaard, K. (1994) Agronomy of white clover. In: Mannetje, L. 't and Frame, J. (eds) *Grassland and Society. Proceedings of the 15th General Meeting of the European Grassland Federation, Wageningen, The Netherlands.* Wageningen Pers, Wageningen, pp. 515–524.

Solangaarachchi, S.M. and Harper, J.L. (1987) The effect of canopy filtered light on the growth of white clover *(Trifolium repens). Oecologia (Berlin)* 72, 372–376.

Soussana, J.F., Vertès, F. and Arregui, M.C. (1995) The regulation of clover shoot growing points density and morphology during short-term clover decline in mixed swards. *European Journal of Agronomy* 4, 205–215.

South, A. (1975) Estimation of slug populations. *Annals of Applied Biology* 53, 255–258.

Sprent, J.I. and Minchin, F.R. (1983) Environmental effects on the physiology of nodulation and nitrogen fixation. In: Jones, D.G. and Davies, D.R. (eds) *Temperate Legumes.* Pitman, London, pp. 269–310.

Steen, R.W.J. and Laidlaw, A.S. (1995) The effect of fertilizer nitrogen input on the stock-carrying capacity of ryegrass/white clover swards continuously grazed by beef cattle. *Irish Journal of Agricultural and Food Research* 34, 123–132.

Stevens, D.R., Drew, K., Laas, F. and Turner, J.D. (1992) Deer production from ryegrass- and tall fescue-based pastures. *Proceedings of the New Zealand Grassland Association* 54, 23–26.

Stevenson, C.A. and Laidlaw, A.S. (1985) The effect of moisture stress on stolon and adventitious root development in white clover *(Trifolium repens* L.). *Plant and Soil* 85, 249–257.

Stewart, T.A. (1984) Utilising white clover in grass based animal production systems. In: Thomson, D.J. (ed.) *Forage Legumes.* Occasional Symposium No. 16, British Grassland Society, Hurley, pp. 93–103.

Stewart, T.A. (1988) *A Decade of Beef from a Grass/White Clover Sward – The Greenmount experience.* Report, Greenmount College of Agriculture and Horticulture, Northern Ireland.

Stewart, T.A. and Haycock, R.E. (1984) Beef production from low N and high N S 24 perennial ryegrass/Blanca white clover swards – a six year farmlet scale comparison. *Research and Development in Agriculture* 1, 103–113.

Suckling, F.E.T. (1951) Results of recent experiments on surface sowing. *Proceedings of the New Zealand Grassland Association* 13, 119–127.

Suckling, F.E.T. (1960) Productivity of pasture species on hill country. *New Zealand Journal of Agricultural Research* 3, 579–591.

Suckling, F.E.T. and Charlton, J.F.L. (1978) A review of the significance of buried legume seeds with particular reference to New Zealand agriculture. *New Zealand Journal of Experimental Agriculture* 6, 211–215.

Sutherland, O.R.W. (1975) A chemical basis for plant resistance to grass grub and black beetle larvae (Coleoptera: Scarabaeidae). *Proceedings of the New Zealand Grassland Association* 37, 126–131.

Suttle, N.F. (1983) The nutritional basis for trace element deficiencies in ruminant livestock. In: Suttle, N.F., Gunn, R.G., Allen, W.M., Linklater, K.A. and Wiener, G. (eds) *Trace Elements in Animal Production and Veterinary Practice.Occasional Publication No. 7,* British Society of Animal Production, pp. 19–25.

Swift, G., Morrison, M.W., Cleland, A.T., Smith-Taylor, C.A.B. and Dickson, J.M. (1992)

Comparison of white clover varieties under cutting and grazing. *Grass and Forage Science* 47, 8–13.

Taylor, J.A. (1976) Upland climates. In: Chandler, T.J. and Gregory, S. (eds) *The Climate of the British Isles*. Longmans, London, pp. 204–287.

Taylor, N.L. (1985) Clovers around the world. In: Taylor, N.L. (ed.) *Clover Science and Technology* ASSA/CSSA/SSSA, Madison, Wisconsin, pp. 2–6.

Taylor, N.L., Quesenberry, K.H. and Anderson, M.K. (1980) Genetic system relationships in the genus *Trifolium*. *Economic Botany* 36, 431–441.

Ten Holte, L. and Van Keulen, H. (1989) Effects of white clover and red clover as a green crop on growth, yield and nitrogen of sugar beet and potatoes. *Developments in Plant and Soil Sciences* 37, 16–24.

Teuber, N. and Laidlaw, A.S. (1995) Effect of herbage rejection by steers on white clover (*Trifolium repens*) branching and development in continuously stocked grass–clover swards. *Journal of Agricultural Science, Cambridge* 124, 205–212.

Teuber, N. and Laidlaw, A.S. (1996) Influence of irradiance on branch growth of white clover stolons in rejected areas within grazed swards. *Grass and Forage Science* 51, 73–80.

Theodorou, M.K., Austin, A.R. and Hitching, S. (1984) A comparison of steers fed on grass and on clover in relation to some microbiological aspects of bloat. In: Thomson, D.J. (ed.) *Forage Legumes*. Occasional Symposium No. 16, British Grassland Society, Hurley, pp. 104–108.

Thomas, H. (1984) Effects of drought on growth and competitive ability of perennial ryegrass and white clover. *Journal of Applied Ecology* 21, 591–602.

Thomas, R.G. (1987) Reproductive development. In: Baker, M.J. and Williams, W.M. (eds) *White Clover*. CAB International, Wallingford, pp. 63–123.

Thomas, R.J., Logan, K.A.B. and Ironside, A.D. (1986) Fate of sheep urine applied to an upland grass sward. *Plant and Soil* 91, 425–427.

Thompson, L. (1993) The influence of radiation environment around the node on morphogenesis and growth of white clover (*Trifolium repens* L.). *Grass and Forage Science* 48, 271–278.

Thompson, L. (1995) Sites of photoperception in white clover. *Grass and Forage Science* 50, 259–262.

Thompson, L. and Harper, J.L. (1988) The effect of grasses on the quality of transmitted radiation and its influence on the growth of white clover (*Trifolium repens*). *Oecologia* 75, 343–347.

Thomson, D.J. (1979) Effect of proportion of legumes in the sward on animal output. In: Charles, A.H. and Haggar, R.J. (eds) *Changes in Sward Composition and Productivity*. Occasional Symposium No. 10, British Grassland Society, Hurley, pp. 101–109.

Thomson, D.J. (1984). The nutritive value of white clover. In: Thomson, D.J. (ed.) *Forage Legumes*. Occasional Symposium No. 16, British Grassland Society, Hurley, pp. 78–92.

Thomson, D.J., Beever, D.E., Haines, M.J., Cammell, S.B., Evans, R.T., Dhanoa, M.S. and Austin, A.R. (1985) Yield and composition of milk from Friesian cows grazing either perennial ryegrass or white clover in early lactation. *Journal of Dairy Research* 52, 17–31.

Thornley, J.H.M., Bergelson, J. and Parsons, A.J. (1995) Complex dynamics in a carbon–nitrogen model of a grass–legume pasture. *Annals of Botany* 75, 79–94.

Tiley, G.E.D. and Frame, J. (1991) Improvement of upland permanent pastures and lowland swards by surface sowing methods. In: *Proceedings of a Conference of the*

European Grassland Federation, Graz, Austria. Federal Research Institute for Agriculture in Alpine Regions, Gumpenstein (BAL), pp. 89–94.

Tiley, G.E.D., Swift, G. and Younie, D. (1986). *Grass and Clover Varieties for the Hills and Uplands*. Scottish Agricultural Colleges Research and Development Note No. 30. SAC, Aberdeen.

Topp, C.F.E. and Doyle, C.J. (1996a) Simulating the impact of global warming on milk and forage production in Scotland. 1. The effects of dry matter yield of grass and grass/clover swards. *Agricultural Systems* 52, 213–242.

Topp, C.F.E. and Doyle, C.J. (1996b) Simulating the impact of global warming on milk and forage production in Scotland. 2. The effects on milk yield and grazing management of dairy herds. *Agricultural Systems* 52, 243–270.

Trolove, S.N., Hedley, M.J., Caradus, J.R. and Mackay, A.D. (1996) Uptake of phosphorus from different sources by *Lotus pedunculatus* and three genotypes of *Trifolium repens* 2. Forms of phosphate used and acidification of the rhizosphere. *Australian Journal of Soil Research* 34, 1027–1040.

Turner, L.B. (1991) The effect of water stress on the vegetative growth of white clover (*Trifolium repens* L.): comparison of long term water deficit and a short term developing water stress. *Journal of Experimental Botany* 42, 311–316.

Tyson, K.C., Roberts, D.H., Clement, C.R. and Garwood, E.A. (1990) Comparison of crop yields and soil conditions during 30 years under annual tillage or grazed pasture. *Journal of Agricultural Science, Cambridge* 115, 29–40.

Ulyatt, M.J. (1981) The feeding value of temperate pastures. In: Morley, F.H.W. (ed.) *Grazing Animals*. Elsevier, Amsterdam, pp. 125–141.

Ulyatt, M.J., Dellow, D.W., John, A., Reid, C.S.W. and Waghorn, G.C. (1986) Contribution of chewing during eating and rumination to the clearance of digesta from the reticulorumen. In: Milligan, L.P., Grovum, W.L. and Dobson, A. (eds) *Control of Digestion and Metabolism in Ruminants. Proceedings of the VIth International Symposium on Ruminant Physiology, Banff, Canada*. Prentice Hall, Englewood Cliffs, New Jersey, pp. 498–515.

Van Bockstaele, E. (1985) Breeding of white clover (*Trifolium repens* L.): objectives and techniques. In: *Proceedings of Workshop of the Commission of the European Communities, Wexford, Ireland*, The Agricultural Institute, Wexford, pp. 81–98.

Varlet-Grancher, C., Moulia, B. and Jacques, R. (1989) Phytochrome mediated effects on white clover morphogenesis. In: *Proceedings of the XVI International Grassland Congress, Nice, France*, Vol. I. Association Française pour la Production Fourragère, Versailles, pp. 477–478.

Vertès, F., LeCorre, L., Simon, J.C. and Rivière, J.M. (1988) Effets du piétinement de printemps sur un peuplement de trèfle blanc pur ou en association. *Fourrages* 116, 347–366.

Vickery, P.J., Wheeler, J.L. and Mulcahy, C. (1987) Factors affecting the hydrogen cyanide potential of white clover (*Trifolium repens* L.). *Australian Journal of Agricultural Research* 38, 1053–1059.

Vidrih, T., Pratnekar, J., Lusin, J. and Drasler, J. (1991) Some experiences on sod seeding of clover and grass into Slovenian grassland. In: *Proceedings of a Conference of the European Grassland Federation, Graz, Austria*. Federal Research Institute for Agriculture in Alpine Regions, Gumpenstein (BAL), pp. 103–106.

Vipond, J.E. and Swift, G. (1992) Developments in legume use in the hills and uplands. In: Hopkins, A. (ed.) *Grass on the Move*. Occasional Symposium No. 20, British Grassland Society, Reading, pp. 54–65.

Voisey, C.R., White, D.W.R., Dudas, B., Appleby, R.D., Ealing, P.M. and Scott, A.G. (1994) *Agrobacterium*-mediated transformation of white clover using direct organogenesis. *Plant Cell Reports* 13, 309–314.

Waghorn, G.C., Jones, W.T., Shelton, I.D. and McNabb, W.C. (1990) Condensed tannins and the nutritive value of herbage. *Proceedings of the New Zealand Grassland Association* 51, 171–176.

Watson, C.J., Jordan, C., Taggart, P.J., Laidlaw, A.S., Garrett, M.K. and Steen, R.W.J. (1992) The leaky N-cycle on grazed grassland. *Aspects of Applied Biology* 30, 215–222.

Watson, R.N., Yeates, G.W., Littler, R.A. and Steele, K.W. (1985) Responses in nitrogen fixation and herbage production following pesticide applications on temperate pastures. In: Chapman, R.B. (ed.) *Proceedings of the 4th Australasian Conference on Grassland Invertebrate Ecology, Hamilton, New Zealand*. Caxton Press, Christchurch. pp. 103–113.

Watt, T.A. (1978) The biology of *Holcus lanatus* L. (Yorkshire fog) and its significance in grassland. *Herbage Abstracts* 48, 195–204.

Webb, K.J. (1996) Opportunities for biotechnology in forage legume breeding. In: Younie, D. (ed.) *Legumes in Sustainable Farming Systems*. Occasional Symposium No. 30, British Grassland Society,Reading, pp. 77–85.

Webb, K.J. and Rhodes, I. (1991) Opportunities for biotechnology in clover breeding. In: den Nijs, A.P.M. and Elgersma, A. (eds) *Proceedings of the 17th Meeting of Eucarpia, Wageningen, The Netherlands*. Pudoc, Wageningen, pp. 111–116.

Weissinger, A.K.II and Parrott, W.A. (1993) Repetitive somatic embryogenesis and plant recovery in white clover. *Plant Cell Reports* 12, 125–128.

Wheeler, J.L. and Vickery, P.J. (1989) Variation in HCN potential among cultivars of white clover (*Trifolium repens*). *Grass and Forage Science* 44, 107–109.

White, J.G.H. (1990) Hill and high country pasture. In: Langer, R.H.M. (ed.) *Pastures: Their Ecology and Management*. Oxford University Press, Auckland, pp. 299–336.

Widdup, K.H. and Turner, J.D. (1983) Performance of 4 white clover populations in monoculture and with ryegrass under grazing. *New Zealand Journal of Experimental Agriculture* 11, 27–31.

Wightman, P.S. and Younie, D. (1994) Responses of grass/white clover miniswards to slurry application. In: Mannetje, L. t' and Frame, J. (eds) *Grassland and Society. Proceedings of the 15th General Meeting of the European Grassland Federation, Wageningen, The Netherlands*. Wageningen Pers, Wageningen, pp. 611–615.

Wilkinson, S.R. and Gross, C.F. (1967) Macro and micronutrient distribution with Ladino clover (*Trifolium repens* L.). *Agronomy Journal* 59, 372–374.

Williams, R.D. (1984) *Crop Protection Handbook – Grass and Clover Swards*. British Council for Crop Protection, Croydon.

Williams, W.M. (1983) White clover. In: Wratt, G.S. and Smith, H.C. (eds) *Plant Breeding in New Zealand*. Butterworths/DSIR, Wellington, pp. 221–229.

Williams, W.M. (1987) Genetics and breeding. In: Baker, M.J. and Williams, W.M. (eds) *White Clover*. CAB International, Wallingford, pp. 343–420.

Wilman, D. and Acuña, P.G.H. (1993) Effects of cutting height on the growth of leaves and stolons in perennial ryegrass–white clover swards. *Journal of Agricultural Science, Cambridge* 121, 39–46.

Wilman, D. and Altimimi, M.A.K. (1984) The *in vitro* digestibility and chemical composition of plant parts in whtie clover, red clover and lucerne during primary growth. *Journal of the Science of Food and Agriculture* 35, 133–138.

Wilman, D. and Asiegbu, J.E. (1982) The effects of variety, cutting interval and nitrogen application on the morphology and development of stolons and leaves of white clover. *Grass and Forage Science* 37, 15–27.

Wilson, J.R. (1993) Organization of plant tissues. In: Jung, H.G., Buxton, D.R., Hatfield, R.D. and Ralph, J. (eds) *Forage Cell Wall Structure and Digestibility*. ASA/CSSA/SSSA, Madison, Wisconsin, pp. 1–32.

Woledge, J. (1971) The effect of light intensity during growth on the subsequent rate of photosynthesis. *Annals of Botany* 35, 311–322.

Woledge, J. (1988) Competition between grass and clover in spring as affected by nitrogen fertilizer. *Annals of Applied Biology* 47, 175–186.

Woledge, J. and Dennis, W.D. (1982) The effect of temperature on the photosynthesis of ryegrass and white clover leaves. *Annals of Botany* 50, 25–35.

Woledge, J. and Parsons, A.J. (1986) Temperate grasslands. In: Baker, N.R. and Long, S.P. (eds) *Photosynthesis in Contrasting Environments*. Elsevier Science Publishers, Amsterdam, pp. 173–197.

Woledge, J., Davidson, I.A. and Tewson, V. (1989) Photosynthesis during winter in ryegrass/white clover mixtures in the field. *New Phytologist* 113, 15–27

Woledge, J., Tewson, V. and Davidson, I.A. (1990) Growth of grass/clover mixtures during winter. *Grass and Forage Science* 45, 191–202.

Woledge, J., Reyneri, A., Tewson, V. and Parsons, A.J. (1992) The effect of cutting on the proportions of perennial ryegrass and white clover in mixtures. *Grass and Forage Science* 47, 169–179.

Wolfe, E.C. and Lazenby, A. (1973) Grass–white clover relationships during pasture development: effect of nitrogen fertilizer with superphosphate. *Australian Journal of Experimental Agriculture and Animal Husbandry* 13, 575–580.

Wong, E., Flux, D.S. and Latch, G.C.M. (1971) The oestrogenic activity of white clover. *New Zealand Journal of Agricultural Research* 14, 633–638.

Wu, L. and Huang, Z.Z. (1992) Selenium assimilation and nutrient element uptake in white clover and tall fescue under the influence of sulphate concentrations and selenium tolerance of the plants. *Journal of Experimental Botany* 43, 549–555.

Yang, X., Baligar, V.C., Martens, D.C. and Clark, R.B. (1996a) Plant tolerance to nickel. 2. Nickel effects on influx and transport of mineral nutrients in four plant species. *Journal of Plant Nutrition* 19, 265–279.

Yang, X., Baligar, V.C., Martens, D.C. and Clark, R.B. (1996b) Cadmium effects on influx and transport of mineral nutrients in plant species. *Journal of Plant Nutrition* 19, 643–656.

Yeates, G.W., Healy, W.B., Widdowson, J.P., Thomson, N.A. and MacDairmid, B.N. (1975) Influence of nematodes on growth of plots of white clover on a yellow brown loam. *New Zealand Journal of Agricultural Research* 18, 411–416.

Yoshida, S. and Yatawaza, M. (1977) Nitrogen fixation capacity during the course of cut-recovery in ladino clover. *Journal of the Japanese Society of Grassland Science* 23, 6–13.

Young, N.R. (1989) Review of legume systems for beef and sheep. In: *The Attractions of Forage Legumes: Joint Meeting of the South West and Wessex Forage Legume Groups*. British Grassland Society, Hurley, pp. 3.1–3.16.

Young, N.R. and Mytton, L.R. (1983) The response of white clover to different strains of *Rhizobium trifolii* in hill land reseeding. *Grass and Forage Science* 38, 13–19.

Younie, D. (1989) Eighteen-month beef production: organic and intensive systems compared. In: *Organic Meat Production in the 90s*. Chalcombe Publications, Marlow, pp. 41–52.

Zeven, A.C. (1991) Four hundred years of Dutch white clover landraces. *Euphytica* 54, 93–99.

Zohary, M. and Heller, D. (1984) *The Genus* Trifolium. Israel Academy of Sciences and Humanities, Jerusalem.

Lucerne (syn. Alfalfa) 3

Introduction

Lucerne is the highest-yielding of the temperate forage legumes and is the most widely grown in warm temperate areas. The area of lucerne in the world exceeds 30 Mha, with the USA, the countries comprising the former Soviet Union and Argentina accounting for about 70% of that area. Other countries with large areas of lucerne include Canada, growing about 2.5 Mha, and Italy and China with about 1 Mha each.

There are three main *Medicago* species recognized as lucerne by most literature sources, i.e. common purple lucerne (*M. sativa* L.), yellow lucerne *(M. falcata* L.) and variegated, hybrid or sand lucerne (*M. media* Pers.). In temperate areas, current commercial cultivars are mostly based on *M. sativa* genes but often have varying degrees of *M. falcata* in their ancestry and so are genetically equivalent to *M. media*, which is considered to have developed from spontaneous hybridization between the other two species (discussed below). Nevertheless, these cultivars retain some of the characteristics of *M. sativa*, e.g. its purple flower colour, and so are usually referred to as 'common lucerne'. Common lucerne is also known as purple medick, snail clover, Chilean clover or purple alfalfa.

Prior to the binominal system of nomenclature, lucerne and its relatives were known as *Medica*. Opinions differ on the derivation of the generic name, some believing it to come from 'Media' pertaining to their early use in Mesopotamia, while others consider it to come from the Latin *medicus*, due to their early-recognized curative properties for livestock. However, Linnaeus classified lucerne as *Medicago sativa* in his *Species Plantarum* in 1753.

Lucerne was one of the first forages to be domesticated. Unlike most other forage legumes, lucerne is generally grown alone in pure swards, although it can be mixed with some other legumes and grasses. It is mainly used for conservation as hay or, to a lesser extent, as silage. It is also used as a dried concentrate feed, in which its protein and pigment content are exploited. It does not tolerate continuous stocking but is used in rotational grazing systems in Australasia and Latin America, for example.

Origin and Distribution

The origin of lucerne is considered to be in Vavilov's Near Eastern centre. More specifically, two centres have been identified, i.e. Transcaucasia and central Asia. The Transcaucasian centre is thought to have given rise to the winter-hardy disease-resistant *M. falcata* type, while the *M. sativa* type, adapted to warm, dry conditions, originated in central Asia (Michaud *et al.*, 1988). Hybridization between the two, which produced variegated lucerne (*M. media* syn. *M. varia*), probably took place in eastern Europe. Yellow lucerne, so called because of its yellow florets, has featured in the evolution of modern lucerne. A native of northern Europe and Asia, yellow lucerne is very winter-hardy, can be drought-resistant and has a creeping rhizomatous growth habit.

The spread of lucerne from its centre has been described (Bolton *et al.*, 1972). Briefly, it spread to the eastern Mediterranean, reaching Italy by about 200 BC. At this time, it was also spreading east into China and, by the beginning of the first century AD, was probably making inroads into North Africa, following Roman invasions. It continued its spread, in the early centuries AD, south of the Mediterranean and westwards and northwards to Spain, while the other route in Europe took it through Switzerland to southern France. Its cultivation around Lake Lucerne is thought to have resulted in the crop taking the name of that Swiss lake and this is still used in Europe today, except in the Iberian peninsula where it is referred to as 'alfalfa' – as it is in the Americas.

One theory concerning the early spread of *M. sativa* amd *M. falcata* and their spontaneous hybridization to produce *M. media* suggests that the two came into contact in the early centuries AD, when *M. sativa* was introduced into northern Europe by the advance of the Romans from the south in the second and third centuries AD, and *M. falcata* was brought westwards later by Attila's armies in the fifth century. Alternatively, it is theorized that the two original species made contact much later, i.e. around the 1600s, when *M. sativa* was gradually spreading northwards as it was introduced into farming systems.

By the fourth century AD, *M. sativa* usage had apparently declined drastically in Europe, but it was reintroduced from North Africa via Spain by the Moors in the eighth century and the Arabic name for the crop, 'alfalfa' (probably from 'aspo-asti' meaning 'horse fodder'), was introduced with it. The Spanish took lucerne to the Americas in the sixteenth century and the name 'alfalfa' was adopted across the Atlantic. It was not until the sixteenth century that it spread north and then east, reaching England in the midseventeenth century and Germany by the mideighteenth century. By this time, it had come into contact with yellow lucerne and so variegated lucerne resulted. Meanwhile, lucerne had been taken to Australasia in the very early nineteenth century and later to South Africa and Canada.

By the early nineteenth century, it was grown in the south-west states of the USA, from where it spread north and east, although this spread to colder climes was limited by its lack of cold-hardiness. Lucerne introduced by western European settlers into the eastern states of the USA in the eighteenth century did not thrive. However, Wendelin Grimm introduced the hybrid variegated lucerne from Germany, and this formed the basis of cultivars capable of surviving the cold winters in the northern states of the USA and in southern Canada.

Plant Development and Physiology

The plant

Lucerne is a perennial upright plant, comprising numerous stems which have originated from crown buds. *Medicago sativa* and *M. media* types have a strong tap root, usually reaching 2–4 m in depth, but root penetration can be deeper in well-drained, deep soils and, exceptionally, roots to a depth of 39 m have been recorded (Angelini, 1979). In contrast, *M. falcata* types have thinner, more branched roots.

The kidney-shaped seeds comprise two cotyledons and an embryo axis, enclosed by a yellow to brown testa, varying in permeability to water and gases. The cotyledons emerge hypogeally during germination and are followed by a unifoliate leaf. Thereafter, trifoliate leaves are produced in an alternate arrangement, leaflets being serrated around the upper margins and with the midribs extending a little beyond the margin of the leaflet, giving a mucronate tip. The leaflets are elongated–ovate in shape. At the base of the petiole, i.e. at the node, the stipules are long and pointed, with serrated margins. Axillary buds form in the axils of these leaves as the stem develops. These buds produce stems, which, in turn, build up a crown of basal buds at their base. This crown is the main source of stems that are produced when the plant regrows after defoliation. Buds are produced in the axils of leaves, borne on nodes above ground level, and are able to develop into branches from the stubble after defoliation of an immature stand, but their contribution to regrowth yield is low (Langer, 1973).

Flowers are racemes, borne in the axils of upper leaves, each purple-coloured floret being typically papilionaceous, with the corolla comprising a large 'standard' petal, two 'wing' petals and the remaining two fused to form a 'keel'. Pollination is generally by bees, such as the honey-bee (*Apis mellifera*) or alkali bee (*Nomia melanderi*). As the insect enters the corolla tube, the ten stamens under tension within the keel are released (tripped) and make contact with the insect, which can then pass pollen on to the stigmata of future florets visited. Seeds develop in spirally coiled pods, producing three or four seeds per pod. The 1000 seed weight averages about 2 g. Table 3.1 summarizes lucerne development from the vegetative to seed-pod stages and Fig. 3.1 shows the plant's morphological features.

Lucerne is generally considered to be a drought-resistant forage plant, the depth of its rooting in this respect being a beneficial factor. As the young plant develops, its root may be three times deeper than the height of its shoot. The

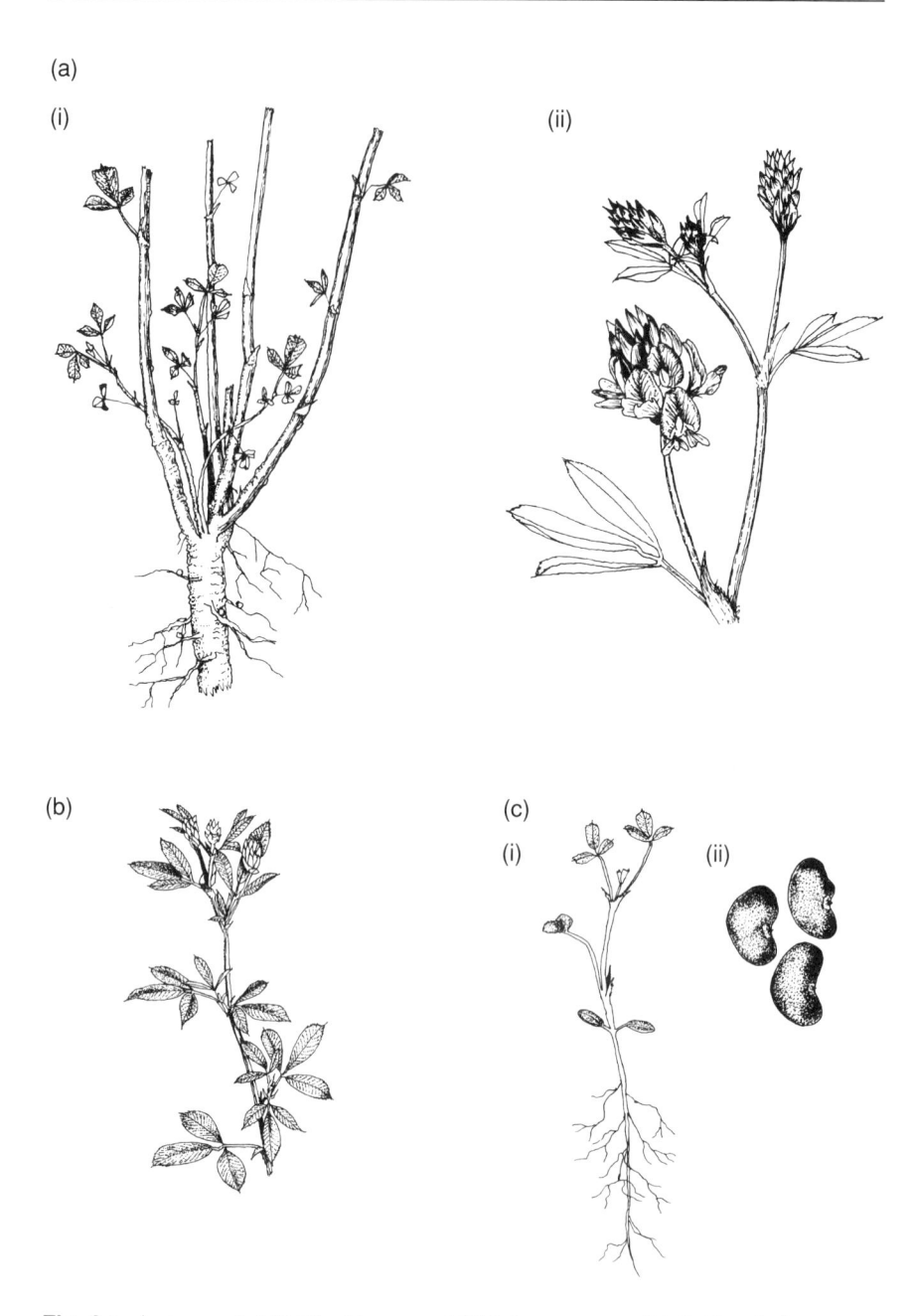

Fig. 3.1. Lucerne. (a) (i) Plant base and (ii) inflorescence. (b) (i) Shoot plant parts at the flowering stage and (ii) leaf. (c) (i) Seedling and (ii) seeds.

Table 3.1. Summary of stages of development of lucerne plants (from Kalu and Fick, 1981).

Stage 0	*Early vegetative:* stem length up to 15 cm; obviously vegetative (no visible buds, flowers or seed pods); axillary buds not easily seen.
Stage 1	*Mid vegetative:* stem length 15–30 cm; obviously vegetative; axillary buds, developing (with 1 or 2 leaves), especially at the mid stem.
Stage 2	*Late vegetative:* stem longer than 30 cm but still vegetative; axillary buds beginning to elongate; inflorescence buds at apex enclosed by young leaves beginning to develop.
Stage 3	*Early bud:* one or two nodes with developing buds near apex on main axis or on branches; no flowers or pods.
Stage 4	*Late bud:* three or more nodes with visible buds; no flowers or pods; clear separation of flower buds in raceme.
Stage 5	*Early flower:* one node with an open flower; no seed pods.
Stage 6	*Late flower:* two or more nodes with open flowers; no seed pods; nodes with flowers spread around mid portion of stem.
Stage 7	*Early seed pod:* one to three nodes with green pods usually on inflorescences at lower nodes initially.
Stage 8	*Late seed pod:* four or more nodes with green seed pods; older stems highly branched; leaves falling off.
Stage 9	*Ripe seed pod:* most pods brown and mature; stem thick and fibrous; seed ready to harvest.

tap root produces branches, but good root growth and strong aerial growth are interdependent. If light conditions are poor, e.g. due to weed competition, shoot and root growths and rhizobial nodulation suffer. Young lucerne plants are then sensitive to competition, due to inadequate nitrogen (N) nutrition because of poor molecular nitrogen (N_2) fixation.

Germination

Germination of lucerne seeds is highly dependent on moisture availability, temperature and salt concentrations in the moisture around the seeds and permeability of the testa (reviewed by Fick *et al.*, 1988). Drought and high temperatures during previous seed development increase the proportion of hard seeds that mature, and these subsequently exhibit slow germination. Uptake of water is minimal at -1.0 to -1.5 MPa soil-water potential, but, even if soil water is apparently present in abundance, salt concentration acting as an osmoticum may limit water availability.

Germination takes place over a wide temperature range, from less than 10°C to about 35°C. Although initial signs of germination are temperature-dependent, differences in germination rate over this range are small, varying within only 1 week. At less than 10°C, germination rate falls sharply with temperature, 75% germination being reached within 2 days at 20°C, compared with about 10 days at 5°C under standard testing conditions (Hampton et al., 1987).

Generally, salt concentration of 350–400 mmol l^{-1} in the germinating medium will inhibit germination, but tolerant cultivars achieve high germination percentages at 400 mmol l^{-1}. Temperature and salt concentration interact, and salt concentration with a negative osmotic potential stronger than −0.7 MPa results in germination inhibition, especially at temperatures above 27°C (Stone et al., 1979). Germination of lucerne seed can be inhibited or delayed by exudates from other lucerne seeds (autotoxicity) with cultivars differing in intensity of effect (Chung and Miller, 1995b).

Temperature

A mean air temperature of 27°C is optimum for seedling growth, but the optimum differs with stage of plant development and regrowth. As shoots develop, the optimum temperature declines by about 5°C (Fick et al., 1988). Leaf development is slower at day/night temperatures of 32°C/26°C than 22°C/16°C, although there are marked cultivar differences (Wilson et al., 1991), *M. sativa* types being the most responsive to temperature. Across a range of cultivars, varying in dormancy characteristics, air temperature of 25°C was optimum for root dry-matter (DM) production, but the optimum for root number per plant was 21°C (Kendall et al., 1994).

Temperature during inflorescence initiation and development influences the rate of maturity, and flowering takes 2 weeks longer at 18°C than at 32°C; however, if temperatures exceed 30°C, the proportion of flowering stems declines. Vernalization of lucerne seeds can advance flowering of the plant by a few days. Seeds subjected to 0.5°C for 2–28 days resulted in a 5-day advance in flowering, compared with seeds not vernalized or vernalized for 1 day (Major et al., 1991).

Excessively high temperatures cause heat injury to lucerne, although adaptive features and mechanisms reduce the impact of temperature greater than 35°C, e.g. possessing leaf characteristics that reflect radiant energy from the canopy or controlling the effective temperature by transpiration.

Nutritive value is influenced by temperature during growth, and, for instance, high temperatures (32°C/26°C day/night temperatures) reduced organic matter digestibility (OMD) by 2–6%, compared with lower temperatures (22°C/16°C), due to a lower concentration of non-structural carbohydrates and a higher content of lignin (Wilson et al., 1991).

Some genotypes of lucerne are extremely winter-hardy, capable of withstanding Alaskan winters. The hardening process is influenced by temperature, commencing when air temperature is about 10°C, intensifying as temperatures approach zero and finalizing when soil temperatures are −1° or −2°C (McKenzie et al., 1988); hardening is retarded if air temperatures exceed about 10°C (Pacquin and Pelletier, 1980). Hardening is associated with induction of dormancy, brought about by shortening days, as well as declining temperatures. However, conditions such as water-saturated soils (Pacquin *et al.*, 1987) or a low plant-N concentration (Vezina and Nadeau, 1991) may prevent the attainment of maximum cold-hardiness.

Cultivars differ in their capability to harden before winter and the less winter-hardy types are slower to harden and quicker to deharden. Winter-hardy varieties can withstand soil temperatures as low as $-20°C$ and compared with less hardy types have higher concentrations of sucrose and total sugars (Duke and Doehlert, 1982), lipids such as polyunsaturated linolenic and linoleic acids and phospholipids (Grenier and Willemot, 1974) and soluble protein, especially proline (Wilding and Smith, 1960). During hardening, concentration of soluble sugars increase in roots and starch concentration declines, cold-hardy dormant types accumulating high concentrations of raffinose and stachyose, rather than sucrose (Castonguay *et al.*, 1995). An indication of the many processes involved in hardening is the wide difference between cold-sensitive and cold tolerant cultivars in activity of some dehydrogenases between hardened and unhardened tissues (Krasnuk *et al.*, 1978).

Exogenously applied abscisic acid (ABA) increased freezing tolerance in seedlings of lucerne and changed protein composition in a manner similar to that which results from cold-hardening induced naturally, suggesting that endogenous ABA may be implicated in gene expression during cold-hardening (Mohapatra *et al.*, 1988).

Light

Light (radiant energy) controls the rate of photosynthesis and influences branching, flowering, cold-hardiness, stem elongation and other processes under photomorphogenic control. Light quality (e.g. relative content of red and far-red light or photoperiod) is particularly important in controlling photomorphogenesis. As some of the experimentation in the past has confounded flux density and photoperiod, interpretation of the effects of light on lucerne growth and development from these studies is not always possible.

The relationship between photosynthesis of individual leaves and photosynthetically active radiation (PAR) shows saturation at an estimated 1200–1400 $\mu mol\ m^{-2}\ s^{-1}$, but the saturation point declines with age, the compensation point occurring at about 0.02 of saturation irradiance, which is about 65 $\mu mol\ m^{-2}\ s^{-1}$ (Sheehy and Popple, 1981). The PAR and canopy photosynthesis relationship is obviously influenced by leaf-area index (LAI), with 95% interception of PAR occurring at an LAI of about 5. With an extinction coefficient of 0.5, saturation of the canopy occurs at about 1600–1800 $\mu mol\ m^{-2}\ s^{-1}$ PAR, producing a net photosynthetic rate of about 5 g carbon dioxide (CO_2) $m^{-2}\ h^{-1}$ (Gosse *et al.*, 1982). The relationship between irradiance and photosynthesis is translated into stem population density, stands growing under 'full-light' conditions having a higher stem density than those subject to shade, if sowing density is low (Cowett and Sprague, 1963). Also, the N concentration of leaves declines faster, in relation to depth below the canopy surface, in dense than in open canopies (Lemaire *et al.*, 1991).

Competition for light intensifies when growth is most rapid, and shoots derived from crown buds gradually become dominant over those from axillary buds. This leads to the death of plants that have a low proportion of stems from crown buds when competition for light becomes important (Gosse *et al.*, 1988); within the plant, the reduction in stems results in redistribution of N from dying stems, which explains the reduction in N content in leaves within dense canopies of lucerne.

Lucerne is a long-day plant, with flowering being mediated by phytochrome, and so a short interruption with red light in the night period in a diurnal sequence can initiate flowering; for example, about 2 µmol m^{-2} s^{-1} PAR for 30 min has been found to advance flowering (Massengall *et al.*, 1971). The minimum photoperiod required to advance flowering to its earliest date varies between cultivars, e.g. out of a group of nine cultivars compared in Canada the required photoperiodic length varied from 17.7 to 19.0 h day^{-1} (Major *et al.*, 1991). Long photoperiods seem to be required for floral development to continue after initiation; otherwise, the developing inflorescences abort. Light also has a quantitative effect on flowering, the number of flowers increasing with irradiance level.

The induction or trigger of cold-hardiness in lucerne is photoperiod-dependent, long nights being required to initiate hardening, which can then be advanced by low temperatures. Stem elongation in autumn in lucerne is also under photoperiodic control. Stem elongation ceases earlier in autumn in winter-dormant, cold-climate types than in less winter-dormant warm-climate cultivars, as the former have a higher critical day length (Guy *et al.*, 1971), and so, generally, winter-dormant types will be more responsive to shortening photoperiods in autumn than the less winter-dormant types (Hesterman and Teuber, 1981). Cultivars that are cold-sensitive do not have the capability to respond to shortening days and thus their hardening is induced by declining temperatures in autumn, but warm spells during hardening may reverse the process (Baldwin, 1990).

Moisture

During the early stages of development, lucerne seedlings acquire sufficient water if soil water potential is higher than −0.6 MPa. Soil type obviously influences the relationship between soil-water potential and yield, but generally yield loss is considerable below potentials of −0.3 to −0.4 MPa. A linear relationship between dry matter (DM) production and available water capacity (AWC) in the soil has been demonstrated (Douglas, 1986). Components of yield adversely affected by drought include leaf size and stem length, followed by leaf and internode number (Vough and Marten, 1971). Generally, leaf growth is affected less than stem elongation, resulting in a higher leaf content in the biomass at lower soil-water potentials down to −2 MPa.

In the early stages of water deficit, shoot production and hence LAI are adversely affected more than roots, and so the main effect of drought is a reduction in light interception, which, in turn, exacerbates the reduction in shoot growth (Durand et al., 1989).

Lucerne physiological functions that are sensitive to drought include photosynthesis, respiration and N_2 fixation, photosynthesis and respiration being almost halved by a soil-water potential of -0.45 MPa (Murata et al., 1966). When subject to moisture stress, N_2 fixation is adversely affected, due to lower root-hair infection rates, lower nodule mass, possibly nodule shedding and reduced specific nitrogenase activity (Sheaffer et al., 1988a); also lucerne nodules are more susceptible to drought than leaves (Irigoyen et al., 1992b). Some aspects of drought stress may also be due to direct physical effects of dehydration.

Lucerne is more drought-tolerant than most other temperate forage legumes, including birdsfoot trefoil and red clover (Peterson et al., 1992), lucerne plants becoming dormant under severe drought conditions. Part of this tolerance is due to deep rooting, although rooting depth is restricted by excess or inadequacy of soil moisture, by certain soil characteristics, such as a compacted soil structure (indurated layer), or by low soil pH, which causes manganese ion (Mn^{2+}) or aluminium ion (Al^{3+}) toxicity to roots. While rooting was best developed in the most irrigated treatment in one trial (Abdul-Jabbar et al., 1982), non-irrigated plots in another trial, with a soil-water potential as strong as -1.5 MPa, had better developed roots in the upper horizon than irrigated plots, indicating that misuse of irrigation can prevent development of the rooting system (Carter and Sheaffer, 1983).

Although lucerne generally has a relatively low stomatal resistance (Kerr et al., 1973), it uses water more efficiently than perennial ryegrass (*Lolium perenne*), for example (Woodward and Sheehy, 1979); this grass used six times more water than lucerne per unit of carbon (C) fixed and, during a dry year, the DM yield from lucerne was three to four times greater than that from perennial ryegrass. At the same AWC, lucerne produced 50% more DM than grass pasture (Douglas, 1986).

Water-use efficiency (WUE), which is an estimate of the quantity of water in cm ha^{-1} transpired during the production of 1 t DM (i.e. evapotranspiration ÷ yield), varies over the season, although it usually ranges between 5 and 9 cm ha^{-1} evapotranspiration t^{-1} of lucerne DM produced (Sheaffer et al., 1988a). Ability to conserve water varies between cultivars and, in Europe, for example, the more northerly Flemish types are less efficient in their use of water than Mediterranean types. The WUE is greatest in spring and early summer in lucerne crops, since growing conditions, especially temperatures, are usually close to optimal and evapotranspiration is relatively low. Under drought conditions, transpiration can be affected more than photosynthesis, since stomata close with reduction in leaf-water potentials, and so WUE may appear to improve (Irigoyen et al., 1992a).

Moisture stress also affects mineral content and nutritional value of lucerne. For example, in Texas, concentrations of calcium (Ca), magnesium

(Mg), zinc (Zn) and phosphorus (P) were higher and potassium (K) concentration lower in non-irrigated than in irrigated treatments (Kidambi *et al.*, 1990) and, if moisture stress is not too severe, it delays maturity, resulting in higher herbage digestibility. Drought has no effect on the degradability of cell walls (Deetz *et al.*, 1996), and so improvement in nutritive value is most likely to be due to a higher leaf : stem ratio in stressed plants (Duru and Langlet, 1993).

Prolonged excess soil moisture seriously affects lucerne yield, root growth and persistence. Anaerobic conditions in saturated soils result in xylem necrosis and production of toxins in the roots, leading to death of organs, and root and crown rots develop in highly humid conditions, e.g. *Phytophthora* root rot (*Phytophthora megasperma*), resulting in a reduction in the plant population (Sheaffer *et al.*, 1988b); cold-hardiness is impaired in saturated soils (Jung and Larson, 1972). Therefore, lucerne is not suited to areas with heavy soils subject to high rainfall or prone to prolonged flooding. Drainage in irrigated crops is also necessary to avoid persistent soil saturation.

Assimilate partitioning

Assimilate partitioning during a growth cycle in lucerne has been reviewed (Heichel *et al.*, 1988). Assimilate source is mostly photosynthetic tissue, and sinks are meristems, nodules and storage organs, such as roots and crowns, but, in early spring or soon after defoliation, the source is storage tissues.

During export from leaves, assimilate takes longer to be translocated from older than from younger leaves at the top of the canopy. There is a cyclical pattern in the concentration of reducing sugars, total sugars, water-soluble carbohydrate (WSC) and total non-structural carbohydrates (TNC) in the leaves, related to the light period. Young leaves tend to supply assimilate to branch buds on the stem, whereas lower leaves supply crown buds as the canopy develops, but both supply roots and nodules.

The supply pattern changes with plant development. In individual plants, at flowering, the shoot apex receives a high proportion of the assimilate from middle leaves on the main stem, changing to the seed pods and axillary buds subtended by the fed leaf, when pod development is under way. Prior to harvest, the main stem and other axillary buds, as well as the apex and the axillary bud associated with the fed leaf, receive a high proportion of the assimilate from the middle leaves. In an establishing lucerne crop about two-thirds of assimilated C is translocated to roots, while in an established stand only about 20% of the C is distributed to roots, increasing to about 45% in autumn regrowth (Khaiti and Lemaire, 1992).

The TNC are concentrated in the crown and roots, particularly in the top 10 cm, but the content varies with stage of regrowth and time of year. Usually, they accumulate in the roots as flowering progresses, reaching a maximum when seed pods develop. After harvesting, some stored TNC from roots and crowns is remobilized for production of new leaves, but much of the decline in reserve carbohydrate is due to respiration losses (Ta *et al.*, 1990). Although there is a net loss from the roots, TNC continues to be translocated

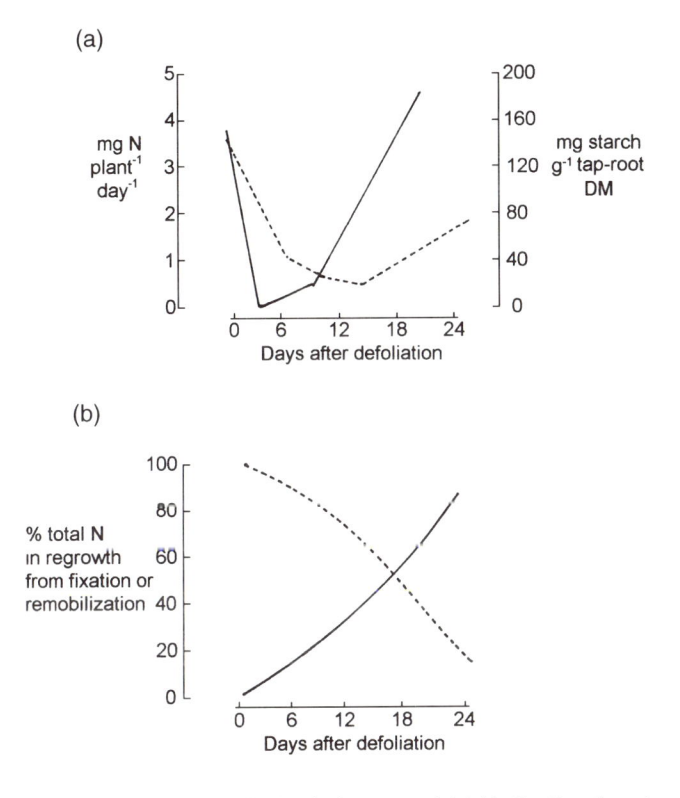

Fig. 3.2. Time course after defoliation in lucerne of (a) N_2 fixation (———) and starch content (-----) in tap root and (b) percentage of N in new growth derived from remobilized N (-----) and N_2 fixation (———) (from Kim et al., 1993).

into storage tissue 1 week after harvesting. The decline in root TNC is accompanied by a decrease in root growth. The reduction in postharvest TNC is greater the more advanced the crop stage of development is at harvesting, and also greater with severe than with lax defoliation. Up to 90% of the storage material is starch. Subsequent to harvest, TNC content is gradually replenished to the preharvest level and usually makes up about 35% of root DM 6–8 weeks after harvest.

During the growing season, TNC levels decline in roots and crowns in response to cutting. Endoamylase activity in tap roots is stimulated by defoliation, and so starch is degraded (Boyce et al., 1992). In addition to carbohydrate decline, reserve N is also lost from the roots subsequent to defoliation, as the shoots are powerful sinks during early regrowth (Kim et al., 1993); N_2 fixation also declines with defoliation, its recovery following a similar pattern to that of shoot regrowth (Fig. 3.2). The concentration of TNC in roots increases under low night temperatures, with water stress or when there is adequate mineral nutrition, especially a sufficient supply of K.

Table 3.2. Effect of frequency of cutting on annual yield (t DM ha^{-1}) and mean ADF, NDF and CP contents (from Hesterman et al., 1993).

Cutting frequency	Harvest year				Mean of 4 years		
	1	2	3	4	ADF	NDF	CP
	(t DM ha^{-1})				(g kg^{-1} DM)		
Four cuts	14.8	13.4	12.6	12.1	310	410	210
Five cuts	15.2	10.9	10.1	9.7	265	373	240

ADF, acid detergent fibre; NDF neutral detergent fibre; CP, crude protein.

Defoliation

The inverse association between cutting interval and DM yield is a general phenomenon, whether the crop is grown in the major growing regions of the USA (Hesterman et al., 1993) or the cool maritime conditions of Scotland (Reid, 1987). In specific examples, three cuts annually yielded more than four cuts, although the forage was of lower nutritive value (Brink and Marten, 1989), as did a four-cut versus a five-cut system (Table 3.2), while, in addition, the yield decrease associated with stand age was less with infrequent harvesting. Often, the harvesting of total nutrients is similar between cutting frequencies, the lower DM yields having higher nutrient concentrations. However, frequent cutting results in reduced persistency, as nutrient reserves do not have an opportunity to be built up between defoliations. Genotypes that develop leaf area rapidly after cutting should be more tolerant of short cutting intervals and so be able to sustain production of herbage with a high nutritive value (Lemaire and Allirand, 1993). Generally, yields will be greater with non-dormant types, if soil moisture is not a limitation, as they produce sufficient herbage late in the season to justify an additional harvest.

Before the advent of cultivars resistant to bacterial wilt, cutting intervals were long and the crops were harvested at full flowering, since plants were considered capable of withstanding attack better when mature. Now, even in northern states of the USA, where the growing season is relatively short, high DM production from lucerne is achieved with three or four harvests in the year (Sheaffer et al., 1988b). Cutting three times a year can produce higher DM production than twice a year, as leaf loss is reduced and net canopy photosynthesis is higher in the more frequently harvested stand (Feuss and Tesar, 1968).

The timing of each harvest is best determined by the stage of development of the primary growth and regrowths. A convenient scale of development within which the stages are easily recognized is vegetative, bud, first flower (10% bloom), half bloom (50%), full bloom and seed-pod set. However, a more detailed, numerate system has been devised, which takes account of developmental stage and stand height, to produce a scale with ten stages (see Table 3.1). This results in cuts being taken every 4–6 weeks in temperate environments (Johnson, 1984; Langer, 1990) and every 4 weeks or so in well irrigated Mediterranean regions.

The DM yield of lucerne increases with advancing maturity but the nutritive value of the herbage is reduced. Maturity results in a decline of the leaf : stem ratio, an increase in lignin content of the stem and leaf loss through senescence or leaf shatter. Cutting intervals need to be sufficiently long to ensure that carbohydrate and N reserves are replenished; otherwise, the stand will become thin and persistence is jeopardized.

While it has long been recognized that carbohydrate reserves in roots and crowns are important for regrowth after defoliation, the association between amount of TNC and regrowth rate is inconsistent (Volenec, 1985). Strong evidence shows that storage N in roots is important in determining the rate of regrowth of lucerne, with much of the N in shoots during the first 2 weeks or so of regrowth being derived from roots, rather than uptake from the soil or N_2 fixation (Kim et al., 1993; Fig. 3.2). Specific pools have been identified as sources of N for regrowth, with young regrowth drawing mainly on the vegetative-storage protein pool, which comprises amino acids and other acetone-soluble N compounds, nitrate and ammonium (Barber et al., 1996). Non-dormant types have a particularly ready supply of protein from this pool, while dormant types have a higher content of insoluble protein, a difference which may explain the faster regrowth of the non-dormant types.

To ensure plant survival over winter, the last harvest in the season should be sufficiently early to allow the plants to build up TNC and N reserves before cessation of growth, but not so early that a heavy canopy develops prior to winter; otherwise, this increases the danger of frost damage to the crown, and reduced plant persistency results. For example, a 6-week period before anticipated cessation of growth is recommended for southern England (Johnson, 1984). Storage of reserves in crowns and roots is promoted by shortening days and lowering temperatures, especially in Flemish and other dormant types, thus providing resources for survival. Defoliation is possible in early winter with cattle without detriment to lucerne persistency, provided its growth has ceased. Defoliation of associated grass or weeds at this time reduces their competitive effects the following spring. However, sheep eat any elongated crown buds in early winter, and this has the effect of reducing yields the following spring. Timing of autumn harvest has little effect on yield the following year in mild Mediterranean climates (Marble et al., 1989), although yields at the first cut can be reduced (Lloveras et al., 1996).

A significant consequence of reduced root and crown reserves, especially over winter, is the reduction in spring of crown buds, which are responsible for stem production in the primary growth and first regrowth each year. At subsequent cuts, axillary buds, originating in the axils of stems, assume importance for regrowth. Stubble height determines the balance between axillary and crown buds, with tall stubbles favouring development of axillary buds (Leach, 1979). Therefore, a high cutting height, e.g. in excess of 12 cm, which increases sites for shoots to be produced and retention of some leaves, may partly compensate for yield depression from short cutting intervals, although the efficiency of utilization of the standing crop is lowered. Conversely, short stubbles can result in higher yields when cutting intervals are long, since the regrowth time, which allows full expression of crown bud growth, is allied to efficient utilization (Smith and Nelson, 1967).

Stubble biomass can ameliorate the effects of frost, reducing its extreme effects on the soil, and helps to retain snow, which insulates lucerne crowns (Barnes and Sheaffer, 1995); thus, a long stubble should be left after the last cut of the season in areas likely to experience hard winters.

The rest interval between grazings is best based on similar principles to cutting defoliation, i.e. first grazing at early bloom stage, followed at 5- to 7-week intervals, depending on rate of growth. As lucerne flowers in regrowths, at least until early autumn, the stage of development of the stems is a guide to suitability for grazing. Potential damage to lucerne by grazing includes overgrazing of the crowns, which results in damage to the crown buds, and possibly axillary buds, which produce the regrowths.

Nitrogen fixation and transfer

Rhizobium meliloti is one of eight cross-inoculation groups in the Rhizobiaceae and is the one that infects and induces nodules on the roots of lucerne. It is a Gram-negative bacterium, which also forms nodules on sweet clover (*Melilotus* spp.) and fenugreek (*Trigonella* spp.). Details of the culture of *R. meliloti* are available (Elkan, 1981).

In lucerne, the release of the flavonoid, luteolin, stimulates production of the *nod* genes in *R. meliloti* to produce a chitin-like lipo-oligosaccharide, which, in turn, induces root hairs to curl and other processes associated with nodule development (Vance, 1997). During infection, pectolytic enzymes degrade the cell walls of the root hair and allow the rhizobia to enter (Munns, 1969). The activity of the enzyme depends on availability of Ca and so, at low soil pH, Ca deficiency may prevent or greatly reduce infection and resultant nodulation. Nodules of lucerne have apical meristems but are cylindrical in shape. Senescent nodules become green coloured at the base, while younger nodules are white or pink, depending on the extent of the zone of the nodule containing leghaemoglobin.

Ineffective nodules produced by *R. meliloti* infection can be either bacterium- or plant-controlled. Irrespective of the source of the ineffectiveness, the common features of such nodules are incomplete development and rapid senescence, excessive starch accumulation, altered membranes and formation of tumour-like nodules (Vance *et al.*, 1988).

Strain × cultivar interactions in the *R. meliloti*/lucerne association exist. Although all strains will generally nodulate any given genotype of lucerne, the degree of effectiveness, as measured by nitrogenase activity and root and shoot DM, may vary (Tan and Tan, 1986). Compatible strains can be introduced at sowing by inoculation, but competition from aggressive indigenous strains may prevent the potential of improved strains being achieved. Certain combinations of *Glomus* spp., vesicular arbuscular mycorrhizal (VAM) fungi, *R. meliloti* strains and lucerne influence N_2 fixation, increased P uptake due to VAM increasing the rate of N_2 fixation (Azcon *et al.*, 1991) and also aiding uptake of soil mineral N.

Estimates of N_2 fixation by lucerne vary widely but are generally higher on an annual basis than for other temperate forage legumes. From a range of sites, annual fixation rates have ranged from 85 to 360 kg N ha^{-1} (Witty *et al.*, 1983; Heichel and Henjum, 1991). Estimates of 700 kg N ha^{-1} from harvested lucerne in New Zealand (Douglas, 1986) did not take into account the proportion of N that was derived from non-fixing sources, including soil mineral N.

Due to the sensitivity of the nitrogenase system to environment, wide variation in N_2 fixation is found from site to site. While soil mineral N or fertilizer N imposes a major inhibition on N_2 fixation, deficiency of certain minerals, such as K, Ca or Mg, or excessive soil acidity also limit fixation. Nodulation occurs at pH less than 4.5, conditional on Ca^{2+} being present, whereas nodulation can be inhibited at a soil pH as high as 5.5 under low Ca^{2+} conditions (Munns, 1970).

One hundred thousand *R. meliloti* cells g^{-1} of soil are considered necessary for nodulation in lucerne seedlings. In areas where lucerne does not occur naturally or has not been grown previously, seeds have to be inoculated with *Rhizobium* to ensure nodulation. Most rhizobia share optimum growth between 29°C and 31°C, but *R. meliloti* strains that have an optimum of 35°C are known. *Rhizobium meliloti* is sensitive to low pH, natural occurrence being markedly reduced in soils below pH 6 (Graham, 1992). Lime pelleting of seed results in increased numbers of *R. meliloti*, but does not affect adhesion to the tap root; however, the number of seedlings with nodules on root branches at the crown is increased by seed pelleting, which, in turn, results in increased N concentration and dry weight of lucerne shoots (Pijnenborg *et al.*, 1991).

Set standards for inoculants have been developed in many countries that grow lucerne, an example being that the culture should have a minimum content of 10^{10} viable rhizobia kg^{-1} of seed to be inoculated at any time up to the viability expiry date, usually 6 months after manufacture of inoculant, be comprised of pure *R. meliloti*, identifiable with the original culture supplied, and be effective for nodulation and N_2 fixation, and the expiry date must be clearly stated.

A major problem with inoculated lucerne seed is survival of rhizobia on the seed. In a survey of new lucerne crops 2 months after sowing, two-thirds had less than one-third of their seedlings nodulated, although nodulation improved with time, two-thirds of mature crops surveyed having 40% or more of their plants nodulated (Blair, 1971). Nodulation of poorly nodulated stands of young plants is improvable by spreading peat inoculant, coated on to calcite granules, after seedling emergence (Close *et al.*, 1971).

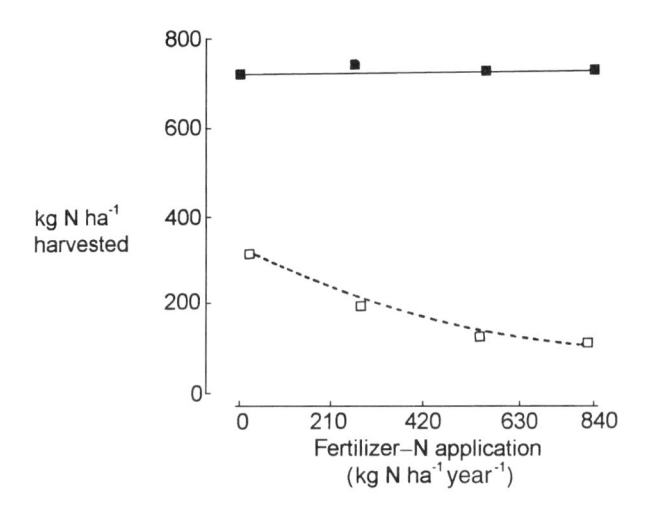

Fig. 3.3. The effect of fertilizer N on total N harvested in lucerne (–■–) and on N_2 fixation (–□–) (total of four harvests, N applied in equal aliquots within an N treatment at the beginning of each growth cycle) (from Lamb *et al.*, 1995).

Generally, infection and, therefore, nodulation and nitrogenase activity in existing nodules are detrimentally affected by the presence of applied mineral-N, the degree of inhibition being related to mineral N concentration in the soil. High root temperatures, e.g. 30°C, low cold-hardening temperatures (5–10°C), moisture stress and any factors, such as low irradiance and low CO_2, which adversely affect photosynthesis, also have detrimental effects on N_2 fixation (MacDowell, 1983).

Defoliation reduces N_2 fixation quite rapidly, and the rate of nitrogenase activity is only about 25% 2–3 days after defoliation, compared with activity prior to defoliation. This effect has been ascribed primarily to reduction in available photosynthate, and so, when leaf expansion commences after defoliation, nitrogenase activity increases (Cralle and Heichel, 1981). However, evidence from defoliation of nodulated white clover plants suggests that the reduction in N_2 fixation is due to feedback inhibition, a consequence of build-up of N products in the nodule, which would have been destined for the shoots in undefoliated plants (Hartwig and Nösberger, 1994). Nitrogenase activity is low at some stages in the annual cycle of crop growth, e.g. prior to winter, particularly in winter-dormant cultivars, with low soil temperatures in autumn probably contributing to the reduced activity.

Nitrogen fixation in lucerne is reduced with application of fertilizer N. However, even at excessively high fertilizer-N applications, such as 210 kg N ha^{-1} per growth cycle, N_2 fixation has been found to account for up to 25% of assimilated N (Fig. 3.3).

The genetic control of nodulation and nodule function by the host is a factor that influences N_2 fixation and is capable of manipulation in breeding programmes to improve N_2 fixation (Viands et al., 1981). The composition of the flavonoids that induce transcription of the *nod* genes differs among genotypes of lucerne, and the chitin-like nodulation factor varies among rhizobia, contributing to an explanation for specificity between legume genotypes and rhizobial strains. Even at germination, cultivars may differ in the relative concentration of flavonoids they secrete; for example, cv. Dohfari excretes twice the amount of quercetin than cv. Saranac (Phillips et al., 1995). Although quercetin does not induce *nod* gene transcription in lucerne, it has a growth-promoting effect on *R. meliloti*, as does luteolin, in addition to its effect on *nod* gene transcription (Hartwig et al., 1991).

In addition to selecting for competitively infective and effective strains of rhizobia specific to a cultivar of lucerne to improve N_2 fixation in the field, there is also scope to breed cultivars that are receptive to a broad spectrum of nodulating factors and so are compatible with natural as well as introduced rhizobia (Barran and Bromfield, 1997). Genotypes that have 'non-*nod*' (no nodulation capability) and 'non-*fix*' (inability to produce effective nodules) capability have been selected and are used in research to investigate the mechanisms associated with the nodulation and N_2-fixation processes.

The energetics of the N_2-fixation process in lucerne has been reviewed (Sheehy et al., 1984). For a crop producing net herbage growth of 105 kg DM ha^{-1} day^{-1}, the equivalent of about 34 kg carbohydrate is expended to support N_2 fixation, i.e. 10 mol of CO_2 lost in respiration per mol N_2 fixed. Lucerne is considered to have a relatively efficient N_2 fixation compared with sainfoin, requiring half of the energy requirement of the latter (Witty et al., 1983). Nevertheless, differences in the efficiency of N_2 fixation between different lucerne/*R. meliloti* associations may not always be reflected in equivalent-sized differences in herbage production (Twary and Heichel, 1991).

Under cutting management, the proportion of total N transferred from lucerne to accompanying species is generally lower than that from red or white clover, and so the increase in grass competition found in other grass/legume associations may be slower in lucerne associations and the proportion of grass less (Heichel and Henjum, 1991). In field studies, only up to 13 kg N ha^{-1} were transferred annually in lucerne/timothy (*Phleum pratense*) mixtures, despite more than 200 kg N ha^{-1} being fixed annually (Ta and Faris, 1987). The amount transferred in another study increased from 5 kg N ha^{-1} in the first year to 19–20 kg N ha^{-1} annually in the next 2 years (Burity et al., 1989,) whereas in a lucerne brome-grass (*Bromus inermis*) association 86 kg N ha^{-1} was fixed, of which 55 kg N ha^{-1} were transferred (Walley et al., 1996). Transfer is due to death and decomposition of below-ground plant parts, especially fine roots and nodules, and net mineralization of the N in plant organic matter (OM) (Dubach and Russelle, 1994). Litter from stubble and shed leaves also contributes to transferable N. From a study of the N content of soil in the rhizosphere and around nodules, it has been concluded that excretion of symbiotically fixed N is not an important component in N transfer from lucerne (Lory *et al.*, 1992), and this has been subsequently confirmed by direct measurement (Russelle *et al.*, 1994). The lower proportion of legume N transferred by lucerne, compared with white clover, for example, is likely to be due to a combination of factors, including the C : N ratio being higher in lucerne roots than in white clover roots and the grass being less able to utilize available N, due to aggressive above-ground growth by lucerne, thus reducing the demand of grass for the available N.

Another aspect of N transfer is the availability of N in lucerne biomass for a following crop, i.e. exploitation of lucerne as a 'plough-down' or green fallow crop or inclusion in a crop rotation. Non-dormant types, even in regions where they would not normally survive the winter, produce more herbage in the sowing year than dormant types, due to herbage growth rates being high in the autumn and N_2 fixation continuing when it would have ceased in dormant types. A non-dormant lucerne produced 30% higher DM yield and had even more significant benefit to N incorporation into the soil, contributing 121 kg N ha^{-1}, compared with 40 kg N ha^{-1} for the dormant-type control (Kelner and Vessey, 1995).

Lucerne has been used traditionally in crop rotations in central Italy. The build-up of N-rich soil OM improves soil fertility for subsequent annual arable crops. When compared with continuous cropping, maximum yields of crops, such as wheat, can be increased by 15–20% when lucerne is included in the rotation (Giardini and Cinti, 1985).

Nitrogen and mineral nutrition

One of the most comprehensive surveys of the literature on the mineral content of lucerne was in the early 1970s (Spedding and Diekmahns, 1972). In summary, the means, in g kg^{-1} DM, reported for the various elements were N 30.8, P 2.9, K 20.9, Ca 16.0, Mg 2.9, sodium (Na) 0.9 and chlorine (Cl) 0.6 and, in mg kg^{-1} DM, iron (Fe) 190, Mn 43, Zn 26, copper (Cu) 8, cobalt (Co) 0.16 and molybolenum (Mo) 0.18, although, in many cases, the range in values was considerable, since management conditions differed widely in the various trials reviewed. Mean concentrations of boron (B) and Al have since been reported to be 30 and 80 mg kg^{-1}, respectively (Hill and Jung, 1975). Most of the minerals of nutritional importance to livestock are in higher concentrations than in grass, Na being a notable exception.

Nutrient requirements for lucerne, referred to below, are based on the critical concentrations, these being the concentration of the specific nutrient in the DM below which a reduction in yield of 5–10% will be experienced. Used correctly, they should be determined for different growth stages and plant parts and account taken of environmental conditions, cultivar and concentrations of other nutrients, but this is not possible in practice. Examples of the impact of these factors on lucerne plant analyses have been cited (Kelling and Matocha, 1990).

Relative to mature stands, herbage from early regrowth tends to have higher concentrations of N, P, K, sulphur (S) and Mg, with Ca and B being unaffected or increased by maturity. Leaves generally have a higher concentration of nutrients than stems, although K can be an exception to this generalization. Glasshouse studies with lucerne tend to result in lower critical values than field studies. The problem of nutrient interaction, in interpreting nutrient analyses, has been partly overcome by the use of nutrient ratios, e.g. by the diagnosis and recommendation integrated system (DRIS) (Russelle and Sheaffer, 1986), although DRIS norms tend to be location specific. Some studies quote sufficiency ranges for nutrients, i.e. the optimum range of concentrations over which DM production will be maximized. To allow comparison with critical levels, the lower limits in the range can be taken as comparable.

While excess liming reduces lucerne growth, due to B deficiency, for example, low soil pH can also be detrimental, particularly if soil Ca concentration is low; Al and Mn toxicity becomes a problem at pH levels less than 5. Aluminium adversely affects root growth, reducing it by 50% if greater than 0.5 mmol kg^{-1} of soil, although some cultivars are more tolerant to soil acidity and associated high soil Al levels than others (Devine et al., 1976). Increasing pH will alleviate Al toxicity but may not reduce Mn sufficiently to avoid leaf damage, and K, Ca, S, Mg and Mo availability increases.

Nitrogen nutrition is complex, due to the impact of N_2 fixation on total N uptake. Critical levels of 25–37 g kg^{-1} DM in the upper horizon of the canopy have been identified, of which 43–64% is derived from fixation (Heichel et al., 1981), the contribution being lower if fertilizer N is applied (Lamb et al., 1995). Inorganic N reduces nodulation, 50 µg ml^{-1} fertilizer N reducing nodulation by 40%. Generally, well-nodulated and N_2-fixing lucerne does not respond to fertilizer N, although the concentration of nitrate N can be higher in N-fertilized lucerne (Schertz and Miller, 1972).

Estimates of critical P levels for whole shoots of lucerne range between 2.1 and 3.0 g kg^{-1} DM, with a higher level (about 3–5 g kg^{-1}) in the young-growth DM of a flowering stand. Lucerne roots take up P mostly in the orthophosphate form. In addition to the effects of P on plant growth generally, P also has a pronounced positive effect on nodule mass and nodule N content. Consequently, P-deficient plants may also be N-deficient, even in plants that have an effective symbiosis with rhizobia.

Stems tend to have about a one-third higher concentration of K in DM than leaves. Critical concentrations for K in the DM of whole shoots range from 8 to 22 g kg^{-1}, with most estimates of K sufficient to maintain growth being at least 12 g kg^{-1}. It is well known that K is implicated in the transfer of assimilate from source to sink, and thus deficiency in K results in lower net photosynthesis, lower concentration of reducing sugars and, as a consequence of low photosynthetic activity, lower N_2 fixation. Due to the generally greater efficiency of companion grasses in utilizing K compared with lucerne, K deficiency may be induced when lucerne is grown in association, unless ameliorative fertilization is undertaken.

Critical levels for Ca in whole-shoot DM is about 15 g kg^{-1}, but estimates in the top 15 cm of foliage in the flowering crop range from 3 to 18 g kg^{-1}, stage of development being important, as Ca levels decline rapidly with advance in maturity. Generally, lucerne has a higher content of Ca than grasses. Calcium availability in the soil is reduced by soil acidity, and in lucerne low soil Ca exacerbates adverse effects of acidity (Munns and Fox, 1977).

Whole shoots of lucerne have critical concentrations for Mg ranging from 2.0 to 3.5 g kg^{-1} DM, and sufficiency levels in the upper horizons of the crop are in the range 3–10 g kg^{-1}. Magnesium concentration declines with crop maturity and, in late-cut hay, may be restricted to levels well below the minimum for animal requirements if soil K levels are high, due to preferential uptake of K by the plants. There is also a seasonal effect on Mg content in lucerne, with Mg content increasing with rising soil temperatures from spring to summer, although this is cultivar-dependent (Gross and Jung, 1978). Since lucerne has a low concentration of Na, sodium chloride (NaCl) supplementation of cattle and sheep grazing lucerne has been beneficial to their health and production (Jagusch, 1982).

Critical levels for S in the DM of whole shoots range from 1.0 to 2.2 g kg^{-1}, and the sufficiency range in the top horizon of the early flowering crop is 3–5 g kg^{-1}. Sulphur deficiency adversely affects N$_2$ fixation, and lucerne has fewer sites in its root mass than grasses to absorb S. Maximum DM production is associated with S levels in shoot DM at harvest of 2.0–3.0 g kg^{-1}, the optima to ensure N concentration being non-limiting for production (Andrews, 1977), although later work has shown no additional effect on N concentration when increasing S above 2 g kg^{-1} (Quigley and Jung, 1984). Critical N : S ratios in the DM of whole shoots have been set at between 11 and 15.

Boron has a sufficiency range of 20–80 mg kg^{-1} DM in the uppermost horizon of the developing canopy, although a critical level of 17–18 mg kg^{-1} in New Zealand has been cited (Douglas, 1986). Boron deficiency can be induced by moisture stress at high soil pH and, since B is relatively immobile, it affects young tissue initially; the upper leaves become yellow and red, and internode and axillary branch elongation is retarded. If plants remain in this condition for several weeks, they become semidormant and do not produce new growth when conditions become suitable for growth unless the plants are defoliated.

Due to the importance of Mo in the functioning of the enzyme nitrogenase in the N$_2$-fixation process, N content in the plant is reduced by Mo deficiency and the symptoms are of N deficiency. Molybdenum availability increases as soil pH rises, and critical levels in the DM are in the range 0.5–0.9 g kg^{-1}. Deficiency is more likely to be manifest in the nodules than in shoots, so herbage analysis is not a good indication of deficiency (Doerge *et al.*, 1985).

Copper deficiency occurs in lucerne, especially in soils high in OM and at a high soil pH. Critical levels are 4–5 mg kg^{-1} in whole-shoot DM and 3–10 mg kg^{-1} for young growth in flowering shoots.

Usually, lucerne has a high tolerance of soils with a high NaCl content and can withstand higher NaCl concentrations in the soil than many other forage species, although cultivars that would be considered 'tolerant' of high NaCl levels are not yet available.

An aspect of nutrient interaction is the impact of uptake of one nutrient on the uptake of others. For example, increasing availability of soil P increases uptake of Na, Ca, Mg and S in lucerne but reduces the concentration of K and some trace elements, while increasing K availability reduces concentration of B, Cu, Mn and Zn (James *et al.*, 1994).

Breeding

Cultivars of lucerne have variable contributions of *M. sativa* ($4n = 32$) and *M. falcata* ($2n = 16$) in their ancestry, the former conferring the capability of rapid growth under well-drained, non-acid soil conditions, and the latter cold-hardiness, disease resistance and some tolerance to grazing.

Lucerne is a cross-fertilizing autotetraploid, relying on bees for pollination. It responds more slowly to selection than a diploid species and equilibrium is not reached within one generation of random mating. Phenotypic mass selection, which may also involve progeny testing, is the usual method for breeding lucerne cultivars, recurrent selection requiring about 75 plants in each cycle (Hill *et al.*, 1969). The additive nature of many of the characters inherited has ensured that this basic system of breeding has usually been successful (Dunbier, 1983).

As there is scope for improvement in most of the agronomic characteristics of lucerne, a breeding programme usually involves selecting for improvement in more than one attribute. To achieve this, a breeding programme either involves 'fixing' each of the attributes in turn within the population (tandem selection) or selecting in sequence within a generation. An example of the latter is in selecting for multiple pest resistance; survivors resistant to a specific pest are then challenged with an infestation of the second pest for which resistance is being sought within the same generation.

Spaced plants in selection of parents or in progeny testing are commonly used in lucerne plant-breeding programmes. However, methods are being developed to allow selections to be made from microswards. Spaced plants continue to have a use in selecting for genotypes that are resistant to a severe environmental stress, where competition with neighbours is of secondary importance. However, selecting for high potential yields for intensive agriculture is more effectively achieved by selecting and evaluating germplasm under sward conditions (Rotili *et al.*, 1996).

Other genetic techniques being developed to improve the prospect of breeding markedly superior cultivars in conventional breeding programmes include selfing to produce parents for controlled crosses and using restriction fragment length polymorphism to determine the degree of heterozygosity in autotetraploid lucerne (Rotili, 1993). Self-crossing is required to concentrate the accumulation of desirable genes and to produce sufficient genetically similar parents when breeding a synthetic cultivar. However, lucerne suffers severe inbreeding depression when selfed (Busbice, 1969), and so an alternative to achieving an adequate number of parents with selected characters is the use of somatic embryogenesis, in which many plants can be vegetatively propagated from parts of one parent plant, under culture conditions (Bowley, 1997). Although in its infancy, the use of somatic embryogenesis to improve variation in lucerne is being intensified (Radeva et al., 1993; Scotti, 1993).

Where incompatibility between lucerne and other species precludes crossing by conventional breeding methods, somatic hybridization has been attempted. This involved fusing protoplasts of *M. sativa* with other *Medicago* species, e.g. *M. borealis* or *M. media*, or with a species from another genus, e.g. sainfoin (*Onobrychis viciifolia*), and overcoming barriers imposed subsequent to fertilization, using embryo rescue techniques (Arcioni et al., 1997).

The use of *in vitro* cell cultures to select for resistance to diseases from genera such as *Fusarium, Verticillium* and *Pseudopeziza* is under investigation, but success depends on the activity of the toxins produced by cell cultures (Arcioni et al., 1993).

Genes from other species have been incorporated into lucerne, mediated by *Agrobacterium tumefaciens*, by particle bombardment or by deoxyribonucleic acid (DNA) uptake (Bowley, 1997). Lucerne has been transformed to produce transgenic plants that contain seed albumin genes from sunflower, in order to increase content of methionine and cysteine, two of the S-containing amino acids that are in limiting supply for wool growth on sheep grazing current lucerne cultivars (Tabe et al., 1995). Other current programmes involving transformation of lucerne include transfer, from a species of *Nicotiana*, of an Mn superoxide dismutase gene, which prevents and repairs damage to cells due to freezing, in order to increase winter-hardiness (Bowley, 1997).

All present-day cultivars bred in the USA and Canada have been derived from nine germplasm bases, namely, *M. falcata*, Ladak, *M. media*, Turkestan, Flemish, Chilean, Peruvian, Indian and African. Although they represent a spectrum in winter dormancy, only the extremes fit into discrete dormancy classess, i.e. *M. falcata*, Ladak, African and Indian (Fairey et al., 1996).

The rate of improvement in yield due to breeding lucerne has been more marked for third-year yields than for first-year yields. Over the period 1965 to 1990, improvement in first-, second- and third-year yields of cultivars approved for registration in Canada has averaged 0.16, 0.27 and 1.07% year^{-1}, respectively, reflecting the advances that have been made in breeding for improved persistence rather than for potential yield (McKersie, 1997). This compares with an estimated mean improvement in DM yield of 0.5–0.6% annually from perennial-ryegrass breeding in Western Europe, albeit over a longer period (Aldrich, 1987; van Wijk and Reheul, 1991).

In addition to breeding for improved yield, cultivars were bred initially for improved winter growth in climates with warm winters, and subsequently for winter-hardiness in regions with hard winters. Winter dormancy has been an important consideration in breeding programmes for lucerne in cooler temperate regions, especially in North America. At registration, a cultivar is given a dormancy index, which is determined from the height of regrowth in October after a defoliation in early September. The indices range from 1 (most dormant) to 9 (least dormant) and each class has a reference cultivar which is a standard for that class.

Breeding for improved N_2 fixation has met with limited success. Improved nodulation rate, nodule and root mass, nitrogenase activity and nitrogen content in lucerne have been achieved by selection in controlled and semicontrolled environments, but the performance of these selections in the field has been disappointing (Jessen et al., 1988; Teuber and Phillips, 1988).

Breeding for disease and pest resistance was introduced into programmes in an attempt to improve overall persistence. Resistance to many of the principal diseases of lucerne has been bred into cultivars in the USA, with resistance to spring black stem (*Phoma medicaginis* var. *medicaginis*), Sclerotinia crown and stem rot (*Sclerotinia trifoliorum*) and Rhizoctonia crown and bud rot being exceptions as yet (Barnes, 1992). Lucerne cultivars are available with resistance to stem nematode (*Ditylenchus dipsaci*) and efforts are being made to breed for resistance to other soil-borne pests, e.g. root-lesion nematode (*Pratylenchus penetrans*) (Christie and Townshend, 1992). Breeding for resistance to *Verticillium* wilt stem and root eelworm has been successful in Western European breeding programmes. These successes have all been achieved via conventional breeding programmes, in contrast to the absence, as yet, from the market of successfully genetically engineered cultivars that have resistance to fungal diseases.

Achievement of adequate winter-hardiness within lucerne, to allow it to withstand harsh winter conditions in regions such as Canada and the northern states of the USA, is strongly associated with low rates of autumn herbage production (Perry et al., 1987). Nevertheless, some cultivars capable of withstanding less extreme winters may have good autumn growth. Out of 251 entries in a study in which autumn growth and winter-hardiness were evaluated, only two had greater winter-hardiness than would have been predicted from autumn herbage production (Schwab et al., 1996). If genetically based, these exceptions suggest that cultivars capable of autumn production and surviving cold winters could be bred.

The development of genotypes of lucerne that are tolerant of grazing has been an objective of plant breeders for many years, but has met with variable success. In the UK, lucerne cultivars are recommended only for conservation, but in the USA, for example, some cultivars have been bred to have broad crowns, thus rendering them more suitable for grazing than narrow-crowned types. In continental Europe, attempts are being made to select and breed for lucerne capable of high and sustained production under frequent-defoliation management (Veronesi et al., 1993). There are also programmes, for example in the USA, France, Italy and Canada, to incorporate tolerance to grazing mainly, but not exclusively, by the incorporation of *M. falcata* genes. In Georgia, USA, some *M. sativa* selections, which were grazing tolerant, have resulted in the cultivar, Alfagraze, being released, and a French Mediterranean type suited to grazing is undergoing registration. In Italy, advantage has been taken of the wide variation in tolerance to grazing, mainly determined by the degree to which the crown is below ground level and found in all of the main groupings of lucerne grown in Europe, i.e. *M. sativa*, *M. falcata*, *M. media* and 'mielga' types, in order to make artificial crosses between them. In particular, high-yielding rhizomatous, rather than creeping, plants have been selected that combine tolerance to grazing resistance and yield (Piano *et al.*, 1996). Selection within the naturally occuring 'mielga' type to take advantage of its creeping growth habit, and hence its high level of persistence, is one objective, but the improvement of its autumn and late-winter growth is a current priority in a breeding programme in the south of France (Prosperi *et al.*, 1996).

Cultivars tolerant of the environmental stresses of cold and drought are available, and selection procedures exist to identify resistance to high salt concentration (Rumbaugh and Pendery, 1990) and tolerance to high Al concentration associated with acid soil conditions (Bouton and Sumner, 1984). As yet, no commercial cultivars are available from these selections and the prospect of breeding a cultivar with high Al tolerance is not imminent. No genotypes in the US 'Alfalfa Core Collection' have been identified as being particularly Al-tolerant (Bouton, 1996). Slow progress towards breeding an Al-tolerant culivar may be partly caused by the slower rate of improvement at each cycle of selection with an autotetraploid, such as lucerne, than with a diploid, and also the extent of inbreeding required to ensure expression of the Al-tolerance genes may be causing inbreeding depression and so mask the Al-tolerance characteristic (Bouton and Parrott, 1997).

Genotypes tolerant of high salt concentrations in soils have been identified and methods developed to select such genotypes for cell batch culture. Although some have been selected that are capable of withstanding NaCl concentrations of more than 300 mmol l^{-1}, characterized by having low concentrations of Na and Cl and a low Na : K ratio, they have not been evaluated in the field (Chaudhary *et al.*, 1996).

Cultivars of lucerne are expected to perform well over a wide range of environments. For example, where variability of yield is due to the prevalence of a given pest over an area, breeding for resistance may overcome that variability. However, other limiting factors may then be exposed, e.g. soil acidity, and so another obstacle can limit the widespread acceptance of a cultivar (Hill and Baylor, 1983). Therefore, cultivars bred from a broad genetic base are most likely to ensure stability over different environments.

Selecting for low lignin content in the herbage, in order to improve nutritive value, generally results in declining yields, due to the resultant higher leaf : stem ratio, and so herbage digestibility is increased at the expense of DM yield. Selection for increased protein content, which is highly heritable, also appears to reduce stem content, with reduction in acid detergent fibre (ADF), but increases the likelihood of lodging. Breeding to reduce mineral imbalances in lucerne to meet ruminant dietary requirements has met with limited success (Hill et al., 1988; Rotili, 1993). For example, an increase in P content by breeding to reduce the Ca : P ratio is only expressed at high soil-P levels.

Another means of attempting to improve nutritive value in lucerne is to increase the number of leaflets per lamina. This is currently under way, although the proportion of plants with the multifoliate characteristic resulting from crosses is variable and a high proportion is required in order to make an impact on crude protein (CP) and digestibility of the harvested herbage. Multifoliate plants do not produce higher yields and they have only slightly lower neutral detergent fibre (NDF) and ADF content than corresponding trifoliate-leaved genotypes (Juan et al., 1993).

The prospects of breeding for resistance to bloat are limited by a lack of potentially important traits within the genus *Medicago*, including insufficient tannin content, low resistance of cells to damage or low saponin content (Hill et al., 1988). Lucerne lines have been selected for thicker cell walls in the leaves, considered to be associated with reduced rate of initial digestion and hence reduced bloating (Goplen et al., 1993), but field-trial results have proved disappointing (Hall et al., 1994). The production of somatic hybrids between lucerne and legumes with a high content of condensed tannins (CT) is another approach towards antibloating characteristics, but, for instance, hybridization with non-bloat-inducing sainfoin has not yet resulted in improved CT content in the hybrid (Li et al., 1993). If successful, the next challenge will be to control the concentration so that it does not reduce digestibility and protect proteins excessively.

Cultivars

Due to the varying contribution of common, yellow and variegated lucernes in the ancestry of the current lucerne cultivars, a wide range of types exist. In Europe, Spanish lucernes, from which African and Peruvian types arose, are believed to be closest to the normal common-lucerne parent. The Australian cultivar, Hunter River, is also of similar ancestry. In contrast, the Grimm form of cultivars used in Canada has a high content of yellow-lucerne genes, conferring winter dormancy, and, hence, winter-hardiness, and a higher tolerance to grazing than those with mainly common-lucerne genes. Other European types, grown mainly in central and southern Europe, are intermediate in their content of common-lucerne and yellow-lucerne genes, including Flamand (Flemish) and Provence (Mediterranean) types, the two principal European groups of lucerne cultivars, the former having more yellow lucerne characteristics than the latter.

The 'mielga' types grown in Spain are also prostrate, due to the production of rhizomes, and are persistent under grazing. Oasis types, e.g. cultivars from Iran, exhibit no winter dormancy and are capable of growth at low temperatures.

The earliest lucerne varieties developed in the USA probably originated from Spanish material, but were not hardy enough for more northerly parts of North America. Hence, introduction of hardier Grimm material from Europe allowed breeders to produce cultivars with the required winter dormancy for cold conditions (Klinowski, 1933). Cultivars have been bred with *M. falcata* genes to gain persistence under grazing, e.g. cvs Rambler and Beaver, which also become dormant in winter and are cold-hardy. However, cv. Alfagraze, which is tolerant of continuous stocking, is not winter-dormant, nor does it have yellow lucerne in its ancestry (Bouton *et al.*, 1991). The cv. SA Standard, which is widely used in the dryland pastures of the southern and western Cape areas of South Africa, is extremely grazing-tolerant (van Heerden, 1993).

Some of the early cultivars bred for resistance to bacterial wilt, such as Ranger or Buffalo, and to spotted alfalfa aphid, e.g. cv. Moapa, are still widely used in North America.

In the USA in recent years, the activity of lucerne breeding has been manifest from the rate of release of new cultivars, 440 having been released between 1962 to 1992, and in the late 1980s only about 5% of the released cultivars originated from publicly funded breeding programmes (Barnes and Sheaffer, 1995). This activity in lucerne breeding has resulted in cultivars being released with resistance to many of the major pests and diseases and with a range of winter-dormancy indices. To facilitate choice of cultivar, databases, such as ALFALFA CATALOG, have been compiled (Table 3.3).

As an illustration of the extent to which lucerne cultivars are resistant to the major pests and diseases, six of the seven registered in the journal *Crop Science* in 1995 were highly resistant to *Fusarium* wilt, the five for northern states and Canada were 'resistant' to *Verticillium* wilt and six of the seven were 'resistant' to bacterial wilt. In contrast, only three had some resistance or were

Table 3.3. Biological information contained in the ALFALFA CATALOG database files for each cultivar (from Townsend *et al.*, 1994).

General information
Dormancy index
Dormancy reference population
Flower colour composition
Origin

Resistance levels	
Spotted alfalfa aphid	Bacterial wilt
Pea aphid	*Fusarium* wilt
Blue alfalfa aphid	Anthracnose
Potato leafhopper	Downy mildew
Stem nematode	*Verticillium* wilt
Root-knot nematode	Common leaf scab
Phytophthora root rot	Scald

highly resistant to at least one of the three aphids considered to be major pests of lucerne (spotted alfalfa, blue alfalfa and pea aphids). Thus, disease resistance seems to have been more readily achieved than pest resistance in the breeding of modern cultivars.

In New Zealand, until the 1960s only two cultivars were predominantly used, namely, Wairau, bred in New Zealand, and Hunter River, from Australia (Iversen and Meijer, 1967), but pest- and disease-tolerant cultivars are now available, such as Grasslands Oranga and Grasslands Otaio. Grasslands Oranga is a general-purpose grazing type selected for resistance to blue-green aphid but is also moderately tolerant to *Verticillium* wilt. Grasslands Otaio was bred for resistance to bacterial wilt, stem nematode and blue-green aphid, with resistance to pea aphid and *Phytophthora* root rot, and is tolerant of *Verticillium* wilt (Charlton, 1995).

In Europe, breeding for lucerne cultivars has been less active than in the USA, with local ecotypes sometimes being grown instead of registered cultivars, partly due to a less stringent registration system in southern Europe, compared with the European Union (EU) generally. For example, in 1989, 85% of the seed used in Italy was from ecotypes, although in France 90% of the seed was of registered cultivars, many of the Flemish type, the most commonly sown being Résis (Guy, 1993). In a comparison of ecotypes used commonly in Italy with modern cultivars bred in North America, ecotypes had higher NDF and cellulose and lower CP contents, with reduced rumen turnover of fibre and DM, compared with the cultivars, demonstrating the benefits of using bred cultivars and the scope for improvement in lucerne in Italy (Andrighetto *et al.*, 1995).

The cultivar Europe is high-yielding and resistant to lodging but is susceptible to *Verticillium* wilt, while cv. Euver is also high-yielding but less susceptible than cv. Europe to *Verticillium* wilt and stem eelworm. The cultivar Vertus has good resistance to *Verticillium* wilt and to stem eelworm but, in their absence, yields less than cvs Europe and Euver.

Breeding programmes in Europe have been successful in increasing resistance to *Verticillium* and *Fusarium*, although increase in potential DM yield over 20 years has been less significant. As the programmes move towards evaluation and selection of genotypes in stands, rather than as spaced plants, and more becomes known about intraspecific competition in lucerne, higher-yielding cultivars may be produced to match the success in breeding for pest and disease resistance (Lorenzetti, 1993).

Seed production

Temperatures of 25–30°C are optimal for seed fertility, pollinator activity and seed filling, and so lucerne seed production is mostly in areas which have high summer temperatures, e.g. western states in the USA and southern European countries. In addition to yield, location can also influence hard-seed content, which is probably inversely related to temperature during seed set (Fairey and Lefkovitch, 1991). Water availability is also critical for seed production and, while irrigation improves seed production, if carefully controlled, overwatering increases bud and flower production but reduces pod number (Steiner *et al.*, 1992). As a rule, moisture stress should be avoided up to the bud stage and again at the green-pod stage, when seeds are filling, but seed yields are adversely affected by wet weather when seeds are ripening (D'Antuono *et al.*, 1988).

Pollination is by bees (honey-bees, short-tongued cutter bees (*Megachile rotundata*), alkali bees or bumble-bees (*Bombus* spp.)). To ensure cultivar purity, seed crops have to be isolated from other crops, areas of less than 2 ha having to be separated further from other crops than larger areas. Proximity to wild flowers enhances pollinator visits (Brookes *et al.*, 1994).

Seed-yield potential is determined by genotype, with yields from cultivars ranging from 20% below to 15% above those from control cultivars in France and with more recently bred cultivars generally having higher seed yields than the older cultivars (Hacquet, 1989). So-called bad years for seed production are associated with high rainfall in the late vegetative stage of growth and this results in lodging, poor pollination and seed fill, while later rainfall may cause sprouting in the pod and difficulties in harvesting the crop.

A low sowing rate – 3–5 kg ha^{-1} – and relatively wide spacing – 30–70 cm – between drills produce high populations of harvestable racemes and 1000-seed weights (Askarian *et al.*, 1995). Seed yield is most highly correlated with number of seeds per pod, followed by number of racemes or pods per unit area. Generally, highest seed yields are achieved in crops with a high population of racemes pollinated over a short period of time, hence ensuring that flowering is confined to a short interval, with crops protected from pests during flowering and seed filling, and with timeous harvesting, so that losses from shedding of ripe seed are avoided. Herbicide use, e.g. glyphosate to control dodder (*Cuscuta* spp.), should be restricted to before bud stage to avoid injury to flowers and subsequent reduced yield and quality (Dawson, 1992).

Guidelines for successful seed production (Dunbier *et al.*, 1983) are as follows:

During establishment:

- Prepare a fine seedbed and use a presowing herbicide.
- Sow at low seed rates in wide rows.
- Sow early to harvest the same season.
- Apply non-acidic fertilizers with the seed to increase early growth.
- Cultivate and rogue between rows to control problem weeds.

Managing established stands:

- Remove winter weed growth by strategic grazing or herbicides.
- Time spring hay cut/grazing to ensure correct crop closure timing.
- Apply irrigation at closing off and later, at bud stage, if dry conditions prevail.
- Monitor fungal leaf diseases and insect pests, and spray, if needed.

- Use leaf-cutter bees at 30,000–50,000 ha^{-1} or short-tongued bumble-bees at 1000–2000 ha^{-1}.
- Desiccate the crop as soon as the first pods start to lose seed in late summer, i.e. when 65–75% of the pods are brown.

In Europe, seed yields increased by more than 50% during the 1980s, to an average of about 400 kg ha^{-1}, mostly due to control of insects (gall midge, mirid bugs and alfalfa weevil) and general improvements in cultural management. In the western states of the USA, i.e. the main lucerne seed-producing region in the country, average yields are about 750 kg ha^{-1} (Barnes and Sheaffer, 1995).

Agronomy and Management

Seed mixtures

While the inclusion of a grass with lucerne may reduce weed invasion, give a more balanced nutrient composition for successful ensiling or for animal feed, or utilize 'transferable' N from lucerne, it may not always improve DM yields relative to a lucerne monoculture. In exceptional circumstances, inclusion of a grass has increased total herbage DM yields over pure-sown lucerne by 70% or even 80% (Douglas, 1986), but, if a benefit is achieved at all, it is normally in the region of 10–15% (Chamblee and Collins, 1988). Maintenance of an acceptable balance between the two components in the long term is usually difficult to achieve. Monocultures are sometimes favoured because selective herbicide control of invasive weeds is simpler, though more costly, to achieve. Competition from weeds is usually most acute in the early stages of lucerne establishment.

In mixtures, defoliation intervals that suit lucerne are likely to be detrimental to the companion grass and vice versa. An example is presented in Fig. 3.4. However, even when perennial ryegrass comprised as little as 18% of the DM after a long cutting interval (40 days), it recovered in the following year when the cutting interval was halved (Jung *et al.*, 1996). The balance between lucerne and companion grass can therefore be controlled by management, but herbage quality also changes, due to the interaction between nutritive value, stage of defoliation and relative contributions of legume and grass.

Although management and environment greatly influence the competitive balance between lucerne and grass, certain species are more competitive towards lucerne than others. Hence, tall fescue (*Festuca arundinacea*) and cocksfoot (*Dactylis glomerata*) are consistently aggressive (Chamblee, 1972), while smooth brome or perennial ryegrass can remain in balance with lucerne (Casler, 1988), ryegrasses persisting with lucerne better than timothy (Jung *et al.*, 1991). In a comparison of five grass species in Scotland, a lucerne/meadow-fescue (*Festuca pratensis*) association produced a higher total DM production than timothy or perennial-ryegrass associations over a 3-year trial (Frame and Harkess, 1987).

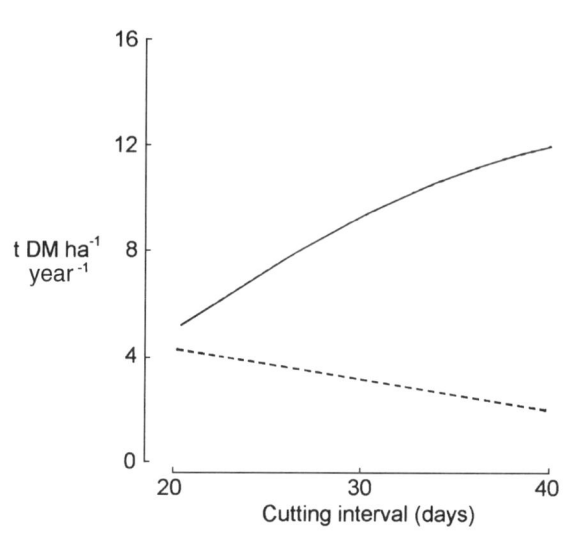

Fig 3.4. Total annual yields of lucerne (——) and perennial ryegrass (-----) in mixture as cutting interval varies in the first harvest year (mean of two trials) (from Jung et al., 1996).

In mixtures, the grass tends to constitute a higher proportion of herbage in spring than in summer. Late maturity in a grass, such as reed canary grass (*Phalaris arundinacea*), puts it at a competitive disadvantage with lucerne, which develops its canopy over the grass early in the season (Jones et al., 1988). In contrast the ability of some grasses to penetrate the canopy of lucerne allows them to maintain a relatively competitive position in the upper layers of the canopy (Jung and Shaffer, 1993). Application of fertilizer N aids the restoration of the grass contribution when it falls below a desired level or, potentially, can maintain total yield when lucerne is declining in a mixture.

The relative root distribution of lucerne and grass through the soil profile influences their competitive abilities. The deep-rooting nature of lucerne puts it at an advantage over accompanying grass during drought. However, tall fescue, with its well-developed rooting system and drought-resistance characteristic, competes vigorously with lucerne, *inter alia* for K (Chamblee and Collins, 1988). Lucerne roots have a higher cation-exchange capacity than grasses and compete vigorously for divalent cations, such as Ca. Lucerne also competes successfully for moisture and some nutrients, P for instance, in the surface layers of soil.

Other factors associated with competition between lucerne and grass include the effect of the companion grass on light quality at the base of the sward canopy. Lucerne crown buds may be prevented from developing when the canopy is dense due to a low red : far-red ratio in irradiance at the canopy base. Thus, at critical times, such as spring, when grass is occupying a high proportion of the canopy, the grass should be defoliated so that the crown buds have a better chance to grow and develop optimally.

Action taken at seeding has met with only limited success in the attainment of a target balance between the species. Varying seed-rate ratios (Jung *et al.*, 1991), the number of rows sown of each (Mooso and Wedin, 1990), sowing time (Chamblee, 1972) and spatially separating the two species in bands to reduce N transfer and prevent grass dominance (Fairey and Lefkovitch, 1990) have all been tried. While delaying sowing to late season aids establishment of grass relative to lucerne, sowing-rate differences result only in short-term effects.

In central USA, sowing annual ryegrass (*Lolium multiflorum*) with lucerne increased yield but reduced quality, compared with monocultures, differences being greater in the sowing year than in the following full harvest year. Early diploid annual ryegrasses were most compatible with lucerne (Sulc and Albrecht, 1996). In semiarid regions in Canada lucerne establishes successfully under irrigation with oats as a nurse crop (Jefferson and Zentner, 1994). In New Zealand, where herbage growth continues over winter, annual ryegrass is sometimes introduced each year into established lucerne swards, with the aim of producing winter herbage when lucerne is relatively dormant and so exploiting complementary growth cycles (Douglas, 1986). However, the system is unpredictable, as vigorous ryegrass growth sometimes severely sets back lucerne growth by as much as 30–40% the following summer. Autumn direct drilling of cereals, such as barley or oats, into lucerne swards has also been successful, but the cereal vegetation accumulated over the winter requires to be grazed off in late winter to avoid adverse effects on lucerne's spring growth.

In investigations of leguminous mixtures, lucerne achieved early dominance over white clover and eventually over red clover (Frame, 1986) and proved too competitive to birdsfoot trefoil (Panciera and Fulkerson, 1981) or, in the semiarid region of western Canada, to sainfoin (Jefferson *et al.*, 1994).

Establishment

Recommended sowing rates for lucerne vary from 10–13 kg ha^{-1} in northern regions of North America to 17–34 kg ha^{-1} in southern regions, where seedling diseases are more prevalent (McKersie, 1997), but elsewhere rates of 6–20 kg ha^{-1} are more common. The seed-bed should be well consolidated to prevent the seed 'settling' too deeply. The optimum depth is about 15 mm, with the number of plants establishing being adversely affected when seed is sown deeper than 25 mm (Sund *et al.*, 1966). For a highly productive stand of lucerne to be established, $3-4 \times 10^6$ seedlings ha^{-1} should emerge. The aim is to establish an ultimate plant population density of 200–400 m^{-2}, although the density of plants declines steadily as the stand develops until an equilibrium is reached (Palmer and Wynn-Williams, 1976); this equilibrium is usually attained at 50–140 plants m^{-2} although maximum yields can be achieved by as low a density as 30 plants m^{-2}. Even at very low seeding rates of 2–3 kg ha^{-1}, yields can be similar to those from crops sown at five or more times this rate, but with a reduced rate of decline in plant density.

If sowing with a companion grass, seeding rates are notionally adjusted to prevent dominance by either component during establishment. High seeding rates of aggressive grasses, such as cocksfoot, reduce lucerne establishment rate, while other less aggressive grasses, such as timothy or meadow fescue, may be adversely affected by high seeding rates of lucerne (reviewed by Chamblee and Collins, 1988). However, sowing-rate effects on species balance are transient, since genotype, environment and postsowing management all determine the long-term composition of the mixture.

Sowing method is important in determining establishment and early yield of lucerne and, particularly, of lucerne/grass mixtures. Drilling seeds over a band of fertilizer gives superior establishment of lucerne compared with broadcasting, particularly under adverse conditions (Tesar and Jackobs, 1972). Yields from stands sown by alternating drills of grass and lucerne are usually inferior, compared with other methods, such as mixing species within drills, particularly if the drills are wide, one factor being the reduced transfer of N from the lucerne, because of less contact between the grass and root rhizospheres. Nevertheless, in Canada, lucerne persistence was as high when sown in drills with brome grass as in alternating rows of the two components, and total yields were similar (Fairey and Lefkovitch, 1994).

Spring-sown crops usually receive sufficient moisture to ensure germination and establishment before water becomes scarce, and spring sowing also usually ensures suitable moisture and temperature conditions for rhizobia survival and infection of the lucerne roots. The sensitivity of young lucerne plants to frost and cold winters, or the possibility of damage due to soil heaving by frost, generally limits late-summer/autumn sowing and so this sowing time is only feasible in areas where winters are mild. Soil moisture may also be limiting in autumn in some regions but can be overcome by irrigation (Janson, 1975).

Inoculation of lucerne seeds with *R. meliloti* is generally recommended, particularly on soils where lucerne has not been sown for the previous 3 or more years. Many countries operate services that supply rhizobia strains to the industry, monitor standards and exercise quality assurance during production of commercial inoculants (see section on N_2 fixation).

High soil temperatures, high soil-nitrate levels, soil-drying winds or extremes in postsowing moisture content reduce the certainty of lucerne establishment (Smith and Stiefel, 1978), as does poor seed-bed preparation and consolidation, inadequate weed control or attack by fungal pathogens. Factors resulting in poor nodulation also reduce lucerne establishment, both directly, due to poor N nutrition, and indirectly, by reducing resistance to fungal attack (Aube and Gagnon, 1970).

Inclusion of a cover crop with spring-sown lucerne results in an economic benefit, since the revenue from the sale of the crop offsets the costs of establishment of lucerne; e.g. wheat, barley, linseed and early-harvested peas are particularly suitable (Wynn-Williams, 1982). While cover crops normally depress yields of lucerne for a few harvests subsequent to its establishment, the effect is minimized by reducing cover-crop seeding rates. A cover crop under dryland conditions is not recommended, since it can be too competitive for limited moisture during establishment.

Lucerne can be successfully established by minimum cultivation, direct drilling or 'no-till' techniques. In some instances, especially if soil moisture is likely to be limiting, lucerne may establish better by direct drilling than by conventional cultivation, as moisture is retained more effectively in undisturbed soil (Brash, 1983). Competition towards the young lucerne plants depends on the nature of the previous crop or pasture, e.g. sowing into the stubble of a grain crop generally presents the seedlings with less competition than into a grass sward. Competition from existing vegetation is controlled by mowing or applying grass-suppressing herbicides, either in bands, within which the seeds are drilled, or as a general application (Byers and Templeton, 1988). The botanical composition of the original sward may influence the ability of lucerne to establish, due to allelopathy, since exudate from tall fescue reduces germination of lucerne seeds and that from cocksfoot reduces the rate of lucerne-seedling growth (Chung and Miller, 1995a).

Irrespective of the previous crop, the probability of success is heightened if soil moisture is not limiting and the drill ensures good control of seed depth and adequate seed–soil contact and soil consolidation. Choice of time of sowing is the most practical means of minimizing risk from inadequate soil moisture, and type of drill plays a large part in ensuring that the soil environment around the seed is optimal. From a comparison of drills, the one most suited to sowing lucerne had independently suspended coulters/spouts sufficiently heavy to cut through litter on the soil surface and hard ground and which pressed the soil around the seed (Waddington, 1992). Control of slugs and some insects may be necessary, as the environment remains relatively undisturbed and moist with direct drilling (Byers and Templeton, 1988).

Fertilizer use

Other than during establishment, when insufficient soil mineral N may be available for the lucerne seedlings, prior to the start of N_2 fixation, fertilizer N need never be applied to lucerne. Generally, N need only be applied at sowing if the soil-nitrate N is less than 15 g kg^{-1} or nodulation is unlikely, due to very low *R. meliloti* counts, and then 25–50 kg N ha^{-1} is a suitable dressing (Hannaway and Shuler, 1993). On light-textured soils, Mg deficiency during lucerne establishment is remedied by the application of Mg limestone or Mg sulphate. Because of its sensitivity to deficiencies in B and Mo, assessment of the status of these elements in the soil at sowing, by soil analysis, and subsequent remedial action, if necessary, are advisable.

When applying organic manures, farmyard manure (FYM) should be ploughed in prior to seedbed preparation and slurry is best applied at a time when damage to the above-ground shoots will be minimal, i.e. preferably outside the growing season and never to establishing crops. Application of manure (slurry) to the seed bed increases soil mineral-N content compared with untreated lucerne, particularly 1–2 months after application, but the consequences are not long-term, the differences between treated and untreated disappearing within 2 years (Schmitt et al., 1994). However, from experiences of farmers, applying high rates of manure before sowing lucerne in a mixture with an aggressive companion grass, e.g. cocksfoot, can severely reduce lucerne persistence

Neither the offtake of nutrients by harvesting successive crops of lucerne nor the requirement for replenishment should be underestimated. Taking a 12 t DM ha^{-1} year^{-1} crop with a P content of 3.3 g kg^{-1} and a K content of 27 g kg^{-1} results in an annual offtake of 40 kg P ha^{-1} and 324 kg K ha^{-1}, and, with an assumed N concentration of 30 g kg^{-1}, the N offtake of 360 kg ha^{-1} underlines the importance of successful N$_2$ fixation in the plants.

The following is a summarized example of the type of advice offered to farmers on the fertilization of lucerne, in this case relevant to the Atlantic provinces in Canada (Agriculture Canada, 1989).

- Soil tests should be performed every 2–3 years.

Postsowing:

- Nitrogen: small amount may be applied to stimulate growth in spring – otherwise avoid its use, as it encourages grass growth.
- Phosphorus: required in early spring. Unless there is a P-deficiency problem, no need to apply after each cut.
- Potassium: apply reasonably high amounts in spring and after first and second cuts.
- Lime: soil pH should be maintained at about 6.5. This can be achieved by liming every 3 years or so.
- Boron: apply at 3- to 4-year intervals.
- Sulphur: apply if signs of deficiency (yellowing of plant with reddening of stems).
- Slurry/FYM applied either in autumn or after cutting, but avoid pollution of waterways.

Irrigation

Lucerne has the ability to tap water deep in the soil profile and to minimize water usage during moisture stress, by reducing its metabolism and becoming semidormant. However, it is sensitive to moisture availability in the upper layers of the soil profile. Experience in the UK shows that it withstands a soil moisture deficit of 150–200 mm.

Moisture stress, as measured by plant water potential (Ψ), severely restricts lucerne growth at Ψ less than -1.0 MPa, while negative growth results at Ψ of about -2.5 MPa; a slow rate of leaf appearance and internode elongation causes reduced growth, while leaf loss on the lower part of the plant results in DM loss exceeding accumulation, i.e. negative growth (Brown and Tanner, 1981).

If drought has not been too severe, response of lucerne to irrigation is rapid, through the remobilization of WSC previously built up in the roots during drought. However, in severe drought conditions, the stand has to be defoliated before it will respond to irrigation. Actual amounts of water required to produce 1 t DM ha^{-1} of lucerne depend on climate, stage of growth and cultural practice. Water use can be as high as 14 mm day^{-1}, this level being confined to the warmer days of summer when a full crop is being achieved, but generally 5–8 cm of water is required to produce 1 t DM (Sheaffer *et al.*, 1988a). Yield improvements from irrigation have ranged from 1.5 to 8.0 t DM ha^{-1}.

Weed control

Lucerne grows slowly during establishment, so weeds readily invade the sward. Apart from competition, weeds in the harvested forage reduce its nutritive value or its sale value as a cash crop, while weed seeds in seed crops may render them unacceptable for certification (Peters and Linscott, 1988). The detrimental effect of weeds in reducing lucerne persistence has also been demonstrated (Latheef *et al.*, 1994). Selecting the time of sowing is one strategy to minimize weed ingress. Autumn sowing may coincide with low germination of weed seed populations but is feasible only in warmer regions, where there is less need to have well developed plants before a probable severe winter than in colder regions.

Conferring nutritional advantages to lucerne growth alleviates weed competition. Banding P and K fertilizer below the sown seed allows lucerne to take advantage of the higher nutrient availability in its rhizosphere. Liming acid soils also helps to tip the soil conditions in lucerne's favour. The effect of N fertilizer varies, in that it confers an advantage to lucerne where the soil is N-deficient but, if weed invasion is heavy, the stimulus of N encourages the growth of weeds and partially negates the effects of herbicides. Thus, N should not be applied if a weed problem is anticipated. The insurance of a high rhizobia infection by inoculation to assist rapid nodulation and N$_2$ fixation becomes all the more essential when N use is inadvisable.

Several cultural or mechanical weed-control measures are available, particularly when the weeds are at a susceptible stage in their life cycle, and these measures can be effectively allied with selected herbicides from a wide range of products. Use of herbicides should conform to government legislation and codes of practice in their handling, storage and application and the manufacturers'-label instructions should be followed carefully.

Cutting the establishing crop to remove annual broad-leaved weeds before they set seed is an effective cultural method of control. However, if cut too early, the young lucerne plants may be damaged and, if cut too late, the annual weeds may have already seeded, thus providing the basis for future problems. Nevertheless, delayed cutting controls grass weeds, so the cutting strategy to be adopted depends on the spectrum of weeds present.

In establishing lucerne swards, the opportunity for weed invasion and colonization is reduced by appropriate presowing management. Stoloniferous or rhizomatous perennials, e.g. couch/quack grass (*Elymus repens*) or black bent (*Agrostis gigantea*), are weakened by continuous cultivation and treatment with a systemic herbicide, such as glyphosate, to destroy the perennating organs. Seedlings from weed seeds released from dormancy are treatable by vegetation-killing herbicides, such as paraquat. Repeating the process of spraying each new cycle of young seedlings, with minimum cultivation prior to sowing, reduces the loading of seeds about to lose their dormancy. This technique has been developed to the extent that lucerne can be drilled directly into the stubble of maize or small-grain crops, following treatment of weeds or volunteer cereals with herbicide – the so-called 'no till' system.

EPTC and trifluralin are used as a pre-emergence herbicides and are used in Italy, for example, to control Johnson grass (*Sorghum halepense*) and fat hen (*Chenopodium album*), respectively. Sethoxydin, fluazifop and haloxyfop are used as effective means of controlling rapid grass-weed ingression in establishing lucerne and can be tank-mixed with 2,4-DB (Peters and Linscott, 1988).

Annual broad-leaved weeds are also controllable by selective 'lucerne-safe' herbicides, eg. 2,4-DB, in undersown stands but also in establishing lucerne monocultures. Although a mixture containing benazolin, 2,4-DB and MCPA is suitable for use in lucerne stands at a relatively low application rate, it is damaging at high rates. Bentazone and linuron have severe effects on lucerne when applied with MCPA and MCPB, unlike their effects on red clover, demonstrating that lucerne can be susceptible to so-called 'clover-safe' herbicides. Bentazone injury is accentuated when applied with oil emulsifier, so, whereas injury is minimal at an application of 0.8 kg active ingredient ha^{-1}, the addition of the oil emulsifier at 1.25% v/v is damaging to lucerne (Harvey, 1991).

Usually, a well-established and managed lucerne crop will be sufficiently aggressive in its growth to compete successfully against weed invasion. However, perennial broad-leaved and grass weeds may become established if the stand becomes thin, due to mismanagement or natural causes, such as ageing of the sward, while winter annuals may ingress during winter dormancy of the lucerne. Advantage can be taken of the dormancy period to apply herbicides which would otherwise be damaging to lucerne if it was actively growing and intercepting the spray. Chlorpropham, if no companion grass is included, MCPA and dinoseb may be used in this way to control broad-leaved weeds. They may also be applied soon after cutting during the growing season, again at a time when there is no lucerne foliage to intercept the spray. Docks (*Rumex* spp.) are controllable by asulam. Herbicide application by a rope-wick applicator allows herbicides to be applied selectively when the weed canopy is above that of the sown crop. Simazine, carbetamide or propyzamide, with metribuzin, controls grass weeds in established lucerne monocultures.

Due to the importance of weeds in determining the persistence of lucerne stands, some long-term studies have been carried out and herbicide strategies evaluated, but with variable results. Paraquat can successfully control annual weeds, such as shepherd's purse (*Capsella bursa-pastoris*) or common chickweed (*Stellaria media*), when applied after the first cut in spring (Foy and Witt, 1993), although rate is critical; applied after the first harvest, at 0.56 kg a.i. ha^{-1}, herbage yield and the carbohydrate content of the lucerne crowns were reduced at the second harvest (Harvey, 1991).

Thiazopyr or trifluorin controls dodder (*Cuscuta indecora*) in lucerne seed crops, but the effect wears off and this allows dodder to germinate late in the season (Cudney *et al.*, 1993). Hexazizone controlled the common weeds in lucerne seed crops in Canada and its use can result in higher-yielding lucerne seed crops, possibly up to 20% more (Mayer *et al.*, 1991).

Pests

Pests of lucerne stunt its growth and development, defoliate plants, hinder or prevent seed development or even kill plants. They are also vectors of viruses or they create entry points for damaging fungi and bacteria. They attack foliage and stems or roots, and even the developing seed pods. While some are a problem in all lucerne-growing regions of the world, others are restricted in distribution, due to environmental constraints, or sometimes they have not had an opportunity to build up their populations in areas to which lucerne has been introduced comparatively recently.

The principal pests of lucerne in temperate regions have been well described: insects and mites (Manglitz and Ratcliffe, 1988) and nematodes (Leath *et al.*, 1988). These accounts are written from an American perspective, but many of the pests described (Table 3.4) and control measures advocated below are of general relevance in lucerne crops grown in other countries.

The foliage of lucerne can be attacked directly by pests or else its growth can be adversely affected by pests attacking other parts of the plant. Leaves are directly consumed by the larvae of alfalfa weevils, cutworms, caterpillars, the larvae of leaf-miners and clover-weevil adults. Biological control by the introduction of viruses and bacteria, e.g. *Bacillus thuringiensis*, complements natural control; for example, natural predators usually keep leaf-miners and caterpillars in check. Generally, early harvesting of an infected crop prevents an excessive build-up of some of the leaf-eating pests by preventing them from completing their life cycle.

Chemical control of leaf eaters is not usually economic, but a broad-spectrum insecticide is effective against alfalfa and clover weevils when applied in autumn, before they enter their reproductive phase. The larvae of alfalfa weevils cause most damage in the first growth cycle, due to the reduction in the plant's photosynthetic area and in N concentration in the herbage (Peterson *et al.*, 1993). However, the larvae that are not controlled by the first cut then feed on young leaves developing from the stubble during the first 10 days of regrowth and depress second-cut yields (Buntin and Pedigo, 1985). The larvae are controllable by the application to the crop of insecticide, usually in spring. In comparison, targeting the insecticide at field margins, where

Table 3.4. Pests of lucerne.

Foliage	
Alfalfa weevils (larvae)	*Hypera* spp.
Caterpillars	*Colias eurythene*
Cutworms/army worms	*Euxoa auxiliaris*
Blister beetles	*Epicauta* spp.
Aphids	
Spotted alfalfa	*Therioaphis maculata/trifolii*
Pea	*Acyrthosiphon pisum*
Blue alfalfa	*Acyrthosiphon kondoi*
Clover weevils	*Sitona* spp.
Grasshoppers	*Melanophus* spp
Leafhoppers	*Agromyza frontella*
Potato leafhoppers	*Empoacea fabae*
Spittlebugs (e.g. meadow)	*Philaenus spumarius*
Roots/crowns	
Nematodes:	
Root knot	*Meloidogyne hapla*
Root lesion	*Pratylenchus penetrans*
Alfalfa stem	*Ditylenchus dipsaci*
Clover-root curculio (larvae)	*Sitona hispidulus*
Snout beetle (larvae)	*Otiorhynchus ligustici*
Leatherjackets	*Tipula* spp.
Seed pods	
Alfalfa seed chalcid (larvae)	*Bruchophagus roddi*
Mirids	*Lygus* spp.

weevils re-enter the field in autumn, has resulted in successful control the following spring, with reduced labour and chemical costs and less environmental damage (Roberts *et al.*, 1987). Biological control of the alfalfa weevil with the wasp *Bathyplectes curculionis*, in combination with spraying and harvesting strategies to implement an integrated pest control (IPC) programme, has been described (Fleming, 1988). Natural parasites of the clover weevil, e.g. the baraconid *Microtonus aethiopoides*, control the weevil population build-up. The parasites are effective when allied with other means of chemical control or with other parasites, since the effectiveness of these other means is not dependent on the population density of clover weevils (Barlow and Goldson, 1993). Predators and pathogens of eggs and larvae are being developed for biological control. Clover weevils can also be controlled by severe grazing, but the treatment may directly damage the crop (Wynn-Williams *et al.*, 1991).

The aphids, leafhoppers and spittlebugs listed in Table 3.4 all suck sap from the leaves and stems of lucerne. In addition to the loss of leaves, due to spotted and blue alfalfa aphids, and loss of winter-hardiness, due to pea aphid, potato leafhopper and spittlebugs, and general stunted growth or delayed maturity, these vectors transmit viruses, which further debilitate the plant

(Hutchins *et al.*, 1991). Defoliation controls aphid numbers per plant, at least temporarily, although the spotted aphid can avoid its removal during grazing by dropping to the ground and reinfecting the regrowth. Early cutting to avoid further aphid or weevil damage may be more damaging to total yield than delaying the first harvest (Latheef *et al.*, 1988). Nevertheless, an early harvest reduces egg population and early life stages of the weevil *Apion pisi* (Lykouressis *et al.*, 1991). Foliage-applied insecticides control aphids, but the remedy may be short-lived, especially if natural predators have also been reduced. Control of the pea and blue aphid has been successfully achieved with the predator *Aphidius ervi* in autumn. Breeding cultivars for resistance to aphid attack has been a major success in the fight against pea and spotted alfalfa aphids. Resistance in these cultivars tends to be confined to local biotypes of the pest, but the principle offers hope for the approach to be successfully repeated for biotypes in other regions.

Potato leafhoppers can be controlled by delaying harvest, a strategy contrary to the control of most other insects. Growing oats with lucerne reduces leafhopper colonization of establishing lucerne in proportion to oat-plant density (Lamp and Zhao, 1994).

Stem nematodes, such as *D. dipsaci*, are a major pest of lucerne throughout the world. They are easily transmitted in the harvested forage, in harvesting equipment and in inadequately cleaned and dressed seed. The effect of infection on the long-term yield of cultivars is considerable; for example, a reduction in yield of 30% over 4 years has been reported. The use of clean seed or seed fumigated with methyl bromide, for example, is essential. In a number of countries, resistance has been bred into some cultivars, but they are not resistant to all nematode biotypes. However, rapid screening methods are now available for selection of lucerne phenotypes resistant to local races of stem nematodes.

Lucerne root damage by root-lesion nematodes (*Pratylenchus* spp.), results in stunted plant growth, while northern root-knot nematode (*Meloidogyne hapla*), although not a major pest of adult plants, kills seedlings. Cultivars resistant to *M. hapla* have been bred. The damage from both nematode species encourages infection by fungi and bacteria, including fungal-wilt and bacterial-wilt agents. Due to the ability of these nematodes to remain viable in the absence of lucerne for a number of years, crop rotation is not very effective in controlling infestations.

The alfalfa snout-beetle larvae, a pest prevalent in Europe, tunnel into lucerne roots and reduce the plant's ability to take up water, but are controllable by crop rotation. Leatherjackets (*Tipula* spp.) are prevalent in heavy textured soils, e.g. in the UK, and damage roots and crowns, but, if the infestation is severe, cost-effective control measures are available

In addition to being adversely affected by root- and foliage-consuming pests, the seed-producing capability of lucerne plants is reduced by insects whose larvae feed on pods, e.g. the larvae of the alfalfa seed chalcid, or else pierce tissues and suck sap around the developing flowers, thus causing flower drop. The larvae of gall midges, which deposit eggs in flower buds, also reduce seed yields of lucerne, but cultural control, by deep ploughing in autumn, crop rotation and isolation and destruction of infected vegetation, minimizes the effect. Burning stubble has a beneficial effect on soil properties, including increased organic matter, N and K, and on seed yield in lucerne (Dormaar and Schaber, 1992).

Organophosphates, carbonates, carbonyl and carbofuran can be used as insecticides, but with increasing pressure to adopt practices perceived to be environmentally friendly, total reliance on chemical control is not sustainable. However, adopting any one control method is fraught with dangers, as the pest may develop immunity from the specific control measure, be it a chemical, biological or genetic (breeding) strategy. The control may be environmentally dependent, in that the parasite or predator may only proliferate and be effective under specific conditions, or the cultural method may be inadequate to cope with the rapid proliferation of the pest. These and many other factors are required to be taken into account when planning a control strategy.

Control clearly depends on the type of pest, the purpose for which the crop is grown and the severity of infestation. Cultural management and chemical or biological control may be applied individually or, more commonly in recent years, combined in IPC programmes, in which the three components are used in harmony and to varying degrees, with the ultimate aim of achieving economic, permanently effective and environmentally safe control. Of course, selection for tolerance or resistance to pests is an ongoing objective in cultivar breeding programmes.

Integrated pest control in lucerne obviously involves a thorough knowledge of the crop and the pest, as well as the response of each to the alternative strategies under consideration. For example, a programme was devised for control of alfalfa weevil, in the presence of the parasite wasp, *B. curculionis*, and with the possibility of resorting to spraying with pesticide or harvesting (Fleming, 1988). The programme comprises the use of a model with three submodels, i.e. to predict the growth of the crop and the population size of the weevil and the parasite, and aims to reduce spraying to a minimum without reducing overall efficacy of control. However, uptake by farmers was limited. Lack of commitment by extension workers, the effort required and lack of expertise of farmers in sampling parasitoids, weevils and crop height and identifying adult stages of the insects, together with the meagre anticipated benefit of saving only one insecticidal spray, all contributed to the poor uptake of the concept. Nevertheless, with increasing environmental pressure being brought to bear on all forms of agriculture, IPC is likely to become increasingly used for successful lucerne production.

Diseases

Fungi, such as *Pythium* spp., can cause seedling damage and, on occasion, complete crop failure, while fungal, bacterial or viral attacks reduce herbage yield, plant number and plant persistence (Summers and Gilchrist, 1991), lower

Table 3.5. Principal diseases of lucerne and casual agents.

Disease	Agent
Bacterial leaf spot	*Xanthomonas alfalfa*
Common leaf spot	*Pseudopeziza medicaginis*
Yellow leaf blotch	*Leptotrochila medicaginis*
Stemphylium	*Stemphylium botryosum*
Leptosphaerulina leaf spot	*Leptosphaerulina briosianna*
	Leptosphaerulina trifolii
Downy mildew	*Peronospora trifoliorum*
Spring blackstem	*Phoma medicaginis* var. *medicaginis*
Alternaria	*Alternaria solani*
Bacterial stem blight	*Pseudomonas medicaginis* or *syringae*
Stagonospora leaf spot	*Stagonospora meliloti*
Rust	*Uromyces striatus*
Summer black stem	*Cercospora medicaginis*
Spring black stem	*Phoma medicaginis*
Fusarium wilt	*Fusarium oxysporum*
Verticillium wilt	*Verticillium albo-atrum*
Bacterial wilt	*Clavibacter michiganense* subsp. *insidiosum*
Sclerotinia crown and stem rot	*Sclerotinia trifoliorum*
Rhizoctonia	*Rhizoctonia solani*
Phytophthora root rot	*Phytophthora megasperma*
Anthracnose	*Colletotrichum trifolii*
Fusarium root rot	*Fusarium* spp.
Aphanomyces root rot	*Aphanomyces euteiches*

herbage nutritive quality (Lenssen *et al.*, 1991) and increase animal health problems, such as infertility, caused by oestrogenic effects, and may also reduce plant cold tolerance (Richard and Martin, 1993). A list of the main foliar and stem diseases and the fungal or bacterial agents responsible is presented in Table 3.5. While activity of agents is usually highest under humid conditions, they have different temperature optima (Barbetti, 1991; Olanya and Campbell, 1991). Although fungicide treatment is not economic, diseases are often sufficiently detrimental to warrant control by management, when feasible, and, of course, by plant breeding in the long term. Leaf damage is sometimes controllable economically by fungicides, e.g. yield improvements from fungicide application of 16–18% in New Zealand and over 40% in Australia, due to suppression of leaf borne diseases (Douglas, 1986). Resistance to leaf spots and black stem has been successfully bred into specific cultivars.

Phytophthora megasperma, the fungus which causes *Phytophthora* root rot, infects the roots of established crops. Crown rot is a common condition in lucerne, and while *Fusarium* spp. are commonly associated with the disease, *Fusarium* is often a secondary invader, subsequent to wounding or other damage caused by organisms such as nematodes or insects, including clover-root curculio (*Sitona hispidulus*) (Keld *et al.*, 1994).

In the case of the lucerne 'wilts', infection is via roots, but symptoms are expressed in above-ground parts. *Fusarium* wilt (*Fusarium oxysporum*) causes the shoots to assume a yellow-red colour and to wilt. The fungus remains in the soil for years, but there are many resistant cultivars. *Verticillium* wilt (*Verticillium albo-atrum*) is the most important disease of lucerne in Western Europe, although it has only recently become a problem in North America. Symptoms are more pronounced after the first year and include chlorosis and streakiness in the expanded leaves, which die back at the tip, while the young leaves curl. European cultivars are resistant and the trait is being bred into cultivars in other countries with a *Verticillium* wilt problem. Fortunately breeding for resistance is additive (Miller and Christie, 1991) and stable, even under drought conditions (Pennypacker *et al.*, 1991). Selection for resistance can be successfully achieved under laboratory conditions, since field and laboratory resistance are correlated (Grau *et al.*, 1991).

Bacterial wilt caused by *Corynebacterium insidiosum* (syn. *Clavibacter michiganense*) is widespread. The vascular system becomes brown, stems are spindly and regrowth is slow. It persists within plant debris in the soil and in seeds and enters the plants via wounds, particularly those caused by root-infecting nematodes. Again, as with many diseases (and pests), the use of resistant cultivars, where available, and management strategies are the most effective means of control, since fungicide application is not economic. Although breeding for disease resistance is expensive, the cost effectiveness of using cultivars highly resistant to *Verticillium* wilt in western Canada has been demonstrated (Smith *et al.*, 1995).

Diseases have different seasonal periods of prevalence and affect plants at different stages of shoot development; for example, pepper spot is more prevalent in recent regrowths, compared with summer black stem, which is found mostly in older shoots (Inch *et al.*, 1993). Cultivars resistant to one specific disease may become susceptible to the development of another, thus elevating its importance.

The impact of a disease caused by fungi or bacteria may be alleviated by management, for example, by maintaining a soil pH higher than 6.5 and high levels of available P and K in the soil and ensuring good drainage, or by including lucerne in rotation with arable crops. Early harvesting and a relatively frequent cutting regime thereafter contain invasion of most of the leaf spots and also spring and summer black stem, although early cutting may be more detrimental to yield than the disease (Duthie and Campbell, 1991). Use of rotations and avoidance of reintroduction from previous host crops are also strategies to minimize the risk of black stem. *Sclerotinia* crown rot can be avoided by using clean seed, and rotation and deep ploughing will reduce the likelihood of serious reinfestation. Leaf damage from *Alternaria* leaf spot, bacterial stem blight, powdery mildew and *Stemphylium* leaf spot is reduced by a *Pseudomonas* bacterial strain known to be a predator in soil (Caseda and Lukezic, 1992). Therefore, biological control of leaf-spot diseases may be possible by developing bacterial predators. Alfalfa mosaic virus is avoidable by using virus-free seed and controlling both aphid attack and alternative weed hosts. Seedling diseases are minimized by treating seed with fungicides, such as metalaxyl, prior to sowing. As *Phytophthora* root rot is prevalent in wet soils, improved drainage, well-controlled irrigation and growing resistant cultivars are useful control measures.

Generally, the crown and root rots prevail due to injury or stress to the plant. Therefore, avoidance of physiological stress from nutritional and moisture factors and biotic stress from insects and nematodes and minimizing mechanical damage to the stubble by harvesting equipment all contribute to disease control in lucerne.

More than 20 viruses are known to infect lucerne and have the potential to seriously reduce yields. For example, alfalfa mosaic alfavirus has been estimated to cause yield losses of 24–67%. Control of insect vectors and sowing resistant cultivars, if available, are current strategies to minimize virus attack. The availability of virus-resistant cultivars should improve when the products of genetic engineering programmes on this topic reach the market (Forster *et al.*, 1997).

Herbage production

In a maturing crop, DM accumulation rate increases in spring, but stem density declines (Woodward and Sheehy, 1979). During this, time stems will have initiated flowers, and these will develop from the bud through to the seed stage. By the time flowers appear, buds in the crown will have elongated and these form the basis of DM production from regrowths.

Within a crop, shoots are at a range of stages of development, which have been described by Kalu and Fick (1981) (see Table 3.1). A method of quantifying the stage of development of a lucerne crop has been developed, in which each shoot in a sample is assigned to a stage class and weighed, the stage × weight for each shoot is summed and the total is divided by the weight to give the mean stage weight (MSW) (Kalu and Fick, 1983). The benefit of this system is that, rather than describing the maturity of a crop by the stage of development of the most advanced shoots, it takes into account the stage of development of all of the shoots and their contribution to yield. This is particularly important when estimating the nutritive value of a crop (discussed later).

Attempts to predict first-cut yields from temperatures, i.e. accumulated degree days, from autumn to spring have met with limited success. In dry winters, predictions are within 15% above and below actual yields, but yields may be overestimated by 30% or more if it has been a wet autumn or spring (Durling *et al.*, 1995).

Potential yield for lucerne (Y_L) under UK conditions has been predicted from rainfall and potential evapotranspiration at a given site (Doyle and Thomson, 1985) as:

$$Y_L = \left[\left(\frac{0.1\,R_A\ -\ 0.033\,PET_S + 94}{100}\right) PET_S\right] 0.0478 - 6.296$$

where Y_L is in t DM ha^{-1}, R_A is mean annual rainfall (mm) and PET_S is potential cumulative summer evapotranspiration (mm).

However, in the field, there are actual limitations imposed by edaphic and other climatic factors and these, together with exposure to pests and diseases, usually cause yields to fall short of the potential. Under field conditions in the south of England, the potential yield for a temperate crop, such as lucerne, is about 15 t DM ha^{-1} (Sheehy *et al.*, 1984). This yield has been consistently achieved in multisite trials in England (Aldrich, 1984), although on-farm annual yields may be as low as 8–10 t DM ha^{-1}, some 30% lower than the predicted potential (Doyle and Thomson, 1985). Yields ranging from 9.4 to 17.7 t ha^{-1} have been recorded in trials in the northern UK (Table 3.6). In France, with generally higher irradiance levels than in the UK, yields in cultivar trial plots vary from 14.5 to 19.0 t DM ha^{-1} (Guy, 1993). In the1980s, national yields in France were 7.5–8.5 t ha^{-1}, but they are now above 10 t ha^{-1}, with yields of crops for drying exceeding 12 t ha^{-1}.

Under rain-fed New Zealand conditions, lucerne yield (Y_L in t DM ha^{-1}) has been related to the AWC in mm as:

$$Y_L = 0.312 + 0.082\ \text{AWC}$$

Over a 12-month growing season in New Zealand, mean growth rates of about 70 kg DM ha^{-1} day^{-1} (25 t DM ha^{-1} annually) have been reported under irrigation (Hoglund *et al.*, 1974). Availability of moisture is usually the principal factor controlling yield of lucerne, so yields ranged from 5 to 23 t DM ha^{-1}, with a few falling below the lower value of the range (Douglas, 1986). Under temperate conditions in the USA, yields approaching 20 t DM ha^{-1} in experimental plots have been achieved (Sheaffer *et al.*, 1988b), although this would be considered the upper limit.

Yields at successive cuts decline over a growing season under a management of fixed cutting intervals and in the absence of moisture stress, irrespective of cutting interval (Fig. 3.5). Efficiency of utilization of intercepted radiation by lucerne was approximately constant over the growing season, i.e.

Table 3.6. Dry matter yields (t ha^{-1}) of lucerne cv. Europe in monoculture, from trials in northern regions of the United Kingdom (cutting frequency, three or four cuts per annum)

	Harvest year				
Country	1	2	3	4	Source
Scotland	10.72	14.84			Frame (1986)
Scotland	13.53	9.39			Reid (1987)
Scotland	9.46	15.80	17.7		Frame and Harkess (1987)
Northern Ireland	11.17	11.24	9.86	11.30	Laidlaw (1985)

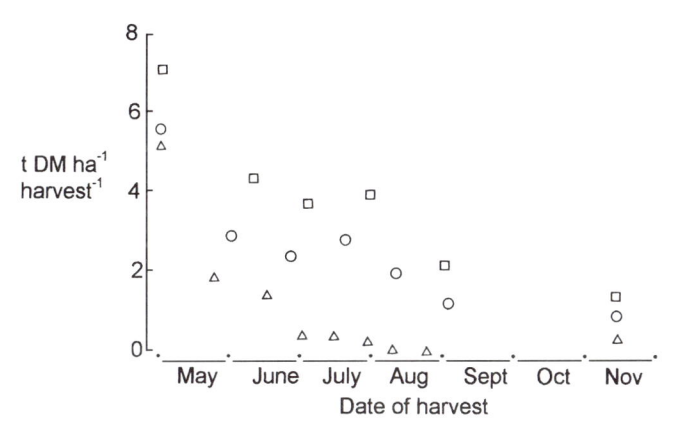

Fig. 3.5. Dry matter yield of irrigated lucerne at successive harvests in southern Italy harvested at 14 day intervals (△), at late bud (○) and late flower (□) stages (from Corletto *et al.*, 1994).

about 2.4 g total biomass DM MJ^{-1}, but when efficiency was related to herbage yields it was lower in the autumn than in summer, due to less assimilate being directed into shoot growth (Khaiti and Lemaire, 1992). While gradually lengthening the interval between cuts reduces the rate of yield decline, herbage quality declines with the lengthening interval.

Progressive decline in lucerne DM yields from year to year has been attributed to many factors: competition from companion grasses, where sown, and from weeds, injury or loss of plants from pests and/or diseases or from winter damage, or adverse management factors, such as uncontrolled grazing, overfrequent cutting, inadequate fertilization or poor drainage. Notwithstanding, lucerne stands decline in yielding ability with age under irrigation and a believed 'optimum' management (Hayman and McBride, 1984).

Lucerne growth and production can be predicted, with varying degrees of accuracy, by simulation models. Seventeen computer models developed to predict lucerne yield were cited in the late 1980s (Fick *et al.*, 1988). Each had limitations and highlighted the absence of data on population dynamics of lucerne plant units (whole plants or stems), thus preventing the general applicability of these models to lucerne crops.

Herbage quality

Compared with perennial ryegrass at similar stages of growth, lucerne has lower cell-wall constituents and digestible fibre but higher cell contents, CP and lignin content (Campling, 1984). Table 3.7 contains a summary of the composition of lucerne, excluding mineral content, which was dealt with earlier, and other aspects of nutritive value are presented in Table 3.2.

As the stems mature, their nutritive value declines and, allied with a decrease in leaf : stem ratio, the nutritive value of total herbage also declines

Table 3.7. Chemical composition of lucerne, excluding minerals (from Spedding and Diekmahns, 1972).

	Range	Mean	No. of studies
g kg⁻¹			
Crude protein*	129–324	193	3
Cellulose	125–362	219	2
Hemicellulose	62–164	124	2
Pectin	76–131	104	1
Non-structural carbohydrates	66–89	75	2
Lignin	40–158	100	2
Organic acids		81	1
Lipids		20.3	1
mg kg⁻¹			
Vitamin A and carotene	9–270	185	2
Oestrogenic compounds	0.4–2.1†	1.25	1
	28–184‡	106	1

* Calculated from N concentration × 6.25
† Uninfected leaves
‡ Leaves infected with pathogenic fungi

steadily. With advancing maturity, NDF, ADF and lignin increased at a rate of about 0.16, 0.06 and 0.03% of DM per day, respectively (Keftassa and Tuvesson, 1993). A 1-week delay in cutting can reduce CP content by 2% and increase cell-wall content by 3% and, while digestibility of the leaf fraction remains high, at between 75 and 80%, stem digestibility declines from over 70% at the vegetative stage to 40% at full bloom; consequently, as stem contribution to DM increases with advancing maturity, digestibility of herbage overall declines from 75% to 60% over the period of maturity (Fig. 3.6). Although the decline in digestibility in lucerne with maturity is faster than in even an early-heading perennial ryegrass during primary growth, the decline in lucerne at the subsequent regrowth is lower than in perennial ryegrass (Demarquilly, 1981). Associations between MSW and nutritive value have been established for a given cultivar grown in the same region but at different sites. For example, during the development of the MSW concept, strong linear relationships were established between MSW, *in vitro* digestibility, ADF and lignin content and quadratic relationships between MSW, CP and NDF (Kalu and Fick, 1983). However, the relationship between MSW and nutritive-value indicators varies with time of year, the decline in nutritive value being slower in summer than in spring for the same change in MSW, and the relationship between MSW and DM digestibility (DMD) and digestible protein also varies between cultivars (Griffin *et al.*, 1994). Therefore, MSW has its limitations in predicting nutritive value, as it is cultivar-, season- and, possibly, region-specific.

Crude-protein content declines in lucerne with advancing maturity, from 250 g kg⁻¹ to less than 200 g kg⁻¹ and the leaves contain two to three times

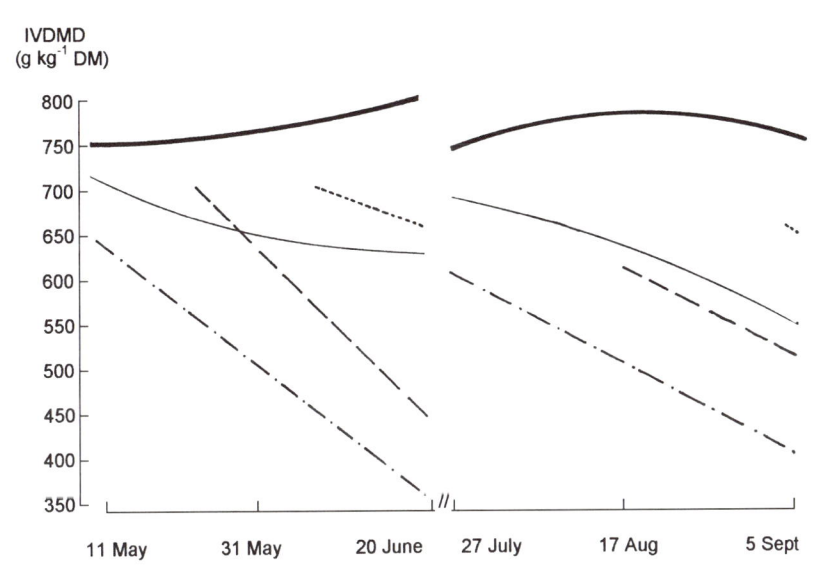

Fig. 3.6. *In vitro* DM digestibility (IVDMD) of total lucerne herbage (——), leaves (——) and of stem up to the sixth node (—·—), sixth to 12th node (– –) and above (-----) during two regrowth periods (from Buxton *et al.*, 1985).

more CP than the stems, but CP yield increases (Marten *et al.*, 1988). Contents of individual amino acids in lucerne and cocksfoot, expressed as a percentage of CP, at a stage at which each species is likely to be conserved, are presented in Table 3.8. Generally, the pattern of amino acid content of protein in both species is similar, although cocksfoot tends to have slightly higher proportions than lucerne and almost a third more essential amino acids than lucerne per unit of CP. In lucerne, aspartic and glutamic acids and leucine are the most prevalent, while the S-containing amino acids each constitute about 1% or less of CP. Stems have a higher content of amides and soluble proteins than leaves, but lower contents of essential amino acids and true proteins (Balde *et al.*, 1993).

A problem with efficiency of utilization of protein in lucerne is that turnover in the rumen is rapid and a high proportion of protein N is lost as ammonia, although ruminal protein degradability declines with plant maturity (Amrane and Michaeletdoreau, 1993). Rate of degradation of protein between genotypes varies, and so reduced protein degradation in the rumen could be achieved by selection (Skinner *et al.*, 1994).

Methods of conservation or processing of lucerne affect nutritive value (Table 3.9). Fibre content does not differ much between herbage preserved at the same stage of growth by ensiling or as hay. Crude protein is generally higher in ensiled lucerne than in hay, due mainly to greater leaf loss in haymaking, but much more of the N comprises non-protein N in silage

Table 3.8. Amino acid content (%) of crude protein in lucerne and cocksfoot (from Balde et al., 1993).

Amino acid	Lucerne (medium bloom)	Cocksfoot (full heading)
Aspartic acid	13.3	16.3
Glutamic acid	9.5	12.8
Leucine*	7.3	9.5
Valine*	6.1	8.5
Alanine	5.9	7.6
Arginine*	4.4	6.2
Phenylalanine*	5.1	6.4
Proline	4.6	5.2
Threonine*	4.7	5.8
Isoleucine*	4.6	6.0
Glycine	4.8	6.2
Serine	4.5	5.3
Lysine*	4.6	5.8
Tyrosine	2.9	3.4
Histidine*	2.0	2.3
Cysteine*	1.2	1.1
Methionine*	0.8	1.1

*Essential amino acid.

Table 3.9. Quality of lucerne hay or silage made from the same second-cut crop (from Broderick, 1995).

Component	Silage (g kg^{-1})	Hay (g kg^{-1})
In fresh weight		
DM	413	580
In DM:		
NDF	354	352
ADF	265	257
CP	212	197
In total N:		
NPN	494	77

(Broderick, 1995). Relative to fresh herbage, barn-dried lucerne hay has slightly higher crude fibre but lower CP, digestibility and net energy, while these differences are accentuated in field-cured hay (Jarrige et al., 1982). The process of haymaking increases the proportion of degradable protein (Broderick et al., 1992). Fraction 1 protein (ribulose 1 : 5 biphosphate carboxylase) is vulnerable during conservation, some being lost during wilting or drying and also by the ensiling process, because of proteolysis (Mangan et al., 1991).

The heat generated during ensiling of wilted, high-DM lucerne forage increases protein protection, but higher respiratory losses than with lower-DM silage cause a reduction in energy content. Application of the silage additive, formic acid, also increases protein protection to advantage, but excessive additive application results in overprotection and subsequent loss (Barry et al., 1978). Short chopping relative to long-chop material influences nutrition directly, by increasing the passage rate of the fed lucerne, since the required particle size is attained quickly in the rumen, and indirectly, by increasing the probability of the desired fermentation.

The feeding of dehydrated lucerne compared with fresh lucerne increases faecal N loss but increases amino acid absorption and N retention, although overheating is detrimental to protein (Goering and Waldo, 1979). Pelleting low-digestibility dried lucerne doubled DM intake, compared with chopping the fresh material, but differences diminished when pelleting and chopping were compared with high digestibility lucerne, since the small particle size within the pellet increased the rate of passage from the rumen (Minson, 1982). Amino acid absorption per unit of energy is higher in pelleted than in fresh lucerne, as drying increases the proportion of rumen-undegradable protein.

Herbage antiquality

Although bloat can be a problem for stock grazing lucerne, especially with cattle, case histories under experimental conditions are scarce (Van Keuren and Matches, 1988). The incidence of bloat is minimized by including a high-fibre component, such as hay or straw, in the diet, grazing a lucerne/grass mixture rather than pure lucerne and having a policy of replacing animals that chronically bloat. Administration of antifoaming agents is short-term in effect, and surfactants, such as poloxalene, have to be available daily. Poloxalene increased the period spent grazing when administered in cracked maize to beef cattle grazing lucerne, suggesting the alleviation of subacute bloat and confirming that subclinical bloat affects animal intake and performance adversely (Dougherty et al., 1992). The development of a controlled-release monensin capsule is a promising method of bloat control (Li et al., 1993). Feeding cattle on lucerne selected for low initial digestion rate, to reduce the rate of gas production, has not resulted in any appreciable reduction in bloat (Hall et al., 1994).

Saponins, although beneficial to lucerne by increasing its resistance to some pests (Tava et al., 1993), interact with rumen bacteria and have a haemolytic effect on animals fed lucerne. Low saponin content can be bred into lucerne, but too low a level may leave the plant vulnerable to pests.

Livestock infertility, due to plant oestrogens, is a hazard when feeding or grazing lucerne. Oestrogen content varies with genotype and is highest at the early-bud growth stage, but oestrogens are also synthesized in response to attacks by aphids and fungal infection on the leaves. Breeding cows or ewes should not be grazed or fed lucerne when oestrogen content is likely to be high and particularly not immediately prior to conception.

Conservation

Lucerne is most commonly cut and conserved as hay or, to a lesser extent, as silage or dried pellets. In summary, under temperate conditions, cutting at 10% bloom stage and then at 5- to 7-week intervals maximizes DM production, provides forage of reasonable nutritive value and helps to maintain sward longevity (Sheaffer *et al.*, 1988b). Obviously, growing conditions determine the interval length between harvests, with recommended intervals varying from 4 to 6 weeks (Aldrich, 1984; Johnson, 1984) to 8 weeks (Griffiths and Poole, 1984). Plant height has been suggested as a suitable guide to determine time to cut (Schmidt, 1993), but, while it may have a use in determining the time to cut for optimum nutritive value, it could result in the crop being cut too frequently and in reduced persistence.

If cut at advanced stages of growth, leaf:stem ratios decrease, with consequent adverse effects on nutritive value. During drying, leaf loss by shatter is a major hazard, due to the rapid loss of moisture from leaves compared with stems; leaf loss may also continue during handling and storage if the hay is not managed carefully, since the DM content of the cured hay can be as high as 90% (Nash, 1985). Leaf shatter may account for losses in DM yield of up to 40% depending on conditions, such losses being double those expected from grass hay.

In haymaking, the major source of loss is during harvesting and baling, whereas the highest losses incurred in making lucerne silage are in the silo. Hay with a high moisture content will also suffer high postharvesting losses, due to overheating and hence loss of carbohydrates. Procedures recommended by the Pennsylvania State University Extension Service to minimize forage loss of lucerne during conservation, in addition to general pointers for good conservation practice, include:

- mowing when there is little chance of rain, but be ready to take risks to avoid cutting schedules slipping;
- laying the cut herbage in as wide a swathe as possible to aid drying;
- when making hay, 'raking' at 60% DM, as slowly as possible and as few times as possible;
- baling at about 82% DM, so that hay is not too dry, resulting in leaf shatter during raking, or too wet, leading to high in-store losses.

Optimum DM content of herbage at chopping for silage is 30% for bunker silos, 35% for concrete-tower silos and 45% for steel-tower silos. Buckmaster *et al.* (1990) present an economic justification for these recommendations, apportioning costs to potential sources of loss. Lucerne can also be made into big-bale silage, baled at 40–50% DM. The shorter wilting time for silage making, compared with haymaking, reduces leaf loss.

Conditioning the hay by crimping and avoiding handling the hay when relative humidity is particularly low, e.g. in the midday sun, reduce leaf shatter. In field trials, application of K carbonate prior to cutting makes the loss of moisture from stems and leaves similar, thus reducing leaf shatter, but the positive effect has not been demonstrated under cool moist conditions. In humid temperate areas, barn drying is usually necessary to complete the curing process. In some countries, chemical preservatives are applied during baling when moisture content is too high to guarantee freedom from microbial spoilage, but uniform application has proved difficult in the field (Douglas, 1986).

Wheel traffic on hayfields has severely reduced stand production and stand life (Sheesley *et al.*, 1974); plant crowns and regrowth shoots were damaged, particularly when field-traffic operations were delayed following initial cutting, and soil compaction limited root development. Thus, as a rule, cutting for hay or silage should be carried out with the lightest of equipment and fewest traffic activities consistent with an efficient system.

During silage making, the low WSC and high CP content, together with a high buffering capacity (twice that of grass silage), of lucerne herbage tends to favour the undesirable clostridial rather than the desirable lactic fermentation. The high buffering capacity due to the high protein and organic-acid content requires 6% lactic acid on a DM basis to be produced during fermentation, compared with 3% for grass silage (Nash, 1985). However, application of acid additives, such as formic acid, lowers the pH of the silage, thus promoting a desirable lactic-acid fermentation of wet herbage, although no benefit results when applied to lucerne wilted with a DM content of more than 35%. When a microbial inoculant additive was tested on lucerne silage, soluble carbohydrate had to be added as a substrate to ensure a satisfactory lactic-acid fermentation (Merry *et al.*, 1989). Lactobacilli or streptococci inoculants reduce cell-wall protein and, if pH falls sufficiently low, result in a decline in sugars, especially arabinose and galactose, in the cell-wall fraction, although glucose and xylose increase (Jones *et al.*, 1992).

Cellulase-enzyme additives lower crude fibre content, pH and ammonium concentration and increase the concentration of CP and WSC (Hristov, 1993). However, they are generally considered to be less effective than bacterial inoculants in aiding ensiling of lucerne.

Precision chopping results in a better quality of lucerne silage than double chopping and improves intake (Dulphy, 1980). Fed to sheep, an increase of more than 20% over double chopped material was achieved, due to both higher-quality silage and the physical benefits of a shorter chop length in enhancing intake, but the chop-length effect on intake was less influential with cattle (Demarquilly and Dulphy, 1977).

The benefits of feeding wilted compared with unwilted lucerne silage have varied from trial to trial. Generally, intake increased progressively as DM content increased, e.g. intake by cattle fed silage at 50% DM was double that from silage at 20% DM, because of improved fermentation rather than as a consequence of higher DM content *per se* (Scales and Barry, 1975). Precision chopping partly offsets the need for full wilting. The application of effective silage additives primarily improves fermentation and, within limits, intake is also improved.

Lucerne is also grown to be dehydrated (so-called green-crop drying), a process requiring a regular supply of herbage with a high protein content. In France, for example, about 100,000 ha are grown for this purpose. Optimum management involves cutting before growth reaches the bud stage. With a similar management, lucerne is also used for zero grazing or, as it is sometimes called, forage feeding or soiling. Fractionation, which is occasionally practised, involves macerating the green crop while still vegetative and extracting juice after the fibre fraction is separated from the macerate. The liquor is heated, the precipitated protein then filtered out and the protein and fibre fractions subsequently dried (Ostrowski-Meissner *et al.*, 1993). The liquor is a rich source of protein, β-carotene and xanthophyll, which are stabilized by introduction of an antioxidant into the liquor (Layug *et al.*, 1996).

Grazing

The same defoliation rules apply to grazed as to cut lucerne, in that sufficient regrowth between defoliations is critical to ensure stand survival, the actual duration between grazings depending on the conditions determining rate of regrowth. For example, in south-east Australia, the minimum regrowth period required to ensure stand survival was between 5 and 6 weeks in summer but 8 weeks in winter (McKinney, 1974). Stocking at a sufficient rate to graze down stands over short grazing periods benefits stand persistence, since young regrowths are not grazed (Janson, 1982). Stocking rate also influences stand longevity and, in one trial, lucerne persisted at stocking rates less than 10 ewes ha^{-1} but declined at over 12 ha^{-1}, probably due to oversevere defoliation and damage to crown buds (Reeve and Sharkey, 1980). Lucerne is not tolerant of continuous stocking and, generally, stands will not persist under this management, unless defoliation is lax. This was demonstrated in the sensitivity of some cultivars to stocking rate when continuously stocked with steers, where stocking at an allocation at 2.7 kg herbage DM kg^{-1} liveweight (LW) gave twice the lucerne-plant density in stands in the second year of the trial than in stands at the higher stocking rate of 0.9 kg DM kg^{-1} LW (Bates *et al.*, 1996).

When lucerne is grown with a companion grass, the interval between grazings has to be adjusted according to prevailing weather conditions. If the grass growth responds particularly well to wet weather, for instance, the short defoliation intervals needed to control it may put the lucerne at a disadvantage and allow the grass to become too dominant (Wilman, 1977). If stocking is at a low density or the period of grazing very short, selective grazing of lucerne leaves by sheep tilts the competitive balance in favour of the grass cospecies or less acceptable weeds (Leach, 1983). In mixed swards, the companion grass should be a highly acceptable species to grazing animals; otherwise, lucerne may be selectively overgrazed. The preference of grazing cattle for grass or lucerne changes with season, grass being preferred in spring while, during summer, lucerne is more attractive to them (Mace, 1982). Nevertheless, lucerne can persist under grazing with a companion grass, e.g. under dry conditions grown with cocksfoot (Charrier *et al.*, 1993). A combination of a hay crop followed by rotational grazing benefited lucerne compared with rotational grazing during the whole growing season (Van Keuren and Matches, 1988).

In New Zealand, a recommended system of grazing lucerne by sheep includes delaying grazing in spring to allow herbage to accumulate, rotationally grazing until closed off for hay, rotationally grazing aftermaths with lambs and then with ewes to clean up the herbage remaining, including weeds, but avoiding excess defoliation and damage to crowns. In autumn, when mating breeding ewes, lucerne should not be grazed, since leaf diseases are prevalent at this time of the year and so oestrogen levels, which can reduce conception rates, may be high (White, 1982).

Animal intake and performance

Compared with grass, lucerne in the vegetative state has higher intake characteristics and a higher animal production response per unit of DM ingested. The likely reasons for these advantages are a rapid passage of digesta out of the rumen, which stimulates appetite, a high concentration of soluble protein, which assists in microbial synthesis in the rumen, the stimulation of cellulose digestion, a low concentration of cell wall in the DM and an adequate supply of minerals and vitamins (Conrad and Klopfenstein, 1988). When cut lucerne and perennial-ryegrass forages of similar maturity were fed to dairy cows, rumen digestion and particle breakdown were enhanced by the lucerne diet, but protein content was higher and the cellulose and hemicellulose contents lower in the lucerne (Waghorn *et al.*, 1989).

In reviewing intake comparisons between lucerne, grazed or fed fresh, and grass, at similar digestibilities, lucerne intakes for sheep, dairy cows and beef cattle were 44, 20 and 28% higher, respectively (Campling, 1984). Intakes of ensiled lucerne by beef cattle, compared with those of ensiled grass of similar digestibility, were superior, by 10–20% (Doyle and Thomson, 1985) to 34% (Pocknee and Campling, 1981). Beef cattle fed lucerne silage produce more LW gain than intake suggests in comparison with grass silage, e.g an advantage in intake of 28% over cocksfoot silage but an LW gain benefit of 58%, because of differences in intake characteristics and nutritive value (Thomson *et al.*, 1991). Intake of lucerne in dried chopped form can be 10% higher than that of grass prepared similarly and of equal digestibility. The introduction of 3 kg of dried lucerne pellets into the diet of dairy cows increased milk protein, without reducing milk-fat content, and intake was increased, compared with adding lucerne silage of similar quality to the diet (Peyraud and Delaby, 1994).

Physical characteristics of leaves and stems differ between cultivars and can potentially affect intake, as these characteristics are related to energy required during mastication and comminution to break down particles to a sufficiently small state to pass through the reticulorumen–abomasum orifice. Cultivars of lucerne differ in the force required to shear their stems; for example, the cv. Vernal, though having thin stems, requires a greater force to shear its stems than cvs Barrier or Anchor (Iwaasa et al., 1996). So thickness of stem need not be an indication of difficulty in being broken down into small enough particles to pass out of the rumen. This is reinforced by the observation that lucerne stems are as thick as those of some tropical grasses but are consumed at twice the rate, suggesting that they require less energy during mastication, rumination and comminution than the stems of these grasses (Mtengeti et al., 1996).

While short-chopped lucerne silage will generally increase intake by dairy cows in total-mix rations over long-chopped silage (Fischer et al., 1994), milk production is increased with a longer chop length, e.g. 10 cm compared with 5 cm, if NDF in the forage is lower than recommended for the diet of dairy cows (Beauchemin et al., 1994). Comparing lucerne and grass silages at equal intakes of metabolizable energy (ME) and available N, cattle fed lucerne silage gained more tissue energy and similar amounts of tissue protein (Tyrrell et al., 1992); higher NDF and a higher content of soluble carbohydrate in the digested OM and other digested nutrients may have conferred an advantage on the lucerne. For a given level of total NDF, higher digestibility of NDF results in higher intakes (Dado and Allen, 1996).

Low energy content is the suggested cause of the limitation to milk production from lucerne diets. Although lucerne has a high CP content, degradability in the rumen is rapid (Balde et al., 1993) and so inadequate amounts of protein reach the small intestine, limiting milk and milk-protein production (Dhiman et al., 1993). The content of either fermentable carbohydrate or non-degradable (bypass) protein in the rumen needs to be increased in lucerne-based diets. The importance of supplementing lucerne silage with protein has been demonstrated in beef-cattle diets, inclusion of rumen non-degradable protein improving feed-conversion efficiency, even with reduced intake (Nicholson et al., 1992).

Comparing milk production and composition from cows fed the same crop of lucerne preserved as silage or as hay, the silage resulted in slightly lower milk yields, with lower protein contents, than hay, but the fat content was higher (Broderick, 1995); the higher yield and protein content of milk from hay-fed cows was most likely to be due to the higher intake of DM (24.0 vs. 22.3 kg DM day^{-1}) and higher content of protected protein, resulting from field-curing of the hay, as these cows responded less to protected-protein supplementation than the silage-fed cows (100 g vs. 30 g milk protein day^{-1}). Higher fibre digestibility in silage than in hay treatments would account for higher milk-fat contents from cows fed silage (Broderick, 1995). Other studies have also shown a benefit to intake by feeding lucerne hay rather than silage to dairy cows (Broderick et al., 1992; Nelson and Satter, 1992).

In the USA, where lucerne is the principal conserved forage fed to dairy cows, this high-protein forage is often complemented with energy-rich maize. A lucerne-silage/grain mixture, 1 : 1 on a DM basis, in comparison with a silage-only diet, provided an additional 11 kg milk cow^{-1} day^{-1}, representing a 38% increase (Cadorniga and Satter, 1993). Supplementing suckler-cow diets grazing low-quality, tall-grass prairie with protein-rich lucerne hay reduced the time taken to conceive and increased the weight of the calf at birth (Vanzant and Cochran, 1994).

Under grazing, higher intake by sheep on lucerne pasture than on grass translates into higher lamb production, sometimes in excess of the relative increase in intake. In New Zealand, lamb production from grazed lucerne averaged 82% higher than that from grazed grass (Douglas, 1986). Feeding an energy supplement of maize grain to lambs grazing lucerne pasture improved lamb performance; for example, 0.25 kg DM day^{-1} increased daily LW gain by 13%, since lucerne protein was used more efficiently (Karnezos *et al.*, 1994). Milk-production levels from dairy cows were similar from grazed lucerne and grazed grass (Bryant, 1978; Sanchez and Campling, 1982), although higher milk production and milk-protein content have also been reported (Conrad *et al.*, 1983).

Alternative Uses

Lucerne's potential as a 'break crop' to improve soil fertility in alternate-husbandry systems is well documented. It can also be used as a green-manure crop, for which its N_2-fixing capacity in the establishment year, especially by non-winter-dormant cultivars, is an advantage.

In Scandanavia, lucerne has been evaluated as a fibre crop, especially for paper-making. Limitations imposed by agricultural cultivars could conceivably be overcome by breeding cultivars with a higher stem content, higher cellulose content and fewer thin-walled cells, characteristics which would improve the yield of pulp and reduce moisture retention (Berggren, 1992).

Other industrial uses of lucerne, due to its high growth rate, include its potential as a biomass crop for energy, a raw product for the production of biodegradable plastics or a substrate for production of fermentation products, e.g. alcohol or lactic acid (Talamucci, 1994). Saponins have insecticidal properties and so could be extracted and purified from cultivars with high saponin content (Tava *et al.*, 1993). Protein, carotene and xanthophyll in extracted juice in the fractionation process may also be considered as industrial raw products.

French studies have suggested that lucerne is a suitable cover for wild life, especially birds. It is one of the few cultivated plants that can produce high levels of biomass with minimum inputs and has been found to be a suitable breeding habitat for some threatened bird species. Thus, lucerne has been proposed as a set-aside cover crop for endangered birds (Boutin *et al.*, 1994).

References

Abdul-Jabbar, A.S., Sammis, T.W. and Lugg, D.G. (1982) Effect of moisture level on the root pattern of alfalfa. *Irrigation Science* 3, 197–207.

Agriculture Canada (1989) *Alfalfa Production and Management in Atlantic Canada*. Publication No. 1833/E, Agriculture Canada, Ottawa. 37 pp.

Aldrich, D.T.A. (1984) Lucerne, red clover and sainfoin: herbage production. In: Thomson, D.J. (ed.) *Forage Legumes*. Occasional Symposium No. 16, British Grassland Society, Hurley, pp. 126–131.

Aldrich, D.T.A. (1987) Developments and procedures in the assessment of grass varieties at NIAB 1950–1987. *Journal of the National Institute of Agricultural Botany* 17, 313–327.

Amrane, R. and Michaeletdoreau, B. (1993) Effect of maturity stage of Italian ryegrass and lucerne on ruminal nitrogen degradability. *Annales de Zootechnic* 42, 31–37.

Andrews, C.S. (1977) The effect of sulphur on the growth, sulphur and nitrogen concentrations of some tropical and temperate pasture legumes. *Australian Journal of Agricultural Research* 11, 1026–1033.

Andrighetto, I., Cozzi, G., Magni, G., Hartman, B., Hinds, B. and Sapienza, D. (1995) Comparison of *in situ* degradation kinetics of lucerne germplasm by ANOVA of nonlinear models. *Animal Feed Science and Technology* 54, 287–299.

Arcioni, S., Damiani, F., Piccirilli, M. and Pupilli, F. (1993) *In vitro* approach to the production of *Medicago sativa* plants resistant to fungi. In: Rotili, P. and Zannone, L. (eds) *The Future of Lucerne Biotechnology, Breeding and Variety Constitution. Proceedings of the X International Conference of* EUCARPIA Medicago *spp. Group*. Instituto Sperimentale per le Colture Foraggere, Lodi, pp. 181–186.

Arcioni, S., Damiani, F., Mariani, A. and Pupilli, F. (1997) Somatic hybridization and embryo rescue for the introduction of wild germplasm. In: McKersie, B.D. and Brown, D.C.W. (eds) *Biotechnology and the Improvement of Forage Legumes*. CAB International, Wallingford, pp. 61–90.

Askarian, M., Hampton, J.G. and Hill, M.J. (1995) Effect of row spacing and sowing rate on seed production of lucerne (*Medicago sativa* L.) cv. Grasslands Oranga. *New Zealand Journal of Agricultural Research* 38, 289–295.

Aube, C. and Gagnon, C. (1970) Influence of certain soil fungi on alfalfa. *Canadian Journal of Plant Science* 50, 159–162.

Angelini, F. (1979) *Coltivacioni Erbacee*: Medicago sativa L, Vol. II. Società Grafica Romana-Roma, pp. 555–587.

Azcon, R., Rubeo, R. and Barea, J.M. (1991) Selective interaction between different species of mycorrhizal fungi and *Rhizobium meliloti* strains and their effect on growth, N fixation (^{15}N) and initiation of *Medicago sativa* L. *New Phytologist* 117, 399–404.

Balde, A.T., Vandersall, J.H., Erdman, R.A., Reeves, J.B. and Glenn, B.P. (1993) Effect of stage of maturity of alfalfa and orchardgrass on *in situ* dry matter and crude protein degradability and amino acid composition. *Animal Feed Science and Technology* 44, 29–43.

Baldwin, B.S. (1990) Evaluation of temperature, photoperiod and divergent selection for the fall dormancy response in alfalfa. *Dissertation Abstracts International* 5, 385A–386A.

Barber, L.D., Joern, B.C., Volenec, J.J. and Cunningham, S.N. (1996) Supplemental nitrogen effects on alfalfa regrowth and nitrogen mobilization from roots. *Crop Science* 36, 1217–1223.

Barbetti, M.J. (1991) Effects of temperature and humidity on diseases caused by *Phoma medicaginis* and *Leptosphaerulina trifolii* in lucerne (*Medicago sativa*). *Plant Pathology* 40, 296–301.

Barlow, N.D. and Goldson, S.L. (1993) A modelling analysis of the successful biological control of *Sitona discoideus* (Coleoptera, Curculionidae) by *Microctonus aethiopoides* (Hymenoptera, Braconidae) in New Zealand. *Journal of Applied Ecology* 30, 165–178.

Barnes, D.K. (1992) Forage legume breeding past, present and future. In: *Proceedings of the 14th General Meeting of the European Grassland Federation, Lahti, Finland*. European Grassland Federation, pp. 78–86.

Barnes, D.K. and Sheaffer, C.C. (1995) Alfalfa. In: Barnes, R.F., Miller, D.A. and Nelson, C.J. (eds) *Forages*, 5th edn, Vol. 1 *An Introduction to Grassland Agriculture*. Iowa State University Press, Ames, Iowa, pp. 205–216

Barran, L.R. and Bromfield, E.P.S. (1997) Competition among rhizobia for nodulation of legumes. In: McKersie, B.D. and Brown, D.C.W. (eds) *Biotechnology and the Improvement of Forage Legumes*. CAB International, Wallingford, pp. 343–374

Barry, T.N., Cook, J.E. and Wilkins, R.J. (1978) The influence of formic acid and formaldehyde additives and type of harvesting machine on the utilisation of nitrogen in lucerne silages. 1. The voluntary intake and nitrogen retention of young sheep consuming the silages with and without intra-peritoneal supplements of DL-methionine. *Journal of Agricultural Science, Cambridge* 91, 701–715.

Bates, G.E., Hoveland, C.S., McCann, M.A., Bouton, J.H. and Hill, N.S. (1996) Plant persistence and animal performance for continuously stocked alfalfa pastures at three forage allowances. *Journal of Production Agriculture* 9, 418–423.

Beauchemin, K.A., Farr, B.I., Rode, L.M. and Schaalje, G.B. (1994) Effects of silage chop length and supplementary long hay on chewing and milk production of dairy cows. *Journal of Dairy Science* 77, 1326–1339.

Berggren, H. (1992) Project Agro-Fibre Industrial use for pulp and paper and related products. In: Rotili, P. and Zannone, L. (eds) *The Future of Lucerne – Biotechnology, Breeding and Variety Constitution. Proceedings of the X International Conference of EUCARPIA, Medicago spp. Group*. Istituto Sperimentale per le Culture Foraggere, Lodi, pp. 55–66.

Blair, I.D. (1971) Lucerne nodulation failures. *Proceedings of the Lincoln Farmers' Conference* 21, 58–66.

Bolton, J.L., Goplen, B.P. and Baenziger, H. (1972) World distribution and historical developments. In: Hanson, C.H. (ed.) *Alfalfa Science and Technology*. American Society of Agronomy, Madison, Wisconsin, pp. 1–34.

Boutin, J.M., Bretagnolle, V. and Moreau, C. (1994) Use of lucerne by wildlife with special reference to birds. *FAO/REUR Technical Series* 36, 47–49.

Bouton, J.H. (1996) Screening the alfalfa core collection for acid soil tolerance. *Crop Science* 36, 198–200.

Bouton, J.H. and Parrott, W.A. (1997) Salinity and aluminium stress. In: McKersie, B.D. and Brown, D.C.W., (eds) *Biotechnology and the Improvement of Forage Legumes*. CAB International, Wallingford, pp. 203–228.

Bouton, J.H. and Sumner, M.E. (1984) Alfalfa (*Medicago sativa* L.) in highly weathered acid soils. V. Field performance of alfalfa selected for acid tolerance. *Plant and Soil* 74, 431–436.

Bouton, J.H., Smith, S.R., Wood, D.T., Hoveland, C.S. and Brummer, E.C. (1991) Registration of 'Alfagraze' alfalfa. *Crop Science* 31, 497.

Bowley, S.R. (1997) Breeding methods for forage legumes. In: McKersie, B.D. and Brown, D.C.W. (eds) *Biotechnology and the Improvement of Forage Legumes.* CAB International, Wallingford, pp. 25–42.

Boyce, P.J., Penalosa, E. and Volenec, J.J. (1992) Amylase activity in tap roots of *Medicago sativa* L. and *Lotus corniculatus* L. following defoliation. *Journal of Experimental Botany* 43, 1053–1059.

Brash, D.L. (1983) Dryland lucerne establishment by overdrilling in central Otago. *Proceedings of the New Zealand Grassland Association* 44, 164–169.

Brink, G.E. and Marten, G.C. (1989) Harvest management of alfalfa – nutrient yield vs. forage quality and relationship to persistence. *Journal of Production Agriculture* 2, 32–36.

Broderick, G.A. (1995) Performance of lactating dairy cows fed either alfalfa silage or alfalfa hay as the sole forage. *Journal of Dairy Science* 78, 320–329.

Broderick, G.A., Abrams, A.M. and Rotz, C.A. (1992) Ruminal *in vitro* degradability of protein in alfalfa harvested as standing forage or baled hay. *Journal of Dairy Science* 75, 2440–2446.

Brookes, B., Small, E., Lefkovitch, L.P., Dammon, H. and Fairey, D.T. (1994) Attractiveness of alfalfa (*Medicago sativa* L.) to wild pollinators in relation to wild flowers. *Canadian Journal of Plant Science* 74, 779–783.

Brown, P.W. and Tanner, C.B. (1981) Alfalfa water potential measurement: a comparison of the pressure chamber and leaf dew point hygrometer. *Crop Science* 21, 240–244.

Bryant, A.M. (1978) Milk yield and composition from cows grazing lucerne. *Proceedings of the New Zealand Society of Animal Production* 38, 185–190.

Buckmaster, D.R., Rotz, C.A. and Black, J.R. (1990) Value of alfalfa losses on dairy farms. *Transactions of the American Society of Agricultural Engineers* 33, 351–360.

Buntin, G.D. and Pedigo, L.P. (1985) Alfalfa development, dry matter accumulation and partitioning after surrogate insect defoliation of stubble. *Crop Science* 25, 1035–1040.

Burity, H.A., Ta, T.C., Faris, M.A. and Coulman, B.E. (1989) Estimation of nitrogen fixation and transfer from alfalfa to associated grasses in mixed swards under field conditions. *Plant and Soil* 114, 249–255.

Busbice, T.H. (1969) Inbreeding in synthetic varieties. *Crop Science* 9, 601–604.

Buxton, D.R., Hornstein, J.S., Wedin, W.F. and Marten, G.C. (1985) Forage quality in stratified canopies of alfalfa, birdsfoot trefoil and red clover. *Crop Science* 25, 273–279.

Byers, R.A. and Templeton, W.C. (1988) Effects of sowing date, placement of seed, vegetation suppression, slugs and insects upon establishment of no-till alfalfa in orchardgrass sod. *Grass and Forage Science* 43, 279–287.

Cadorniga, C. and Satter, L.D. (1993) Protein versus energy supplementation of high alfalfa silage diets for early lactation cows. *Journal of Dairy Science* 76, 1972–1977.

Campling, R.C. (1984) Lucerne, red clover and other forage legumes: feeding value and animal production. In: Thomson, D.J. (ed.) *Forage Legumes.* Occasional Symposium No. 16, British Grassland Society, Hurley, pp. 140–146.

Carter, P.R. and Sheaffer, C.C. (1983) Alfalfa response to soil water deficits. 1. Growth, forage quality, yield, water use and water use efficiency. *Crop Science* 23, 669–675.

Caseda, L.E. and Lukezic, F.L. (1992) Control of leaf spot diseases of alfalfa and tomato with applications of the bacterial predator *Pseudomonas* strain 679–2. *Plant Disease* 76, 1217–1220.

Casler, M.D. (1988) Performance of orchardgrass, smooth bromegrass and ryegrass in binary mixtures with alfalfa. *Agronomy Journal* 80, 509–514.

Castonguay, Y., Nadeau, P., Lechasseur, P. and Chouinard, L. (1995) Differential accumulation of carbohydrates in alfalfa cultivars of contrasting winterhardiness. *Crop Science* 35, 509–516.

Chamblee, D.S. (1972) Relationships with other species in a mixture. In: Hanson, C.H. (ed.) *Alfalfa Science and Technology*. American Society of Agronomy, Madison, Wisconsin, pp. 211–228.

Chamblee, D.S. and Collins, M. (1988) Relationships with other species. In: Hanson, A.A., Barnes, D.K. and Hill, R.R., Jr (eds) *Alfalfa and Alfalfa Improvement*. Agronomy Monograph No. 29, CSSA/ASSA/SSSA, Madison, Wisconsin, pp. 439–461.

Charlton, J.F.L. (ed.) (1995) *The Grasslands Range of Forage and Conservation Plants*. AgResearch, Palmerston North New Zealand, 76 pp.

Charrier, X., Emile, J.C. and Guy, P. (1993) Recherche de génotypes de luzerne adaptés au pâturage. *Fourrages* 135, 507–510.

Chaudhary, M.T., Wainwright, S.J. and Merrett, M.J. (1996) Comparative NaCl tolerance of lucerne plants regenerated from salt-selected suspension cultivars. *Plant Science* 114, 221–232.

Christie, B.R. and Townshend, J.L. (1992) Selection for resistance to the root lesion nematodes in alfalfa. *Canadian Journal of Plant Science* 72, 593–598.

Chung, I.M. and Miller, D.A. (1995a) Differences in autotoxicity among seven alfalfa cultivars. *Agronomy Journal* 87, 596–600.

Chung, I.M. and Miller, D.A. (1995b) Allelopathic influence of nine forage grass extracts on germination and seedling growth of alfalfa. *Agronomy Journal* 87, 767–772.

Close, R.C., Whitelaw, J.S. and Taylor, G.G. (1971) Studies on the nodulation of lucerne in the field. 1. After inoculation with *Rhizobium*. 2. After using certain fungicides as seed treatments. In: *Proceedings of the 4th Australasian Legume Nodulation Conference*, Paper 12, p. 4.

Conrad, H.R. and Klopfenstein, T.J. (1988) Role in livestock feeding – greenchop, silage, hay and dehy. In: Hanson, A.A., Barnes, D.K. and Hill, R.R., Jr (eds) *Alfalfa and Alfalfa Improvement*. Agronomy Monograph No. 29, ASA/CSSA/SSSA, Madison, Wisconsin, pp. 539–566.

Conrad, H.R., van Keuren, R.W. and Dehority, B.A. (1983) Top-grazing high-protein forages with lactating cows. In: Smith, J.A. and Hayes, V.W. (eds) *Proceedings of the XIV International Grassland Congress, Lexington, USA*. Westview Press, Boulder, Colorado, pp. 690–692

Corletto, A., Cazzato, E. and Ventricelli, P. (1994) The effect of cutting management systems on survival, DMY and protein content on alfalfa (*Medicago sativa* L.). *FAO/REUR Technical Series* 36, 93–98.

Cowett, E.R. and Sprague, M.A. (1963) Effect of stand density and light intensity on the microenvironment and stem production of alfalfa. *Agronomy Journal* 55, 432–434.

Cralle, H.T. and Heichel, G.H. (1981) Nitrogen fixation and vegetative regrowth of alfalfa and birdsfoot trefoil after successive harvests or floral debudding. *Plant Physiology* 67, 895–905.

Cudney, D.W., Orloff, S.B. and Demason, D.A. (1993) Effects of thiazopyr and trifluorin on dodder (*Cuscutata indecora*) in alfalfa (*Medicago sativa* L.). *Weed Technology* 7, 860–864.

Dado, R.G. and Allen, M.S. (1996) Enhanced intake and production of cows offered ensiled alfalfa with higher NDF digestibility. *Journal of Dairy Science* 79, 418–428.

D'Antuono, C.F., Lovato, A. and Rossi Pisa, P. (1988) Effetti di alcuni fattori meteorologici sulla produzione di seme di erba medica (*Medicago sativa* L.). *Rivista di Agronomia* 22, 137–148.

Dawson, J.H. (1992) Response of alfalfa (*Medicago sativa*) grown for seed production to glyphosate and SC-0224. *Weed Technology* 7, 860–864.

Deetz, D.A., Jung, H.G. and Buxton, D.R. (1996) Water deficit effects on cell wall composition and *in vitro* degradability of structural polysaccharides. *Crop Science* 36, 383–388.

Demarquilly, C. (ed.) (1981) *Prévision de la Valeur Nutritive des Aliments des Ruminants.* Centre de Recherches Zootechniques et Vétérinaires de Theix, INRA, Beaumont.

Demarquilly, C. and Dulphy, J.P. (1977) Effect of ensiling on feed intake and animal performance. In: *Proceedings of the International Meeting on Animal Production from Temperate Grassland, Dublin, Ireland.* Irish Grassland and Animal Production Association and An Foras Taluntais, Dublin, pp. 53–61.

Devine, T.E., Fry, C.D., Fleming, A.L., Hanson, C.H., Campbell, T.A., McMurtey, J.E. and Schwartz, J.W. (1976) Development of alfalfa strains with differential tolerance to aluminium toxicity. *Plant and Soil* 44, 73–79.

Dhiman, T.R., Cadorniga, C. and Satter, L.D. (1993) Protein and energy supplementation of high alfalfa silage diets during early lactation. *Journal of Dairy Science* 76, 1945–1959.

Doerge, T.A., Bottomley, P.J. and Gardner, E.H. (1985) Molybdenum limitations to alfalfa growth and nitrogen content on a moderately acid, high phosphorus soil. *Agronomy Journal* 77, 895–901.

Dormaar, J.F. and Schaber, B.D. (1992) Burning of alfalfa stubble for insect control as it affects soil chemical properties. *Canadian Journal of Soil Science* 72, 169–175.

Dougherty, C.T., Collins, M., Bradley, N.W., Lauriault, L.M. and Cornelius, P.L. (1992) The effects of poloxalene on ingestion by cattle grazing lucerne. *Grass and Forage Science* 47, 180–188.

Douglas, J.A. (1986) The production and utilization of lucerne in New Zealand. Review paper. *Grass and Forage Science* 41, 81–128.

Doyle, C.J. and Thomson, D.J. (1985) Potential of lucerne in British Agriculture: an economic assessment. *Grass and Forage Science* 40, 57–68.

Dubach, M. and Russelle, M.P. (1994) Forage legume roots and nodules and their role in nitrogen transfer. *Agronomy Journal* 86, 259–266.

Duke, S.H. and Doehlert, D.C. (1982) Root respiration, nodulation and enzyme activities in alfalfa during cold acclimation. *Crop Science* 29, 489–495.

Dulphy, J.P. (1980) The intake of conserved forages. In: Thomas, C. (ed.) *Forage Conservation in the 80s.* Occasional Symposium No. 11, British Grassland Society, Hurley, pp. 107–121.

Dunbier, M.W. (1983) Lucerne. In: Wratt, G.S. and Smith, H.C. (eds) *Plant Breeding in New Zealand.* Butterworths/DSIR, Wellington, pp. 243–252.

Dunbier, M.W., Wynn-Williams, R.B. and Purves, R.G. (1983) Lucerne seed production in New Zealand – achievements and potential. *Proceedings of the New Zealand Grassland Association* 44, 30–35.

Durand, J.L., Lemaire, G., Gosse, G. and Charter, M. (1989) Analyse de la conversion de l'énergie solaire en matière sèche par un peuplement de lucerne (*Medicago sativa* L) soumis à un déficit hydrique. *Agronomie* 9, 599–607.

Durling, J.C., Hesterman, O.B. and Rotz, C.A. (1995) Predicting first-cut alfalfa yields from preceding winter weather. *Journal of Production Agriculture* 8, 254–259.

Duru, M. and Langlet, A. (1993) Effets de la compétition pour la lumière et du déficit en eau sur l'évolution du rapport feuille/tige de la luzerne. *Fourrages* 134, 199–204.

Duthie, J.A. and Campbell, C.L. (1991) Effects of plant debris on intensity of crop spot diseases, incidence of pathogens and growth of alfalfa. *Phytopathology* 81, 511–517.

Elkan, G.H. (1981) The taxonomy of the Rhizobiaceae. In: Giles, K.L. and Atherly, A.G. (eds) *Biology of the Rhizobiaceae*. International Review of Cytology, Academic Press, New York, pp. 1–14.

Fairey, D.T. and Lefkovitch, L.P. (1990) Herbage production from a conventional mixture versus alternative strips of grass and legume. *Agronomy Journal* 82, 737–744.

Fairey, D.T. and Lefkovitch, L.P. (1991) Hard seed content of alfalfa grown in Canada. *Canadian Journal of Plant Science* 71, 437–444.

Fairey, D.T. and Lefkovitch, L.P. (1994) Alternating strips of grass and legume and nitrogen fertilization strategy for long term herbage production from brome grass. *Canadian Journal of Plant Science* 75, 649–654.

Fairey, D.T., Lefkovitch, L.P. and Fairey, N.A. (1996) The relationship between fall dormancy and germplasm source in North American alfalfa cultivars. *Canadian Journal of Plant Science* 76, 429–433.

Feuss, R.W. and Tesar, M.B. (1968) Photosynthetic efficiency, yields and leaf loss in alfalfa. *Crop Science* 8, 159–163.

Fick, G.W., Holt, D.A. and Lugg, D.G. (1988) Environmental physiology and crop growth. In: Hanson, A.A., Barnes, D.K. and Hill, R.R., Jr (eds) *Alfalfa and Alfalfa Improvement*. Agronomy Monograph No 29, ASA/CSSA/SSSA, Madison. Wisconsin, pp. 163–191.

Fischer, J.M., Buchanan-Smith, J.G., Campbell, C., Grieve, D.G. and Allen, O.B. (1994) Effects of forage particle size and long hay for cows fed total mixed rations based on alfalfa and corn. *Journal of Dairy Science* 77, 217–229.

Fleming, R.A. (1988) Difficulties implementing a modelling-based integrated pest management program for alfalfa. *Memoirs of the Entomology Society of Canada* 143, 47–59.

Forster, R.L.S., Beck, D.L. and Lough, T.J. (1997) Engineering for resistance to virus diseases. In: McKersie, B.D. and Brown, D.C.W. (eds) *Biotechnology and the Improvement of Forage Legumes*. CAB International, Wallingford, pp. 291–318.

Foy, C.L. and Witt, H.L. (1993) Effects of paraquat on weed-control and yield of alfalfa. *Weed Technology* 1, 495–506.

Frame, J. (1986) The production and quality potential of four forage legumes sown alone and combined in various associations. *Crop Research* 25, 103–122.

Frame, J. and Harkess, R.D. (1987) The productivity of four forage legumes sown alone and with each of five companion grasses. *Grass and Forage Science* 42, 213–223.

Giardini, A. and Cinti, F. (1985) Praticoltura da vicenda nell'Italia centrale. [Rotated meadows in central Italy] *Rivista d'Agronomia* 19, 104–121.

Goering, H.K. and Waldo, D.R. (1979) The effects of dehydration on protein utilisation in ruminants. In: Howarth, R.E. (ed.) *Proceedings of the Second International Green Crop Dryers' Congress, Saskatoon, Canada*. University of Saskatchewan Press, Saskatoon, pp. 277–291.

Goplen, B.P., Howarth, R.E. and Lees, G.L. (1993) Selection of alfalfa for a lower initial rate of digestion and corresponding changes in epidermal and mesophyll cell wall thickness. *Canadian Journal of Plant Science* 73, 111–122.

Gosse, G., Chartier, M., Varlet-Grancher, C. and Bonhomme, R. (1982) Photosynthetically active radiation by alfalfa – variations and modelling. *Agronomie* 2, 583–588.

Gosse, G., Lemaire, G., Chartier, M. and Balfourier, F. (1988) Structure of a lucerne population (*Medicago sativa* L.) and dynamics of stem competition for light during regrowth. *Journal of Applied Ecology* 25, 609–617.

Graham, P.H. (1992) Stress tolerance in *Rhizobium* and *Bradyrhizobium* and nodulation under adverse soil conditions. *Canadian Journal of Microbiology*, 38, 475–484.

Grau, C.R., Nygaard, S.L., Arny, D.C. and Delwiche, P.A. (1991) Comparison of methods to evaluate alfalfa cultivars for reaction to *Verticillium alba-atrum*. *Plant Disease* 75, 82–85.

Grenier, G. and Willemot, C. (1974) Lipid changes in roots of frost hardy and less hardy alfalfa varieties under hardening conditions. *Cryobiology* 11, 324–331.

Griffin, T.S., Cassida, K.A., Hesterman, O.B. and Rust, S.R. (1994) Alfalfa maturity and cultivar effects on chemical and *in situ* estimates of protein degradability. *Crop Science* 34, 1654–1661.

Griffiths, T.W. and Poole, D.A. (1984) Effects of cutting dates and frequency of cutting on yield and quality of lucerne. *Research and Development in Agriculture* 1, 177–179.

Gross, C.F. and Jung, G.A. (1978) Magnesium, Ca and K concentration in temperate origin forage species as affected by temperature and Mg fertilisation. *Agronomy Journal* 70, 397–403.

Guy, P. (1993) Lucerne in Europe: statistical elements. In: Rotili, P. and Zannone, L. (eds) *The Future of Lucerne Biotechnology Breeding and Variety Constitution. Proceedings of the X International Conference of EUCARPIA*, Medicago *spp.* Group. Istituto Sperimentale per le Colture Fouraggère, Lodi, pp. 13–17.

Guy, P., Blondon, F. and Durand, J. (1971) Action de la température et de la durée d'éclairement sur la croissance et la floraison de deux types éloignes de luzerne cultivée. *Medicago sativa* L. *Annales d'Amélioration des Plantes Fourragères*, 21, 409–422.

Hacquet, J. (1989) Genetic variability and climatic factors affecting lucerne seed production. Workshop A. In: *Proceedings of the XVI International Grassland Congress, Nice, France*. Association Française pour la Production Fourragère, Versailles, p. 1844.

Hall, J.W., Majat, W., Stout, D.G., Cheng, K.J., Goplen, B.P. and Howarth, R.E. (1994) Bloat in cattle fed alfalfa method selected for a low initial rate of digestion. *Canadian Journal of Animal Science* 74, 451–456.

Hampton, J.G., Charlton, J.F.L., Bell, D.D. and Scott, D.J. (1987) Temperature effects on germination of New Zealand legumes. *Proceedings of the New Zealand Grassland Association* 48, 177–184.

Hannaway, D.B. and Shuler, P.E. (1993) Nitrogen fertilisation in alfalfa production. *Journal of Production Agriculture* 6, 80–85.

Hartwig, U.A. and Nösberger, J. (1994) What triggers the regulation of nitrogenase activity in forage legume nodules after defoliation. *Plant and Soil* 161, 109–114.

Hartwig, U.A., Joseph, C.M. and Phillips, D.A. (1991) Flavonoids released naturally from alfalfa seeds enhance growth rate of *Rhizobium meliloti*. *Plant Physiology* 95, 797–803.

Harvey, R.G. (1991) Bentazone for annual weed control in newly seeded alfalfa. *Weed Technology* 5, 154–158.

Hayman, J.M. and McBride, S.D. (1984) *The Response of Pasture and Lucerne to Irrigation.* Technical Report 17, Winchmore Irrigation Research Station, Ministry of Agriculture and Fisheries, p. 79. (Cited in Douglas,1986.)

Heichel, G.H. and Henjum, K.I. (1991) Dinitrogen fixation, nitrogen transfer and productivity of forage legume–grass communities. *Crop Science* 31, 202–208.

Heichel, G.H., Barnes, D.K. and Vance, C.P. (1981) Nitrogen fixation of alfalfa in the seedling year. *Crop Science* 21, 330–335.

Heichel, G.H., Delaney, R.H. and Cralle, H.T. (1988) Carbon assimilation, partitioning and utilization. In: Hanson, A.A., Barnes, D.K. and Hill, R.R., Jr (eds) *Alfalfa and Alfalfa Improvement.* Agronomy Monograph No. 29, ASSA/CSSA/SSSA, Madison, Wisconsin, pp. 195–228.

Hesterman, O.B. and Teuber, L.R. (1981) Effect of photoperiod and irradiance on fall dormancy in alfalfa. In: *Agronomy Abstracts, 73rd Annual Meeting, American Society of Agronomy.* American Society of Agronomy, Madison, Wisconsin, p. 87.

Hesterman, O.B., Kells, J.J. and Tiffin, P.L. (1993) Interaction among harvest frequency, fertilizer and herbicide use with intensively manged alfalfa in the north-central USA. In: Baker, M.J. (ed.) *Proceedings of the XVII International Grassland Congress, New Zealand and Australia,* Vol. I. New Zealand Grassland Association et al., Palmerston North, pp. 885–887.

Hill, R.R., Jr and Baylor, J.E. (1983) Genotype × environment interaction analysis for yield in alfalfa. *Crop Science* 23, 811–815.

Hill, R.R., Jr and Jung, G.A. (1975) Genetic variability for chemical composition of alfalfa. 1. Mineral elements. *Crop Science* 15, 652–657.

Hill, R.R., Jr, Hanson, C.H. and Bisbice, T.H. (1969) Effect of four recurrent selection programs on two alfalfa populations. *Crop Science* 9, 363–365.

Hill, R.R., Jr, Shenk, J.S. and Barnes, R.F. (1988) Breeding for yield and quality. In: Hanson, A.A., Barnes, D.K. and Hill, R.R., Jr (eds) *Alfalfa and Alfalfa Improvement.* Agronomy Monograph No 29, ASA/CSSA/SSSA, Madison, Wisconsin, pp. 809–825.

Hoglund, J.H., Dougherty, C.T. and Langer, R.H.M. (1974) Response of irrigated lucerne to defoliation and nitrogen fertilizer. *New Zealand Journal of Experimental Agriculture* 2, 7–11.

Hristov, A.N. (1993) Effect of a commercial enzyme preparation on alfalfa silage fermentation and protein degradability. *Animal Feed Science and Technology* 42, 273–282.

Hutchins, S.H., Buntin, G.D. and Pedigo, L.P. (1991) Impact of insect feeding on alfalfa regrowth: a review of physiological responses and economic consequences. *Agronomy Journal* 82, 1035–1044.

Inch, M.J., Irwin, J.A.G. and Bray, R.A. (1993) Seasonal variation in lucerne foliar diseases and cultivar reaction to leaf spot pathogens in the field in Southern Queensland. *Australian Journal of Experimental Agriculture* 33, 343–348.

Irigoyen, J.J., Emerich, D.D. and Sanchez-Diaz, M. (1992a) Alfalfa leaf senescence induced by drought stress. photosynthesis, hydrogen peroxide metabolism, lipid peroxidation and ethylene evolution. *Physiologia Plantarum* 84, 67–72.

Irigoyen, J.J., Sanchez-Diaz, M. and Emerich, D.D. (1992b) Transient increase of anaerobically-induced enzymes during short-term drought of alfalfa root nodules. *Journal of Plant Physiology* 139, 397–402.

Iversen, C.E. and Meijer, G. (1967) Types and varieties of lucerne. In: Langer, R.H.M. (ed.) *The Lucerne Crop.* Reed, Wellington, pp. 74–84.

Iwaasa, A.D., Beauchemin, K.A., Buchanan-Smith, J.G. and Acharya, S.N. (1996) Effect of stage of maturity and growth cycle on shearing force and cell wall chemical constituents of alfalfa stems. *Canadian Journal of Animal Science* 76, 321–328.

Jagusch, K.T. (1982) Nutrition of ruminants grazing lucerne. In: Wynn-Williams, R.B. (ed.) *Lucerne for the 1980s.* Special Publication No. 1, Agronomy Society of New Zealand, pp. 73–78.

James, D.W., Hurst, C.J. and Tintal, T.A. (1994) Alfalfa cultivar response to phosphorus and potassium deficiency – elemental composition of the herbage. *Journal of Plant Nutrition* 18, 2447–2464.

Janson, C.G. (1975) Irrigation of lucerne in its establishment season. *New Zealand Journal of Experimental Agriculture* 3, 223–228.

Janson, C.G. (1982) Lucerne grazing management research. In: Wynn-Williams, R.B. (ed.) *Lucerne for the 1980s.* Special Publication No. 1, Agronomy Society of New Zealand, pp. 85–90.

Jarrige, R., Dermarquilly, C. and Dulphy J.P., (1982) Forage conservation. In: Hacker, J.B. (ed.) *Nutritional Limits to Animal Production from Pasture.* Commonwealth Agricultural Bureaux, Farnham Royal, pp. 367–387.

Jefferson, P.G. and Zentner, R.P. (1994) Effect of an oat companion crop on irrigated alfalfa yield and economic returns in south west Saskatchewan. *Canadian Journal of Plant Science* 74, 465–470.

Jefferson, P.G., Lawrence, T., Irvine, R.B. and Kelly, G.A. (1994) Evaluation of sainfoin–alfalfa mixtures for forage production and compatibility at a semi-arid location in Southern Saskatchewan. *Canadian Journal of Plant Science* 74, 785–791.

Jessen, D.L., Barnes, D.K. and Vance, C.P. (1988) Bidirectional selection in alfalfa for activity of nodule nitrogen and carbon assimilating enzymes. *Crop Science*, 28 18–22.

Johnson, J. (1984) Syndicate report. Lucerne. In: Thomson, D.J. (ed.) *Forage Legumes.* Occasional Symposium No. 16, British Grassland Society, Hurley, pp. 225–227.

Jones, T.A., Carlson, I.T. and Buxton, D.R. (1988) Reed canary grass binary mixtures with alfalfa and birdsfoot trefoil in comparison to monocultures. *Agronomy Journal* 80, 49–55.

Jones, B.A., Hatfield, R.D. and Muck, R.E. (1992) Effect of fermentation and bacterial inoculation of lucerne cell walls. *Journal of the Science of Food and Agriculture* 60, 147–153.

Juan, N.A., Sheaffer, C.C., Barnes, D.K., Swanson, D.R. and Halgerson, J.H. (1993) Leaf and stem traits and herbage quality of multifoliate alfalfa. *Agronomy Journal* 85, 1121–1127.

Jung, G.A. and Larson, K.L. (1972) Cold, drought and heat tolerance. In: Hanson, C.H. (ed.) *Alfalfa Science and Technology.* American Society of Agronomy, Madison, Wisconsin, pp. 185–209.

Jung, G.A. and Shaffer, J.A. (1993) Component yields and quality of binary mixtures of lucerne and perennial, Italian or short rotation hybrid ryegrass. *Grass and Forage Science* 48, 118–125.

Jung, G.A., Shaffer, J.A. and Rosenberger, J.L. (1991) Sward dynamics and herbage nutritional value of alfalfa–ryegrass mixtures. *Agronomy Journal* 83, 786–794.

Jung, G.A., Shaffer, J.A. and Everhart, J.H. (1996) Harvest frequency and cultivar influence on yield and protein of alfalfa–ryegrass mixtures. *Agronomy Journal* 88, 817–822

Kalu, B.A. and Fick, G.W. (1981) Quantifying morphological development of alfalfa for studies of herbage quality. *Crop Science* 21, 267–271.

Kalu, B.A. and Fick, G.W. (1983) Morphological stage of development as a predictor of alfalfa herbage quality. *Crop Science* 6, 1167–1172.

Karnezos, T.P., Matches, A.G., Preston, R.L. and Brown, C.P. (1994) Corn supplementation of lambs grazing alfalfa. *Journal of Animal Science* 72, 783–789.

Keftassa, D. and Tuvesson, M. (1993) The nutritive value of lucerne (*Medicago sativa*, L.) in different development stages. *Swedish Journal of Agricultural Research* 23, 153–159.

Keld, D.W., Bergstrom, G.C. and Shields, E.J. (1994) Prevalence, severity and association of fungal crown and root rots with injury by the clover root curculio in New York alfalfa. *Plant Diseases* 78, 491–495.

Kelling, K.A. and Matocha, J.E. (1990) Plant analysis as an aid in fertilizing forage crops. In: Kelling, K.A., Matocha, J.E. and Westerman, R.L. (eds) *Soil Testing and Plant Analysis*, (3rd Edn). Soil Science Society of America, Madison, Wisconsin, pp. 603–643.

Kelner, D.J. and Vessey, J.K. (1995) Nitrogen fixation and growth of one year stands of nondormant alfalfa in minitubes. *Canadian Journal of Plant Science* 75, 655–665.

Kendall, W.A., Shaffer, J.A. and Hill, R.R. (1994) Effect of temperature and water variables on the juvenile growth of lucerne and red clover. *Grass and Forage Science* 49, 264–269.

Kerr, J.P., McPherson, H.G. and Talbot, J.S. (1973) Comparative evapotranspiration rates of lucerne, paspalum and maize. In: *Proceedings of the First Australasian Conference on Heat and Mass Transfer*, Monash University, Melbourne, Section 3, pp. 1–8.

Khaiti, M. and Lemaire, G. (1992) Dynamics of shoot and root growth of lucerne after seeding and after cutting. *European Journal of Agronomy* 1, 241–247.

Kidambi, S.P., Matches, A.G. and Bolger, T.P. (1990) Mineral concentrations in alfalfa and sainfoin as influenced by soil moisture level. *Agronomy Journal* 82, 229–236.

Kim, T.H., Ourry, A., Boucaud, J. and Lemaire, G. (1993) Partitioning of nitrogen derived from N_2-fixation and reserves in nodulated *Medicago sativa* L. during regrowth. *Journal of Experimental Botany* 44, 555–562.

Klinowski, M. (1933) *Lucerne – Its Ecological Position and Distribution in the World*. Herbage Plants, Bulletin 12, Imperial Bureau of Plant Genetics, Aberystwyth, Wales.

Krasnuk, M., Jung, G.A. and Witham, F.H. (1978) Dehydrogenase levels in cold-tolerant and cold sensitive alfalfa. *Agronomy Journal* 70, 605–613.

Laidlaw, A.S. (1985) A note on the DM production of lucerne (*Medicago sativa*) in a 'weed-free' soil in Northern Ireland. *Record of Agricultural Research, Department of Agriculture for Northern Ireland* 33, 1–4.

Lamb, J.F.S., Barnes, D.K., Russelle, M.P., Vance, C.P., Heichel, G.H. and Henjum, K.I. (1995) Ineffectively and effectively nodulated alfalfa demonstrate biological nitrogen fixation continues with high nitrogen fertilization. *Crop Science* 35, 153–157.

Lamp, W.O. and Zhao, L.M. (1994) Prediction and manipulation of movement by polyphagous, highly mobile pests. *Journal of Agricultural Entomology* 10, 267–281.

Langer, R.H.M. (1973) Lucerne. In: Langer, R.H.M. (ed.) *Pastures and Pasture Plants*. Reed, Wellington, pp. 347–363.

Langer, R.H.M. (1990) Pasture plants. In: Langer, R.H.M. (ed.) *Pastures: Their Ecology and Management*. Oxford University Press, Auckland, pp. 39–74

Latheef, M.A., Caddel, J.L., Berberet, R.C. and Stritzke, J.F. (1988) Alfalfa production as

influenced by pest stress and early first harvest in Oklahoma. *Crop Protection* 1, 190–197.
Latheef, M.A., Berbert, R.C., Stritzke, J.F. Caddel, J.L. and McNew, R.W. (1994) Productivity and persistence of declining alfalfa stands as influenced by the alfalfa weevil, weeds and early first harvest in Oklahoma. *Canadian Entomologist* 124, 135–144.
Layug, D.V., Ohshima, M., Yokota, H., Nagatomo, T. and Ostrowski-Meissner, H.T. (1996) Effect of maturity stage on the protein and carotenoid yields of alfalfa leaf extract and press cake. *Grassland Science* 41, 287–293.
Leach, G.J. (1979) Regrowth characteristics of lucerne under different systems of grazing management. *Australian Journal of Agricultural Research* 30, 445–465.
Leach, G.J. (1983) Influence of rest interval, grazing duration and mowing on the growth, mineral content and utilisation of a lucerne pasture in a subtropical environment. *Journal of Agricultural Science, Cambridge* 101, 169–183.
Leath, K.T., Ervin, D.C. and Griffin, G.D. (1988) Diseases and nematodes. In: Hanson, A.A., Barnes, D.K. and Hill, R.R., Jr (eds) *Alfalfa and Alfalfa Improvement*. Agronomy Monograph No. 29, ASA/CSSA/SSSA, Madison, Wisconsin, pp 621–670.
Lemaire, G. and Allirand, J.M. (1993) Relation entre croissance et qualité de la luzerne: interaction génotype mode d'exploitation. *Fourrages* 134, 183–198.
Lemaire, G., Orillon, B., Gosse, G., Chartier, M. and Allirand, J.M. (1991) Nitrogen distribution within a lucerne canopy during regrowth – relation with light distribution. *Annals of Botany* 68, 483–488.
Lenssen, A.W., Sorensen, E.L., Posler, G.L. and Blodgett, S.L. (1991) Depression of forage quality of alfalfa leaves and stems by *Acyrthosiphon kondoi* (Homoptera, Aphididae). *Environmental Entomology* 20, 71–76.
Li, Y.G., Tanner, G.J., Delves, A.C. and Larkin, P.J. (1993) Assymetric somatic hybrid plants between *Medicago sativa* L. (alfalfa, lucerne) and *Onobrychis viciifolia* Scop. (sainfoin). *Theoretical and Applied Genetics* 87, 455–463.
Lloveras, J., Ferrán, X., Muñoz, F., Torres, L. and Alsina, J. (1996) Effect of autumn harvest management on alfalfa (*Medicago sativa* L.) production and quality in Mediterranean areas. In: Parente, G., Frame, J. and Orsi, S. (eds) *Grassland Science in Europe. Proceedings of the 16th General Meeting of the European Grassland Federation, Grado, Italy*, Vol. 1, ERSA, Gorizia, pp. 99–101.
Lorenzetti, F. (1993) Methods and procedures in variety constitution. In: Rotili, P. and Zannone, L. (eds) *The Future of Lucerne Biotechnology, Breeding and Variety Constitution: Proceedings of the X International Conference of EUCARPIA, Medicago spp. Group*. Istituto Sperimentale per le Colture Foraggere, Lodi, pp. 494–498.
Lory, J.A., Russelle, M.P. and Heichel, G.H. (1992) Quantification of symbiotically fixed nitrogen in soil surrounding alfalfa roots and nodules. *Agronomy Journal* 84, 1033–1040.
Lykouressis, D.P., Emmanouel, N.G. and Parentis, A.A. (1991) Studies on biology and population structure of three curculionid pests of lucerne in Greece. *Journal of Applied Entomology* 112, 317–320.
MacDowell, F.D.H. (1983) Effects of light intensity and CO_2 concentration on the kinetics of first month growth and nitrogen fixation of alfalfa. *Canadian Journal of Botany* 61, 731–740.
McKenzie, J.S., Pacquin, R. and Duke, S.H. (1988) Cold and heat tolerance. In: Hanson, A.A., Barnes, D.K. and Hill, R.R., Jr (eds) *Alfalfa and Alfalfa Improvement*. Agronomy Monograph No. 29, ASA/CSSA/SSSA, Madison, Wisconsin, pp. 259–302.

McKersie, B.D. (1997) Improving forage production systems using biotechnology. In McKersie, B.D. and Brown, D.C.W. (eds) *Biotechnology and the Improvement of Forage Legumes*. CAB International, Wallingford, pp. 3–24.

McKinney, G.T. (1974) Management of lucerne for sheep grazing on the southern tablelands of New South Wales. *Australian Journal of Experimental Agriculture and Animal Husbandry* 71, 726–734.

Mace, M.J. (1982) Grazing management in practice – North Island dairying. In: Wynn-Williams R.B. (ed.) *Lucerne for the 1980s*. Agronomy Society of New Zealand, pp. 91–96.

Major, D.J., Hanna, M.R. and Beasley, B.W. (1991) Photoperiod response characteristics of alfalfa (*Medicago sativa* L.) cultivars. *Canadian Journal of Plant Science* 71, 87–93.

Mangan, J.L., Harrison, F.A. and Vetter, R.L. (1991) Immunoreactive fraction 1 leaf protein and dry matter content during wilting and ensiling of ryegrass and alfalfa. *Journal of Dairy Science* 74, 2186–2199.

Manglitz, G.R. and Ratcliffe, R.H. (1988) Insects and mites. In: Hanson, A.A., Barnes, D.K. and Hill, R.R., Jr (eds) *Alfalfa and Alfalfa Improvement*. Agronomy Monograph 29, ASA/CSSA/SSSA, Madison, Wisconsin, pp. 671–704.

Marble, V.L., Shauner, C.A., Jr and Peterson, G. (1989) Effect of fall/winter harvest on alfalfa (*Medicago sativa* L.) productivity and persistence in a Mediterranean climate. In: *Proceedings of the XVI International Grassland Congress, Nice, France*. Association Française pour la Production Fourragère, Versailles, pp. 597–598.

Marten, G.C., Buxton, D.R. and Barnes, R.F. (1988) In: Hanson, A.A., Barnes, D.K. and Hill, R.R., Jr (eds) *Alfalfa and Alfalfa Improvement*. Agronomy Monograph No. 29, ASA/CSSA/SSSA, Madison, Wisconsin, pp. 463–491.

Massengall, M.A., Dobrenz, A.K., Brubaker, H.A. and Bard, A.E. (1971) Response of alfalfa (*Medicago* spp.) to light interruption of the dark period. *Crop Science* 11, 9–12.

Mayer, J.R., Richards, K.W. and Schaalje, G.B. (1991) Effect of plant density and herbicide application on alfalfa seed and weed yields. *Canadian Journal of Botany* 61, 731–740.

Merry, J.R., Beever, D.E. and Theodorou, M.K. (1989) Additives – their potential, for improving the nutritive value of silages. In: *Proceedings of Technology and Management Workshop, Truro, Nova Scotia, Canada*, pp. 44–73.

Michaud, R., Lehman, W.F. and Rumbaugh, M.D. (1988) World distribution and historical development. In: Hanson, A.A., Barnes, D.K. and Hill, R.R., Jr (eds) *Alfalfa and Alfalfa Improvement*. Agronomy Monograph No. 29, ASA/CSSA/SSSA, Madison, Wisconsin, pp. 25–91.

Miller, P.R. and Christie, B.R. (1991) Genetics of resistance to verticillium wilt in Vertus alfalfa. *Crop Science* 31, 1492–1495.

Minson, D.J. (1982) Effect of chemical and physical composition of herbage eaten upon intake. In: Hacker, J.B. (ed.) *Nutritional Limits to Animal Production from Pastures*. Commonwealth Agricultural Bureaux, Farnham Royal, pp. 167–182.

Mohapatra, S.S., Poole, R.J. and Dhinsa, R.S. (1988) Abscisic acid regulated gene expression in relation to freezing tolerance in alfalfa. *Plant Physiology* 87, 468–473.

Mooso, G.D. and Wedin, W.F. (1990) Yield dynamics of canopy components in alfalfa–grass mixtures. *Agronomy Journal* 82, 696–701.

Mtengeti, E.J., Wilman, D. and Moseley, G. (1996) Differences between twelve forage species in physical breakdown when eaten. *Journal of Agricultural Science, Cambridge* 126, 287–293.

Munns, D.N. (1969) Enzymic breakdown of pectin and acid-inhibition of the infection of *Medicago* roots by *Rhizobium. Plant and Soil* 30, 117–119.

Munns, D.N. (1970) Nodulation of *Medicago sativa* in solution culture. 5. Calcium and pH requirements during infection. *Plant and Soil* 32, 90–102.

Munns, D.N. and Fox, R.L. (1977) Comparative lime requirements of tropical and temperate legumes. *Plant and Soil* 46, 533–548.

Murata, Y., Lyama, J. and Honma, T. (1966) Studies on the photosynthesis of forage crops: the influence of soil moisture content on the photosynthesis and respiration of seedlings in various forage crops. *Proceedings of Crop Science Society of Japan* 34, 385–390. (Cited in Sheaffer *et al.*, 1988a.)

Nash, M.J. (1985) *Crop Conservation and Storage in Cool Temperate Climates.* Pergamon Press, Oxford, 286 pp.

Nelson, W.F. and Satter, L.D. (1992) Impact of stage of maturity and method of preservation of alfalfa on digestion in lactating dairy cows. *Journal of Dairy Science* 75, 1571–1580.

Nicholson, J.W.G., Charmley, E. and Bush, R.S. (1992) The effect of supplemental protein source on ammonia levels in rumen fluid and blood and intake of alfalfa silage by beef cattle. *Canadian Journal of Animal Science* 72, 853–862.

Olanya, O.M. and Campbell, C.L. (1991) Isolate characteristics and epidemic components of *Leptosphaerulina* leaf spots on alfalfa and white clover. *Phytopathology* 80, 1278–1282.

Ostrowski-Meissner, H.T., Pearson, C.J. and Shiels, L.M. (1993) Effect of seasonal variation in the fractionation pattern of lucerne selected for commercial crop processing in Australia. In: Baker, M.J. (ed.) *Proceedings of the XVIII International Grassland Congress, New Zealand and Australia,* Vol. II. New Zealand Grassland Association *et al.,* Palmerston North, pp. 1518–1520.

Pacquin, R. and Pelletier, G. (1980) Acclimatation naturelle de la lucerne (*Medicago media* Pers) au froid. 10 variations de la teneur en proline libre des familles et des collets. *Physiologie Végétal* 19, 103–117.

Pacquin, R., Berniercardon, M. and Castronguay, Y. (1987) Influence of soil moisture, temperature and length of feeding on alfalfa survival. *Canadian Journal of Plant Science* 67, 765–775.

Palmer, T.P. and Wynn-Williams, R.B. (1976) Relationships between density and yield of lucerne. *New Zealand Journal of Experimental Agriculture* 4, 71–77.

Panciera, M.T. and Fulkerson, R.S. (1981) Some responses of alfalfa–birdsfoot trefoil mixtures to seeding year management. *Canadian Journal of Plant Science* 61, 929–938.

Pennypacker, B.W., Leath, K.T. and Hill, R.R. (1991) Impact of drought stress on the expression of resistance to *Verticillium albo-atrum* in alfalfa. *Phytopathology* 81, 1014–1024.

Perry, M.C., McIntosh, M.S., Weibold, W.J. and Welterlen, M. (1987) Genetic analysis of cold hardiness and dormancy in alfalfa. *Genome* 29, 144–149.

Peters, E.J. and Linscott, D.L. (1988) Weeds and weed control. In: Hanson, A.A., Barnes, D.K. and Hill, R.R., Jr (eds) *Alfalfa and Alfalfa Improvement.* Agronomy Monograph No. 29, ASA/CSSA/SSSA, Madison, Wisconsin, pp. 705–735.

Peterson, P.R., Sheaffer, C.C. and Hall, M.W. (1992) Drought effects on perennial forage legume yield and quality. *Agronomy Journal* 84, 774–779.

Peterson, R.K.D., Danielson, S.D. and Higley, L.G. (1993) Yield response of alfalfa to simulated alfalfa weevil injury and development of economic injury levels. *Agronomy Journal* 85, 595-601.
Peyraud, J.L. and Delaby, L. (1994) Use of high quality dried lucerne in diets for dairy cows. *Productions Animales* 7, 125-134.
Phillips, D.A., Wery, J., Joseph, C.M., Jones, A.D. and Teuber, L.R. (1995) Release of flavonoids and betaines from seeds of seven *Medicago* species. *Crop Science* 35, 805-808.
Piano, E., Valentine, P., Percetti, L. and Romani, M. (1996) Evaluation of a lucerne germplasm collection in relation to traits conferrring grazing tolerance. *Euphytica* 89, 279-288.
Pijnenborg, J.W.M., Lie, T.A. and Zehnder, A.J.B. (1991) Nodulation of lucerne (*Medicago sativa*) in an acid soil – effects of inoculation size and lime pelleting. *Plant and Soil* 131, 1-10.
Pocknee, B.R. and Campling, R.C. (1981) Voluntary intake and digestibility of lucerne silages by steers. *Grass and Forage Science* 36, 141-142.
Prosperi, J.M., Angevain, M., Bonnin, I., Chaulet, E., Genier, G. and Jenczewski, E., Olivieri, I. and Ronfort, J. (1996) Genetic diversity, preservation and use of genetic resources of Mediterranean legumes: alfalfa and medics. In: Genier, G. and Prosperi, J.M. (eds) *The Genus Medicago in the Mediterranean Region: Current Situation and Prospects in Research.* Cahiers Options Méditerranéennes, Vol. 18, FAO/CIHEAM, Zaragoza, pp. 71-90.
Quigley, E.H. and Jung, G.A. (1984) Alfalfa and corn response to sulphur fertilization on three Pennsylvania soils. *Communications in Soil Science and Plant Analysis* 15, 213-226.
Radeva, V., Ilieva, A., Alanassov, A., Denchev, P. and Dragiyska, R. (1993) Study of develoment and adaptation abilities of regenerant lucerne plants obtained by the method of somatic embryogenesis under conditions of physiological drought. In: Rotili, P. and Zannone, L. (eds) *The Future of Lucerne Biotechnology, Breeding and Variety Constitution. Proceedings of the X International Conference of EUCARPIA, Medicago spp. Group.* Istituto Sperimentale per le Colture Foraggere, Lodi, pp. 223-229.
Reeve, J.L. and Sharkey, M.J. (1980) Effect of stocking rate, time of lambing and inclusion of lucerne on prime lamb production in north eastern Victoria. *Australian Journal of Experimental Agriculture* 20, 637-653.
Reid, D. (1987) A study of the potential of lucerne as a silage crop in a high-rainfall area. *Grass and Forage Science* 42, 185-190.
Richard, C. and Martin, J.G. (1993) The influence of *Fusarium oxysporum* F sp. *medicaginis* on total soluble sugar concentration of infected alfalfa roots. *Canadian Journal of Plant Science* 73, 647-649.
Roberts, S.J., Zavaleta, L.R., Grube, A.H., Armbrust, E.J. and Pansch, R.D. (1987) Evaluation of a pest control technique – fall spray control of alfalfa weevil (Coleoptera, Curculionidae) in alfalfa fields. *Journal of Economic Entomology* 80, 859-866.
Rotili, P. (1993) Methods and procedures in variety constitution. In: Rotili, P. and Zannone, L. (eds) *The Future of Lucerne Biotechnology, Breeding and Variety Constitution. Proceedings of the X International Conference of the EUCARPIA, Medicago spp. Group.* Istituto Sperimentale per le Colture Foraggere, Lodi, pp. 499-508.
Rotili, P., Berardo., N., Gnocchi, G., Pecetti, L., Piano, E. and Scotti, C. (1996) Research

Activity on *Medicago* spp. at Istituto Sperimentale per la Colture Foraggere Lodi, Italy. In: Genier, G. and Prosperi, J.M. (eds) *The Genus* Medicago *in the Mediterranean Region: Current Situation and Prospects in Research*. Cahiers Options Méditerranéennes, Vol. 18, FAO/CIHEAM, Zaragoza, pp. 11–22.

Rumbaugh, M.D. and Pendery, B.M. (1990) Germination salt resistance of alfalfa (*Medicago sativa* L.) in relation to subspecies and centres of diversity. *Plant and Soil* 124, 47–51.

Russelle, M.P. and Sheaffer, C.C. (1986) Use of the diagnosis and recommendation integrated system with alfalfa. *Agronomy Journal* 78, 557–560.

Russelle, M.P., Allan, D.L. and Gourley, C.J.P. (1994) Direct assessment of symbiotically fixed nitrogen in the rhizosphere of alfalfa. *Plant and Soil* 159, 233–243.

Sanchez, O.L. and Campling, R.C. (1982) Comparison of milk production from cows grazing lucerne and ryegrass pastures. *Grass and Forage Science* 37, 172–173.

Scales, G.H. and Barry, T.N. (1975) Winter performance of beef weaners fed untreated and formaldehyde-treated wilted lucerne silages and hay. *Proceedings of the New Zealand Society of Animal Production* 35, 160–167.

Schertz, D.L. and Miller, D.A. (1972) Nitrate N accumulation in soil profile under alfalfa. *Agronomy Journal* 64, 660–664.

Schmidt, L. (1993) Use of plant height for determining the nutritive value, yield and optimal use span of lucerne. In: Baker, M.J. (ed.) *Proceedings of the XVII International Grassland Congress, New Zealand and Australia*, vol. I. New Zealand Grassland Association *et al.*, Palmerston North, pp. 894–895.

Schmitt, M.A., Sheaffer, C.C. and Randall, G.W. (1994) Manure and fertiliser effects on alfalfa plant nitrogen and soil nitrogen. *Journal of Production Agriculture* 7, 104–109.

Schwab, P.M., Barnes, D.K., Sheaffer, C.C. and Li, P.H. (1996) Factors affecting a laboratory evaluation of alfalfa cold tolerance. *Crop Science* 36, 318–324.

Scotti, C. (1993) Pod fertility at different levels of selfing in lucerne plants derived from *in vitro* culture. In: Rotili, P and Zannone, L. (eds) *The Future of Lucerne Biotechnology, Breeding and Variety Constitution. Proceedings of the X International Conference of the EUCARPIA,* Medicago *spp. Group*. Istituto Sperimentale per le Colture Foraggere, Lodi, pp. 230–235.

Sheaffer, C.C., Tanner, C.B. and Kirkham, M.B. (1988a) Alfalfa water relations and irrigation. In: Hanson, A.A., Barnes, D.K. and Hill, R.R., Jr (eds) *Alfalfa and Alfalfa Improvement*. Agronomy Monograph No. 29, ASA/CSSA/SSSA, Madison, Wisconsin, pp. 373–409.

Sheaffer, C.C., Lacefield, G.D. and Marble, V.I. (1988b) Cutting schedules and stands. In: Hanson, A.A., Barnes, D.K. and Hill, R.R., Jr (eds) *Alfalfa and Alfalfa Improvement*. Agronomy Monograph No. 29, ASA/CSSA/SSSA, Madison, Wisconsin, pp. 411–437.

Sheehy, J.E. and Popple, S.C. (1981) Photosynthesis, water relations, temperature and canopy structure as factors influencing the growth of sainfoin (*Onobrychis viciifolia* Scop.) and lucerne (*Medicago sativa* L.). *Annals of Botany* 48, 113–128.

Sheehy, J.E., Minchin, F.R. and McNeill, A. (1984) Physiological principles governing the growth and development of lucerne, sainfoin and red clover. In: Thomson, D.J. (ed.) *Forage Legumes*. Occasional Symposium No. 16, British Grassland Society, Hurley, pp. 112–125.

Sheesley, R., Grimes, D.W., McClennan, W.D., Summers, C.G. and Marble, V. (1974) Influence of wheel traffic on yield and stand longevity of alfalfa. *California Agriculture* 28, 6–8.

Skinner, D.Z., Fritz, J.O. and Klocke, L.L. (1994) Protein degradability in a diverse array of alfalfa germplasm sources. *Crop Science* 34, 1396–1399.

Smith, D. and Nelson, C.J. (1967) Growth of birdsfoot trefoil and alfalfa. 1. Responses to height and frequency of cutting. *Crop Science* 7, 130–133.

Smith, E.G., Acharya, S.N. and Huang, H.C. (1995) Economics of growing *Verticillium* wilt resistant and adapted alfalfa cultivars in western Canada. *Agronomy Journal* 87, 1206–1210.

Smith, R.G. and Stiefel, W. (1978) Pasture and lucerne research on sand country. *Proceedings of the New Zealand Grassland Association* 39, 61–69.

Spedding, C.R.W. and Diekmahns, E.C. (1972) *Grasses and Legumes in British Agriculture*. Bulletin No. 49, Commonwealth Bureau of Pastures and Field Crops, Farnham Royal, 511 pp.

Steiner, J.J., Hutmacher, R.B., Gamble, S.D., Ayars, J.E. and Vail, S.S. (1992) Alfalfa seed water management. 1. Crop reproductive development and seed yield. *Crop Science* 32, 476–481.

Stone, J.E., Marx, D.B. and Dobrenz, A.K. (1979) Interaction of sodium chloride and temperature on germination of two alfalfa cultivars. *Agronomy Journal* 71, 425–427.

Sulc, R.M. and Albrecht, K.A. (1996) Alfalfa establishment with diverse annual ryegrass cultivars. *Agronomy Journal* 88, 442–447.

Summers, C.G. and Gilchrist, D.G. (1991) Temporal changes in forage alfalfa associated with insect and disease stress. *Journal of Economic Entomology* 84, 1353–1363.

Sund, J.M., Barrington, G.P. and Scholl, J.M. (1966) Methods and depth of sowing forage grasses and legumes. In: Hill, A.G.G. (ed.) *Proceedings of the X International Grassland Congress, Helsinki, Finland*. Finnish Grassland Association, Helsinki, pp. 319–323.

Ta, T.C. and Faris, M.A. (1987) Effects of alfalfa proportions and clipping frequencies on timothy alfalfa mixtures. 2. Nitrogen fixation and transfer. *Agronomy Journal* 79, 820–824.

Ta, T.C., MacDowall, F.D. and Faris, M.A. (1990) Utilization of carbon and nitrogenous reserves of alfalfa roots in supporting N_2-fixation and shoot regrowth. *Plant and Soil* 127, 231–236.

Tabe, L.M., Wardley-Richardson, T., Ceriotti, A., Arayan, A., McNabb, W., Moore, A. and Higgins, T.J.V. (1995) A biotechnological approach to improving the nutritive value of alfalfa. *Journal of Animal Science* 73, 2752–2759.

Talamucci, P. (1994) Lucerne's role in farming systems, technical itineraries and managements for different uses in diverse physical and socio-economic environments. *FAO/REUR Technical Series* 36, pp. 6–17.

Tan, G.Y. and Tan, W.K. (1986) Interaction between alfalfa cultivars and rhizobium strains for nitrogen fixation. *Theoretical and Applied Genetics* 71, 724–729.

Tava, A., Forti, D. and Odoardi, M. (1993) Alfalfa saponins: isolation, chemical characterisation and biological activity against insects. In: Rotili, P. and Zannone, L. (eds) *The Future of Lucerne – Biotechnology, Breeding and Variety Constitution. Proceedings of the X International Conference of the EUCARPIA, Medicago spp. Group, Lodi, Italy.* Istituto Sperimentale per le Colture Foraggere, Lodi, pp. 283–288.

Tesar, M.B. and Jackobs, J.A. (1972) Establishing the stand. In: Hanson, C.H. (ed.) *Alfalfa Science and Technology*. American Society of Agronomy, Madison, Wisconsin, pp. 415–445.

Teuber, L.R. and Phillips, D.A. (1988) Influences of selection method and N environment on breeding alfalfa for increased forage yield and quality. *Crop Science* 28, 599–604.

Thomson, D.J., Waldo, J.R., Goering, H.K. and Tyrell, H.F. (1991) Voluntary intake, growth rate and tissue retention by Holstein steers fed formaldehyde treated and formic treated alfalfa and orchardgrass silages. *Journal of Animal Science* 69, 4644–4659.

Townsend, M.S., Henning, J.A. and Currier, C.G. (1994) The ALFALFA CATALOG software package. *Agronomy Journal* 86, 337–339.

Twary, S.N. and Heichel, G.H. (1991) Carbon cost of dinitrogen fixation associated with dry matter accummulation in alfalfa. *Crop Science* 31, 985–992.

Tyrrell, H.F., Thomson, D.J., Waldo, D.R., Goering, H.K. and Haaland, G.L. (1992) Utilisation of energy and nitrogen by yearling Holstein cattle fed direct cut alfalfa or orchardgrass ensiled with formic acid plus formaldehyde. *Journal of Animal Science* 69, 3163–3177.

Vance, C.P. (1997) Nitrogen fixation capacity. In: McKersie, B.D. and Brown, D.C.W. (eds) *Biotechnology and the Improvement of Forage Legumes*. CAB International, Wallingford, pp. 375–407.

Vance, C.P., Heichel, G.H. and Phillips, D.A. (1988) Nodulation and symbiotic dinitrogen fixation. In: Hanson, A.A., Barnes, D.K. and Hill, R.R., Jr (eds) *Alfalfa and Alfalfa Improvement*. Agronomy Monograph No. 29, ASA/CSSA/SSSA, Madison, Wisconsin, pp. 229–257.

van Heerden, J.M. (1993) Production of dryland lucerne in the western and southern Cape regions of South Africa. In: Baker, M.J. (ed.) *Proceedings of the XVII International Grassland Congress, New Zealand and Australia*, Vol. I. New Zealand Grassland Association, et al., Palmerston North, pp. 773–774.

Van Keuren, R.W. and Matches, A.G. (1988) Pasture production and utilisation. In: Hanson, A.A., Barnes, D.K. and Hill, R.R., Jr (eds) *Alfalfa and Alfalfa Improvement*. Agronomy Monograph No. 29, ASA/CSSA/SSSA, Madison, Wisconsin, pp. 515–538.

van Wijk, A.J.P. and Rehuel, D. (1991) Achievements in fodder crops breeding in maritime Europe. In: den Nijs, A.P.M. and Elgersman, A. (eds) *Proceedings 16th Meeting of the Fodder Crop Section, EUCARPIA*. Pudoc, Wageningen, pp. 13–18.

Vanzant, E.S. and Cochran, R.C. (1994) Performance and forage utilisation by beef cattle receiving increasing amounts of alfalfa hay as a supplement to low quality, tall grass-prairie forage. *Journal of Animal Science* 72, 1059–1067.

Veronesi, F., Mariani, A. and Tavoletti, S. (1993) Breeding for lucerne tolerant to frequent cutting: results and perspectives. In: Rotili, P. and Zannone, L. *The Future of Lucerne Biotechnology, Breeding and Variety Constitution. Proceedings of the X International Conference of EUCARPIA, Medicago spp. Group*. Istituto Sperimentale per le Colture Foraggere, Lodi, pp. 289–294.

Vezina, L.P. and Nadeau, P. (1991) The combined effects of rhizobial inoculation and nitrogen fertilization on growth and cold acclimation of alfalfa (*Medicago sativa* cv Saranac). *Annals of Botany* 68, 359–363.

Viands, D.R., Barnes, D.K. and Heichel, B.H. (1981). *Nitrogen Fixation in Alfalfa: Response to Bidirectional Selection for Associated Characteristics*. USDA Technical Bulletin 1643, US Government Printing Office, Washington, DC.

Volenec, J.J. (1985) Leaf area expansion and shoot elongation of diverse alfalfa germplasm. *Crop Science* 25, 822–827.

Vough, L.E. and Marten, G.C. (1971) Influence of soil moisture and ambient temperature on yield and quality of alfalfa forage. *Agronomy Journal* 63, 40–42.

Waddington, J. (1992) A comparison of drills for direct seeding alfalfa into established grassland. *Journal of Range Management* 45, 483–487.

Waghorn, G.C., Shelton, I.D. and Thomas, V.J. (1989) Particle breakdown and rumen digestion of fresh ryegrass (*Lolium perenne* L) and lucerne (*Medicago sativa* L.) fed to cows during a restricted feeding period. *British Journal of Nutrition* 61, 409–423.

Walley, F.L., Tomm, G.O., Matus, A., Slinkard, A.E. and Vankessell, C. (1996) Allocation and cycling of nitrogen in an alfalfa–bromegrass sward. *Agronomy Journal* 88, 834–843.

White, J.G.H. (1982) Lucerne grazing management for the 80s. In: Wynn-Williams, R.B. (ed.) *Lucerne for the 80s*. Special Publication No. 1, Agronomy Society of New Zealand, pp. 111–114.

Wilding, M.D. and Smith, D. (1960) Free amino acids in alfalfa as related to coldhardiness. *Plant Physiology* 35, 726–732.

Wilman, D. (1977) The effect of grazing compared with cutting at different frequencies on a lucerne–cocksfoot ley. *Journal of Agricultural Science, Cambridge* 88, 483–492.

Wilson, J.R., Deinum, B. and Engels, F.M. (1991) Temperature effects on anatomy and digestibility of leaf and stem of tropical and temperate forage species. *Netherlands Journal of Agricultural Science* 39, 31–48.

Witty, J.F., Minchin, F.R. and Sheehy, J.E. (1983) Carbon costs of nitrogenase activity in legume root nodules determined using acetylene and oxygen. *Journal of Experimental Botany* 34, 951–963.

Woodward, F.I. and Sheehy, J.E. (1979) Microclimate, photosynthesis and growth of lucerne (*Medicago sativa* L.) 2. Canopy structure and growth. *Annals of Botany* 44, 709–719.

Wynn-Williams, R.B. (1982) Lucerne establishment – conventional. In: Wynn-Williams, R.B. (ed.) *Lucerne for the 80s*. Special Publication No. 1, Agronomy Society of New Zealand, pp. 11–20.

Wynn-Williams, R.B., Rea, M.B., Purves, R.G. and Hawthorne, B.T. (1991) Influence of winter treading on lucerne growth and survival. *New Zealand Journal of Agricultural Research* 34, 271–275.

Red Clover 4

Introduction

Red clover (*Trifolium pratense* L.) is one of the most important herbage legumes in temperate agriculture, particularly in Europe and North America. The 4.5 Mha in the USA make it the second most important legume after lucerne, although it has declined in recent years from a previous 5–6 Mha (Taylor and Smith, 1995). In some regions, such as Scandinavia, it is the most widely used forage legume. It is a short-lived perennial species, highly productive for 2 and, sometimes, 3 years. The upright growth habit of red clover makes it eminently suitable for hay or silage cropping, but it is also a useful component of grazed swards, where it has a high acceptability to stock, except if allowed to become stemmy with advancing maturity.

Originating in south-eastern Europe and Asia Minor (Fergus and Hollowell, 1960; Smith *et al.*, 1985), red clover was cultivated in southern Europe in the third and fourth centuries AD and introduced into Spain in the sixteenth century and, through The Netherlands and German states, into England in the midseventeenth century (Evans, 1976). Settlement in the Americas and Australasia by Europeans resulted in widespread use and distribution in these regions. In North America, it is grown in the humid north-east and in the Pacific north-west of the USA under irrigation and is used as an annual in south-east USA (Taylor and Smith, 1995).

Historically, its success has been due to its adaptation to a wide range of soil and environmental conditions, its molecular nitrogen (N_2)-fixing ability and its high nutritive value for ruminants. It was also valued as a break crop in arable farming, *inter alia* because of the residual nitrogen (N) benefit to subsequent crops in the rotation and as a source of organic matter (OM).

Despite these virtues, red-clover usage has declined in some European countries. The factors that have contributed to this decline include increased reliance upon fertilizer N, simplification of modern seed mixtures, continuous instead of rotational arable cropping, ready accessibility of protein concentrates and variable performance of red clover in swards. Nevertheless, the recent growing interest in environmentally friendly systems of animal production, with the grassland being less intensively fertilized with N than hitherto, is leading to reappraisal of red clover, as a forage plant generally (Frame, 1990) or as a component of species-rich 'wild-flower' mixtures for extensive grassland systems (Frame et al., 1994).

There is also renewed interest in its role as a green manure or 'ploughdown' crop, because of its low-cost seed and rapid early growth (Ten Holte and Van Keulen, 1988; Christie et al., 1992), and as a break crop in organic arable-farming systems (Lampkin, 1990). Novelty uses, in common with white clover, include the possibility of making a special wine from the flowerheads and the development of 'lucky' four-leaved cultivars as garden or pot plants.

Plant Development and Physiology

Germination

After sowing, the first indication of germination is the appearance below the soil surface of the radicle, which will eventually develop into a tap root. Above the surface, the first evidence is the emergence of two oval cotyledons, followed by a single unifoliate leaf 5–10 days after germination commences; thereafter, the leaves which develop are trifoliate. Germination is restricted at solute pressures of -0.8 MPa (Hegarty and Ross, 1979). Low temperatures also restrict germination in red clover, imbibition rate being faster at 25°C than at 7°C.

Mature red-clover seeds often contain a high proportion of hard seeds. This arises from the ripening phase, following fertilization of the florets by bumble-bees, when the moisture content of the seed declines to about 25% and then, more slowly, to 14%, as the impermeable layer in the epidermis develops. Hard-seededness in red clover is inversely related to temperature during seed set, although the relationship varies between cultivars (Puri and Laidlaw, 1984).

Experimentally, the seed-coat impermeability can be broken by high temperatures of over 100°C for a few minutes or by soaking in concentrated sulphuric acid. Radiofrequency electrical treatment is also effective (Nelson et al., 1976). In practice, it can be broken by physical impact during harvesting or gradual abrasion after sowing by 'weathering' from fluctuating conditions in the soil. The hard seeds in sown mixtures represent a reservoir from which some seeds germinate over a period of time, thus assisting the stability of the clover-plant population in the sward.

Seedling growth

Subsequent to germination, the young seedling is capable of growth at water potentials as low as -1.5 MPa (Hegarty and Ross, 1979). The most powerful sinks for assimilate in the young red-clover plant, based on the proportion of assimilate claimed, are axillary branches, followed by roots, with less than 10% being apportioned to the terminal apex of the main axis (Ryle et al., 1981).

During seedling development, the young tap root produces branches at short regular intervals along its length, and root hairs of about 0.2 mm are also formed. The root : shoot ratio remains more or less constant with development through the seedling stage.

Red-clover seedlings grow at 7°C or above but become chlorotic and die at above 40°C (Fergus and Hollowell, 1960). Young seedlings are susceptible to freezing temperatures, but cold-hardinesss increases with seedling age up to the four- to six-leaf stage (Kilpatrick et al., 1966). After freezing at temperatures of $-9°C$, survival of plants improved when they were over 12 weeks old (Choo and Suzuki, 1983). Net growth rate of seedlings reaches a maximum at 520 Wm^{-2} (about 2200 µmol m^{-2} s^{-1}) PAR, although, as discussed below, canopy photosynthesis increases beyond this level.

Plant development

The main axis of red clover above ground remains in a rosette form, as the branch buds, borne in the axils of cotyledons, and true leaves expand and produce branches; the internodes of the branches remain short during establishment but are subsequently capable of elongation. An accumulation of axillary buds formed at the base of the shoot produces a crown at the soil surface (Fig. 4.1). During primary growth, the basal buds produce upright stems, but later the plant overwinters as a rosette with a central crown. Late-flowering types develop more stems than early-heading types. The plant has a deep tap root, which confers a moderate to high degree of resistance to drought, but nevertheless it is less deep rooting than lucerne or sainfoin. Plants that survive for 3 or 4 years have mainly adventitious and lateral roots growing from the crown (Spedding and Diekmahns, 1972; Montpetit and Coulman, 1991a).

Inflorescences, which are ovoid racemes of pink or purple florets, are axillary. Pollination of the florets is by insects, and the pods that result from fertilization contain either one or two seeds, varying in colour but generally yellow, purple or brown. The 1000-seed weight is 1.8 g for diploids and 3.4 g for tetraploids, brown seeds in a seed lot generally being lighter than purple seeds.

Temperature

The optimum temperature for red clover growth is 20–25°C but it grows between 7°C and up to 40°C (Ruiz, 1973). Increasing temperature above the optimum to more than 30°C adversely affects roots more than shoots.

Total available carbohydrate (TAC) decreases in response to elevated temperatures, respiration increasing with temperature; following prolonged periods of high temperatures, usually associated with drought, plants are weakened and subsequent winter survival may be compromised. There may also be an interaction between plants weakened by low TAC and soil-borne root-rotting organisms (Fergus and Hollowell, 1960).

(a)

Fig. 4.1. Red clover. (a) Plant at stem elongation stage. (b) (i) inflorescence, (ii) seed and (iii) seedling.

Fig. 4.1 continued

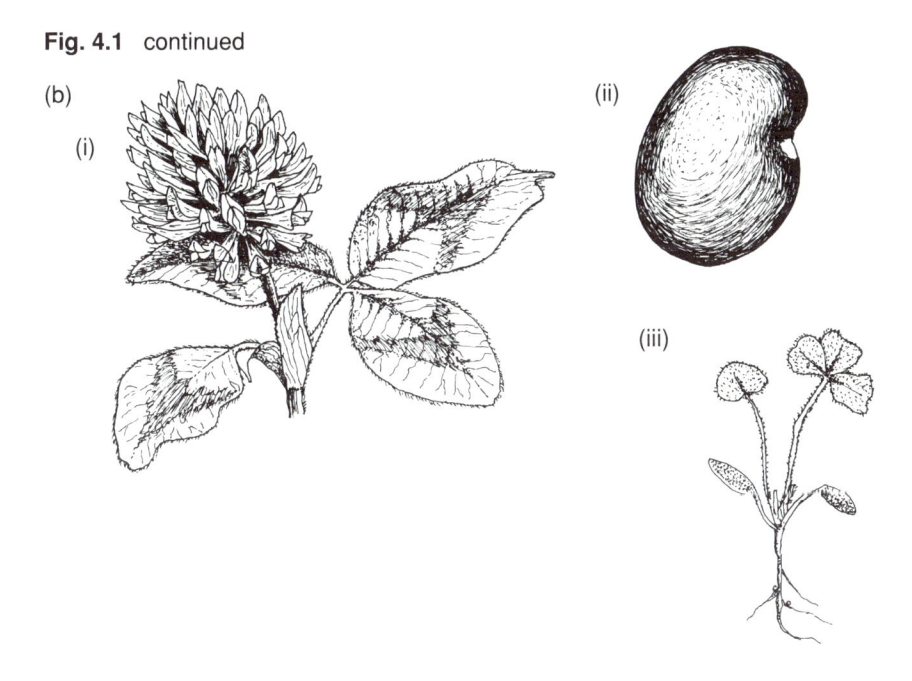

(b)

(i)

(ii)

(iii)

Flowering in red clover is initiated without prior exposure to low temperatures, i.e. it does not have a vernalization requirement (Taylor and Quesenberry, 1996), but, in some cultivars and ecotypes, low temperatures (2–10°C) may advance flowering, e.g. by exposing germinating seeds to 3°C (Fejer, 1960). Generally, however, increasing temperatures at the time of floral initiation advance the flower appearance. Temperature can interact with heading type; low temperatures during the early stages of development result in early-heading types achieving higher rates of photosynthesis and dark respiration than late-flowering types when both are subsequently grown under optimum growing conditions (Chugunova and Romanova, 1988).

Ability to withstand subzero temperatures improves with age, and plants at growth stages below the six-leaf stage are particularly vulnerable (Kilpatrick *et al.*, 1966). Low temperatures – slightly above 0°C – induce hardening, sometimes in conjunction with shortening days. Factors that reduce cold hardening include insect or disease damage, excessive accumulation of herbage entering the winter or mineral (calcium (Ca), potassium (K) or Sulphur (S)) deficiency. Cold-hardiness is also associated with physiological state, genotypes that have flowered having lower hardiness than those that have remained vegetative (Christie and Choo, 1991). However, flowering is not detrimental if there is sufficient time afterwards for strong rosettes to develop before winter.

As total non-structural carbohydrate (TNC) concentration in the roots can decline over winter by 50% or more of the autumn level, cold-hardiness is associated with high TNC concentrations at the beginning of winter. Low utilization of TNC during winter is also important for survival (Kendall et al., 1962). Levels of TNC decline with plant age, even in winter-hardy cultivars (Smurygin and Brazhnikova, 1989). Hardened roots relative to unhardened roots have a higher unfreezable-to-freezable water ratio, a higher osmotic pressure, a slightly higher pH and higher contents of total sugars, dextrins, starch and non-protein N, but a lower moisture content and lower specific conductance. Considering the timing of autumn defoliation and its effect on plant persistency and subsequent production, the best management is that which allows adequate accumulation of carbohydrate reserves prior to winter.

Rhizobium strains isolated from red clover growing at far-northern latitudes in Finland had higher N_2-fixation potential than more southerly strains when grown with red clover under simulated cold climatic conditions in a phytotron, indicating adaptation to temperature conditions at source (Lipsanen and Lindstrom, 1986).

Light

Saturation of leaf photosynthesis takes place at photosynthetically active radiation (PAR) of about 1500 µmol m^{-2} s^{-1} (Hesketh and Moss, 1963), with the compensation point, when photosynthesis only balances respiration, at about 140 µmol m^{-2} s^{-1}. Saturation for canopy photosynthesis has been shown to be in excess of 2400 µmol m^{-2} s^{-1}. The photosynthetic efficiency of the sward canopy is dependent on the stage of development. The critical leaf-area index (LAI), i.e. the area of leaf per unit area of ground that intercepts 95% of incident PAR, ranges from 3.0 to 3.5 for vegetative swards in the field or microswards under glasshouse conditions to 5.5 for reproductive swards in the field; presumably, higher irradiance in the latter allows more leaves to photosynthesize above compensation point (Nösberger and Joggi, 1981).

Irradiance levels during leaf development can influence the potential photosynthesis and longevity of individual leaves. The photosynthesis rate of 7-day-old leaves that develop at about 800 µmol m^{-2} s^{-1} is 60% greater than that of those leaves developing under low irradiance (160 µmol m^{-2} s^{-1}) levels, but the latter leaves can sustain a higher rate of photosynthesis after about 22 days of age (Joggi *et al.*, 1983). Irradiance levels also influence the morphology of red clover and imposing shade results in an increase in mean leaf area, petiole length and leaf : stem ratio and a decrease in root : shoot ratio (Bowley *et al.*, 1987). A reduced root system typical of shaded (e.g. undersown) red clover may militate against plant survival, particularly when the nurse crop is harvested and clover plants have to cope with the sudden increase in transpirational load. None the less, red clover is more tolerant of low light intensity than other forage legumes (Taylor and Smith, 1995).

Elongation of the stem and initiation of flowers require long days, flower initiation usually requiring longer days than stem elongation (Taylor and Quesenberry, 1996). The minimum day length required to initiate flowering varies according to maturity group of the cultivar. Early-flowering types require up to 14 h irradiance per day, whereas late-heading types require a longer photoperiod. Late-heading types also have a longer juvenile phase prior to being receptive to long days than early-heading types, the difference being as much as ten leaf-appearance intervals, e.g. 13 as opposed to three leaves (Jones, 1974). Day length is also involved in the induction of cold-hardiness in red clover, with shortening days having a positive effect (Umaerus, 1963).

Moisture

Estimates of water use in red-clover crops are in the range 400–600 kg water kg^{-1} dry matter (DM) (Bowley *et al.*, 1984). This efficient use of soil moisture, compared with many plant species, is a consequence of red clover's ability to tap water deep in the soil profile, potentially as deep as 3 m, although tap roots would normally extend to about 1 m in depth. A water potential of −0.1 MPa can induce the first symptoms of moisture stress in red clover (Dougherty, 1972) and wilting has occurred at a water potential of −0.8 to −1.2 MPa (Hardie and Leyton, 1981); infection with the vesicular arbuscular mycorrhizal (VAM) fungus *Glomus mossae* reduces the water potential at which wilting is induced to the range −1.8 to −2.4 MPa. However, water uptake does not increase due to the presence of mycorrhiza *per se* but, rather, may be due to an associated increase in phosphorus (P) uptake (Fitter, 1988).

Whereas irrigation of herbage crops of red clover is not usually economically feasible, seed crops grown in low-rainfall areas require irrigation. For example, in the drier western states in the USA, 90–140 cm of irrigation water may be required during the growing season. Not only will development of the inflorescence and seed set be adversely affected by low moisture availability but winter survival may also be jeopardized, due to freeze desiccation. When required, irrigation water is applied to red clover seed crops in furrows (Rincker and Rampton, 1985), the optimum time for irrigation being at peak flowering (Oliva *et al.*, 1994a, b).

Nitrogen and minerals

Seedlings depending on N_2 fixation have slower shoot-growth rates than those receiving high rates of mineral N (Maag and Nösberger, 1980), so fertilizer N applied to the seed bed advances clover establishment, especially where there is a shortage of available N in the soil. Abundant soil mineral N results in increased shoot growth. Nitrogen content of red-clover herbage is often in the range 35–45 g kg^{-1} DM, content being lower in mature than in vegetative crops.

The cation exchange capacity (CEC) in red clover roots ranges from 130 to 320 mmol kg^{-1} (Mouat and Anderson, 1974) and, since this range is generally higher than the CEC of grass roots, red clover has a competitive advantage over companion grass for divalent ions, such as Ca^{2+} or magnesium (Mg^{2+}), although not for monovalent ions, such as K^+ or sodium (Na^+). For the same reason, tetraploid red clover has a competitive advantage over diploid clover, if sown together. Also, the root systems of tetraploids are greater than those of diploids and this may result in more effective absorption of mineral elements (Chloupek, 1976); the larger root system will also result in enhanced drought resistance.

During the first 40 days or so after germination, P uptake by red clover is low, due to the young seedlings relying heavily on seed reserves. Because of a high requirement for P in nodules actively fixing N_2, the P content in the roots of established red clover is higher than in roots of some non-N_2-fixing species, e.g. Italian ryegrass (Kemp and Blair,1991). Compared with perennial ryegrass, red clover has the capacity to acidify its rhizosphere, increasing P availability around its roots (Gerke et al., 1995).

Maximum yields are achieved with herbage P concentrations of 2–4 g kg^{-1} DM (Spedding and Diekmahns, 1972), with critical concentration for P in the range 2.5–2.7 g kg^{-1} (Wright et al., 1984). Infection with VAM fungi can benefit P uptake by red clover (Hardie and Leyton, 1981), and the more P is limiting the greater may be the response to VAM fungal infection. Although herbage P concentration usually increases due to VAM infection, increased red-clover growth may not necessarily result, since other factors may be limiting growth (Sainz and Arines, 1988a, b).

The critical level for K in red clover is in the range 18–22.5 g kg^{-1} DM (Kelling and Matocha, 1990). Due to its ability to produce an abundance of lateral and adventitious roots near the soil surface, in comparison with lucerne, for example, red clover seems to be more able to absorb K (Smith and Smith, 1977).

The optimum pH for red-clover growth is 6.0–7.5, but the critical pH for the growth of associated *Rhizobium* is 4.7–4.9, with nodulation usually being totally inhibited at pH 4.2 (Rice et al., 1977). Associated with low pH, the availability of manganese (Mn) and aluminium (Al) increases and this can result in toxicity to the roots and subsequent plant injury. Aluminium impedes root penetration, may increase red clover's susceptibility to drought and, at levels of 40 μmol l^{-1} or more, adversely affects red-clover growth in general (MacLeod and Jackson, 1965). The response of red clover to Al varies among cultivars, so there is scope to breed for increased tolerance (Baligar et al., 1987; Campbell et al., 1990). Infection by VAM fungi may depress Mn uptake by red clover, especially in soil with high Mn levels (Arines et al., 1990), Mn concentrations in the shoots and roots of infected red clover being less than 50% and 25% of those in uninfected shoots and roots, respectively.

Defoliation

Red-clover production per annum and plant persistence are inversely related to defoliation frequency (Sheldrick et al., 1986). Defoliation causes a reduction in root-elongation rate (Evans, 1973), the shedding of nodules and roots – although not as pronounced as in white clover (Butler et al., 1959) – and a reduction in the specific N_2 fixation of the nodules that remain. Three-weekly defoliation reduces the number of live roots, compared with 6-weekly defoliation (Planquaert and Raphalen, 1973). During regrowth, shoot and root grow in proportion and the root : shoot ratio remains fairly constant.

In the early stages of regrowth, TAC is remobilized from the tap roots. As the LAI increases, TAC concentration again increases in the roots. Net assimilation rate (NAR) decreases with length of regrowth period, as LAI increases, as does the proportion of old leaves and also plant respiration (Joggi *et al.*, 1983).

In the early stages of red-clover growth or regrowth, 60–70% of carbon (C) initially remains in leaves. During this time, N_2 fixation is normally high, roots having a high demand for assimilate from the leaves of the vegetative crown branches. However, as internodes on reproductive shoots elongate and branching increases, the proportion of leaves associated with the crown branches declines, as does allocation to the roots (Fernandez and Warembourg, 1987). During this period of decline in availability of C to the roots, N_2 fixation is adversely affected. The distribution of assimilate from different leaves varies, depending on the leaf's relative age. The developing branch in the axis of a mature leaf has a high priority for assimilate from that leaf, whereas roots have a strong claim on photosynthate from young, expanding leaves (Fig. 4.2).

Carbohydrates are important in red clover for the plant's survival over the winter. The polysaccharide starch is the principal storage carbohydrate, which accumulates in the roots during the growing season and is depleted during winter. For example, in Sweden, the absolute amount of starch in red-clover

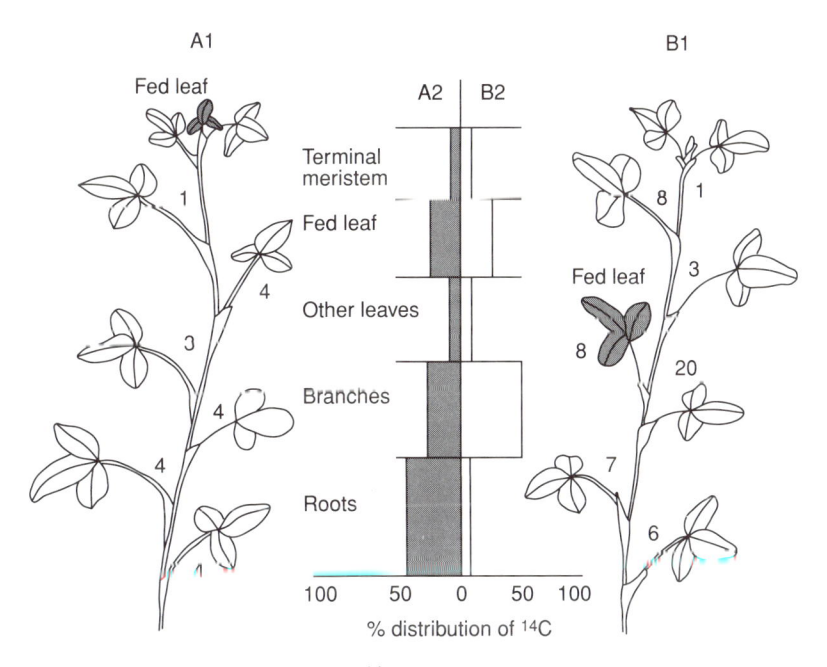

Fig. 4.2. Percentage distribution of ^{14}C, after youngest or fourth youngest leaf is fed, among branches and axillary buds (A1, B1) and throughout the whole plant (A2, B2) (from Ryle *et al.*, 1979).

roots declined from about 50 g m^{-2} of ground in autumn (October) to 1–2 g m^{-2} in the following spring (April) (Halling, 1988). Carbohydrate concentration in autumn has an influence on growth in the following spring, this growth being significantly correlated to TNC and starch concentration in roots in autumn. Taking more than one autumn harvest reduces carbohydrate accumulation, particularly in early-heading types, and, in combination with a reduction in plant numbers, reduces yield at the first harvest in the following year, an effect exacerbated by grazing in autumn (Hay, 1985).

Nitrogen fixation

The total amount of N fixed by red clover and the contribution it makes to N content vary according to a number of factors, including climatic and soil conditions, presence and effectiveness of *Rhizobium leguminosarum* bv. *trifolii*, companion species and plant stage of development.

Nitrogen fixation can contribute up to 80% of total N assimilation in red clover (Heichel *et al.*, 1985). However, the rate of N$_2$ fixation may be greatly reduced due to drought, accumulation of inorganic N in soil (Maag and Nösberger, 1980), soil acidity or plant defoliation. As the proportion of companion grass in the sward increases, so also does the proportion of N in red clover derived from fixation (Mallarino *et al.*, 1990a), presumably due to soil mineral N being depleted by the companion grass. Concomitantly, the total N fixed decreases in accordance with a decreasing clover presence. Annual N$_2$-fixation rates in red-clover crops vary widely, but published rates are mostly in the range 100–250 kg N ha^{-1} (Bowley *et al.*, 1984; Smith *et al.*, 1985), although annual fixation rates as high as 390 kg N ha^{-1} have been recorded in Uruguay (Mallarino *et al.*, 1990b). A monoculture of red clover, producing 10 t DM ha^{-1} with a mean N content of 33 g kg^{-1} in the DM, of which 70% is from N$_2$ fixation, will fix 230 kg N ha^{-1} (excluding the N contribution to stubble and below-ground tissue).

The introduction of *Rhizobium* by inoculation of clover seed can increase red-clover yields in the field. Even with only 20% occupancy of nodules by introduced strains, high yields can result (Martensson, 1990). However, effectiveness is not always associated with competitiveness, and so a highly effective strain might not successfully compete with less effective wild strains (Ames-Gottfred and Christie, 1989). Nevertheless, inoculation of red clover by strains selected for effectiveness under acid conditions, originating from acid sites, may increase yields under acid conditions at pH as low as 4.1, i.e. at a soil pH generally considered to inhibit nodulation, albeit that herbage yields are still low (Lindstrom and Millyniemi, 1987).

At present, inoculation with *Rhizobium* is recommended when introducing red clover into areas that do not have a history of growing red clover. While inoculating seed irrespective of previous crop/vegetation history might be considered an insurance against slow nodulation or nodulation by strains with low effectiveness, the inoculated strain might not occupy many of the nodules formed, due to either indigenous strains being more competitive than the introduced strains or soil conditions not favouring the latter, e.g. low soil pH. More is required to be known about the factors influencing *Rhizobium* competitiveness before inoculants with predictable effectiveness can be developed (Taylor and Quesenberry, 1996).

Nitrogen fixation in red clover, as in all other legumes, is energy-demanding, and about 3.2 mg C is required by roots mg^{-1} N fixed (Fernandez and Warembourg, 1987). Consequently, less C is available for shoot growth for a given rate of C assimilation in N_2-fixing plants than in plants relying on mineral N, and so shoot-growth rates are slower in the N_2-fixing plants than in the latter.

The possibility of selecting red-clover phenotypes for increased N_2-fixing ability, at least when in association with a specific strain of *Rhizobium*, has been demonstrated (Christie *et al.*, 1992). The criterion of nodule number per plant is as good a trait to consider as nitrogenase activity per plant when selecting for increased N_2 fixation in red clover.

Nitrogen transfer

As with many other legumes, the companion grass grown with red clover can benefit from N 'released' from the clover. Estimates of N transferred by red clover have been made in a few studies, in which the N derived from the legume has a lower $^{15}N : ^{14}N$ ratio than that of grass, when the mixture is grown in soil enriched with ^{15}N, due to fixed atmospheric N_2 (which has a low $^{15}N : ^{14}N$ ratio) diluting ^{15}N in the legume, and the data from some of these studies are summarized in Table 4.1. Despite the wide range of conditions under which these experiments were carried out, the proportion of clover-derived N that is transferred in three of the four studies is similar, i.e. 10–13% of total N assimilated by the legume (including N fixed). However, these data obscure the low proportion of legume N transferred in the sowing year, presumably due to very slow decomposition of roots and nodules.

Overall, the amount of N transferred is a relatively small proportion of total legume N, so most of the N is either incorporated into soil OM, an attribute exploited when including red clover in rotations to increase soil N for a subsequent arable crop, or is harvested in the forage. One year's growth of

Table 4.1 Nitrogen fixed and transfer of red clover-derived N to companion grass (kg ha^{-1} year^{-1})

Companion	Year	Red clover (%)	N_2 fixed	Total legume N	N transferred	Source
Italian ryegrass	2	61	224	259	30	Boller and Nösberger (1987)
Tall fescue	2	62	179	204	22	Mallarino *et al.* (1990a,b)
Reed canary grass	4	54	32	81	11	Heichel and Henjum (1991)
Cocksfoot	2	57	123	127	36	Farnham and George (1993)

red clover provided the equivalent of 140–160 kg N ha^{-1} for a subsequent maize (*Zea mays* L.) crop, which produced 16.5 t ha^{-1} DM as whole-crop maize (Collins, 1993).

Breeding

Red clover is a diploid species ($2n = 14$), although some cultivars bred over the past 40 years have been tetraploids ($2n = 28$). Self-incompatible alleles and insect pollination ensure cross-pollination, but tetraploids exhibit a degree of self-fertility.

Red clover was originally marketed as landraces, distinctive ecotypes evolving rapidly due to populations within the species responding readily to selection pressure (Hay, 1985). Now cultivars are bred, usually by phenotypic recurrent selection, in which individuals are evaluated by their phenotypic response, e.g. resistance to a specific disease, and those selected are polycrossed or open-pollinated with a wider population. This may be repeated for a number of cycles.

Selection has tended to be for traits associated with increasing forage yield and persistence, such as resistance to foliar, crown or root diseases or to root or stem nematodes. As *Fusarium* spp., the fungi often associated with root rot in red clover, are only parasitic on red clover that has already been subject to stress or wounded, it is unlikely that breeding for resistance to that complex of diseases will be successful. When tetraploids were first developed in Europe, several cultivars were more productive and persistent than existing diploids, because of resistance to diseases and pests (Åkerberg *et al.*, 1963,) but tetraploidy *per se* is not a guarantee of superiority (Frame, 1976), particularly as a number of improved diploid cultivars were subsequently released. Diploid red clovers were always the favoured option for North American breeders (Taylor and Smith, 1995).

Phenotypic recurrent selection has resulted in improved resistance to northern anthracnose (*Kabatiella caulivora*), clover rot (*Sclerotinia trifoliorum*), powdery mildew (*Erysiphe polygoni*), yellow clover aphid (*Therioaphis trifolii*) and pea aphid (*Acyrthosyphon pisum*) (Taylor and Smith, 1979).

Improved persistence, resistance to stress and to nematodes and increased seed production in diploid red clover have been successfully achieved by maternal line selection, in which the most successful maternal plants (evaluated by progeny) are selected for the next cycle of breeding. However, progeny testing in red clover is constrained by difficulties in maintaining parental clovers during the testing phase.

Mass selection, development of polyploids, hybridization with other *Trifolium* species and induction of self-fertility, to produce inbred lines for hybridization to maximize heterosis, have been applied with varying degrees of success in red-clover breeding programmes. Gametic non-reduction to widen the genetic base of tetraploids (Taylor and Berger, 1989) and the use of leaf-culture-derived protoplast fusion to produce interspecific hybrids (Myers et al., 1989) are also being developed for red-clover breeding. Diploid red clover hybridized with the persistent perennial *Trifolium alpestre* has resulted in F_2 plants by embryo rescue, the hybrid resembling *T. alpestre* (Phillips et al., 1992).

Despite the prolonged and extensive breeding effort that has been devoted to breeding for persistence in red clover, the most long-lived of modern cultivars are still, at best, short-lived perennials. In many countries, there has been little or no red-clover breeding over the past 10–20 years, mainly on account of the increased emphasis on N-fertilized grass swards.

A common seed-production practice in parts of North America is to harvest the second-cut crop for seed, following a first-cut crop for conservation, and so red-clover seed is relatively cheap and available on-farm or among neighbouring farmers. Thus, the rewards for plant breeding are low, and so breeding efforts have remained modest and mainly in the hands of publicly funded institutes, in total contrast to the vibrant and profitable lucerne breeding, which is almost totally in the hands of private seed companies (McKersie, 1997).

Cultivars

Cultivars are grouped into either early- or late-flowering categories. The earlies, formerly known as 'double-cut' types and referred to as medium red clovers in the USA, where they are the most widely grown types, or as cowgrass in New Zealand, are most productive in early season. Because they have two successions of basal bud formation and shoot elongation, two main cuts can be taken in the first harvest year. Persistence and, hence, production are reduced in subsequent years, dependent upon cultivar and stand health. Compared with the early cultivars, the late-flowering or 'single-cut' types start growth later in the spring and flower later, so a higher proportion of the annual production is obtained from the first cut. They have wider crowns, more numerous aerial stems and a better-developed adventitious root system and are more persistent, a major factor being the crop of basal buds produced in autumn, which do not develop inflorescences until the following year (Rumball, 1983; Montpetit and Coulman, 1991b).

Improved plant persistence has been a major objective of red-clover breeders, and the difference between earlies and lates in this characteristic has become less clear-cut. The best cultivars of both types have the potential for high production and persistence over 2–3 harvest years, or even longer in disease- and pest-free environments (McBratney, 1981, 1984), and are suitable as non-dominant components of rotationally grazed pastures or of red-clover-dominant swards cut for hay or silage.

In the past, tolerance to grazing was not a major breeding objective, so the development of such a characteristic is noteworthy. In Tasmania, Australia, a semiprostrate cultivar (Astred), which reproduces by seed and also vegetatively by stolons and daughter plants, has shown long-term persistence under close grazing (Smith and Bishop, 1993).

Under certain management conditions, the oestrogen content in red clover can cause a lowered reproduction rate in breeding ewes. However, cultivars with low contents of the isoflavone formononetin, which is responsible for the oestrogenic effect, have been developed in the past in Australia (Francis and Quinlivan, 1974) and also more recently in New Zealand, e.g. the diploid cultivars Enterprise and Grasslands Colenso.

In the USA, medium red-clover cultivars include Arlington, Kenstar, Marathon and Redland II. The cultivars Florex and Ottawa are examples of Canadian medium types, while Norlac and Altaswede are late-flowering single-cut types. In Sweden, where red clover is a common constituent of seed mixtures, examples of cultivars developed are the tetraploids Sara and Fanny, and the diploids Jasper and Britta. Other European examples are Merviot (Belgium), Marcom (France), Kuhn (the Netherlands), Rajah (Denmark) and Deben (the UK). No cultivars have been released recently in the UK, reflecting its declining use there.

Seed production

The USA and, to a lesser extent, Canada are the major producers of red-clover seed in world terms. Seed production can be a specialist enterprise or, in some regions, subsidiary to forage production.

Red clover is cross-pollinated, mainly by wild, long-tongued bumble-bees (*Bombus* spp.), so rough ground and a suitable static food source near the seed crop are desirable to encourage nesting. It is also possible to supplement the local bee population available for pollination by introducing additional queen bees in spring (Clifford, 1973). Honey-bees (*Apis mellifera*), which can be transported and concentrated in seed-production areas, are also effective pollinators of red-clover cultivars with short corolla tubes (Palmer-Jones *et al.*, 1966).

Seed is harvested from the regrowth of early-flowering clover after a cut in early season, thus avoiding excessive vegetative growth and potential lodging of the seed crop. For late-flowering clovers, seed is harvested from the primary growth and can be taken for 2 or more years, whereas early-flowering types are usually harvested for only 1 year (Kelly, 1988).

Clover seed is harvested when most of the seed heads are brown and the seeds hard, with purple coloration. Sometimes, a desiccant, e.g. diquat, is used before the intended date of combine-harvesting, but the crop can also be threshed after cutting and swathing. The optimum time to harvest red clover is about 1 month after the end of rapid inflorescence appearance, corresponding to when 1–4% of the total inflorescence number is classed as 'unripe'. A profile of the rate of inflorescence appearance and harvest times at which seed yields were highest for two early-flowering cultivars and and a late-flowering cultivar are shown in Fig. 4.3.

Chemical growth regulators, particularly daminozide, can improve the potential yield of red clover through increases in seed setting and density of inflorescence, following shortening of the corolla-tube length, which

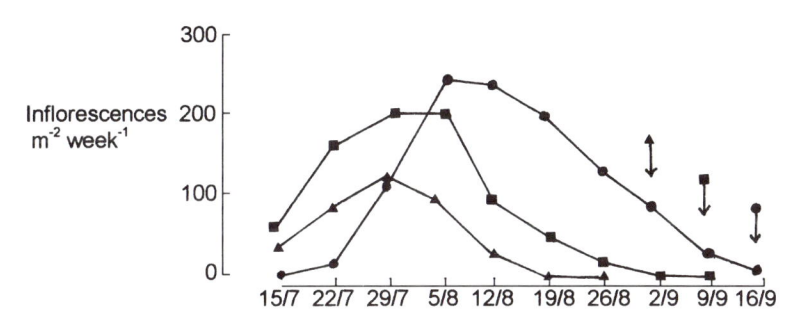

Fig. 4.3. Rate of inflorescence production and time of harvest at which maximum seed yield is produced (arrow) for three cultivars of red clover, Sabtoron (■), Hungaropoly (▲) and S123 (●) (from Puri and Laidlaw, 1984).

improves pollination by bees, and reduced plant height, which alleviates lodging (Mela, 1969; Christie and Choo, 1990); however, seed-yield increases have been very variable, ranging from 0 to over 100%, with about 30% being more typical (Taylor and Quesenberry, 1996).

Potential seed yield under northern European conditions is of the order of 1800 kg ha^{-1}, although actual seed yield is likely to be only 25–50% of this, due to losses during harvesting (Dennis and Haas, 1967). For the USA, an annual production of 11,300–13,600 t from average seed yields of 600–700 kg ha^{-1} was noted (Rincker and Rampton, 1985). World production in 1990 included 12,507 t (USA), 3901 t (Canada), 1165 t (France) and 300 t (New Zealand) (Taylor and Quesenberry, 1996). In eastern Canada, seed production in the seeding year varied from 45 to 580 kg ha^{-1}, yields being affected by soil type, cultivar and different weather conditions among years (Belzile, 1994).

Tetraploid clovers usually produce much lower seed yields than diploids, due to a combination of low inflorescence number per unit area and poor seed setting (Dennis, 1980). Low seed set is in part due to tetraploid red clover being a young plant species, which will require a number of generations before it is stabilized and genetically in balance (Jonsson, 1985).

Agronomy and Management

Seed mixtures

When making up a seed mixture, the red-clover cultivar selected should be on a country's recommended list, this being proof of its value as a superior cultivar. Seed supplies of newly recommended cultivars can be scarce initially, more so with tetraploids, because of their lower seed potential, than with diploids. Individual cultivars for seed mixtures should be chosen on the basis of their specific characteristics, and resistance ratings to diseases and pests are a major consideration.

In traditional general-purpose mixtures for cattle and sheep enterprises, red clover was sown at 3–6 kg ha^{-1}, chiefly to enhance hay yields in the first full harvest year and to improve the feeding value of grazed aftermaths. For example, in Sweden, 4–6 kg ha^{-1} red clover, with 18–20 kg ha^{-1} grass, was a typical seed mixture (Arnemo and Steen, 1972). In the USA, 10–12 kg ha^{-1} are sown in monocultures and about half this rate for mixtures with grass (Taylor and Smith, 1995). Red clover can also improve the quality of autumn-saved forage for out-wintered livestock, where the climate allows this practice, and is also a valuable constituent of seed mixtures for meat production from Angora goats (Stevens et al., 1992) or for venison production from red deer (Hunt and Hay, 1989; Barry et al., 1993).

The role of red-clover-dominant swards for silage production has been intensively researched in the UK. Swards based on monocultures at 12–15 kg ha^{-1} seed or on similar rates of clover, together with 4–5 kg ha^{-1} grass seed have shown considerable short-term potential, although farmer adoption of such swards has been low (Frame, 1990). In several countries where the application rates of fertilizer N on grassland have steadily increased in past decades, the usage of red-clover seed has declined. This has been reflected in the UK, for instance, by the reduced tonnage of seed delivered annually for use on farms by seed merchants – c. 3000 t in the 1960s, 1000 t in the 1970s, 200 t in the 1980s and 100 t or less in the early 1990s, according to Ministry of Agriculture, Fisheries and Food statistics.

Companion grass

In general, the addition of a companion grass gives greater total herbage production with a higher digestibility than red-clover monocultures, but clover content is decreased, resulting in reduced crude protein (CP) concentration of the herbage. The levels of digestibility and CP depend upon the frequency of harvesting and stages of growth of the sward components. The grass companion is usually more dominant in the mixture at the first harvest in June than at later harvests. The production advantage of the mixed association varies according to the production potential of the grass companion and persistency of red clover, and may be most marked only in the later life of the sward (McBratney, 1981; Frame et al., 1985). The companion grass chosen also influences the seasonal distribution of production, rather than the annual total (Connolly, 1970), heading date of the grass being a major influence in this respect. In Canada, red-clover plants survive better in severe winters when sown with a grass, rather than in monoculture (Belzile, 1987).

A range of individual companion grasses have been evaluated, and several species have proved satisfactory, notably perennial ryegrass (*Lolium perenne*) (Frame et al., 1972; Laidlaw and McBratney, 1980), timothy (*Phleum pratense*) (McBratney, 1981; Frame et al., 1985), meadow fescue (*Festuca pratensis*) (Frame and Harkess, 1987) and Italian ryegrass (*Lolium multiflorum*) (Boller, 1996). Intermediate-heading perennial ryegrass is highly suitable, particularly tetraploid cultivars, since they are less competitive than diploids, on account of their more open growth habit (Frame and Harkess, 1987). However, a mixture of the more winter-hardy timothy and meadow fescue is favoured in most Scandinavian countries and tall fescue (*Festuca arundinacea*) is popular in the USA, while timothy is the main companion grass in Atlantic Canada. Tall fescue also proved to be a compatible companion grass in southern New Zealand (Hay and Ryan, 1989).

Establishment

Sowing direct in the early to midseason period gives significantly higher DM production and red-clover contents during the following harvest year than later sowings (Table 4.2) and the depression in production due to later sowings is not compensated for by increased clover seed rate, because of decreased plant survival (Frame *et al.*, 1976b). Well-developed plants from early-season sowing are also more resistant to winter damage, particularly when not allowed to flower. However, winter-hardiness is not impaired if there is sufficient time after flowering for strong basal growth to develop before winter.

Another advantage of early-season sowing is a greater production of herbage in the establishment year, although this may still only be 50–60% of the production obtainable in the first full harvest year. In western USA, cutting late (90 days after germination of red clover) in the seeding year rather than early (40 days) was beneficial not only for establishment-year yield, but also for crop persistency and yield the following harvest year (Hall and Eckert, 1992). It is possible to sow red clover, conventionally or by direct drilling, after an early-harvested winter-barley crop, but only when autumn climatic conditions are favourable to vigorous clover-plant development. In temperate zones with a mild winter climate, red clover can be sown in late season as a winter annual forage.

Red-clover establishment can be satisfactory when a clover monoculture or grass/clover mixture is sown under a grain-cereal cover, and this is practised

Table 4.2. Effect of date of sowing in the establishment year on herbage production from pure-sown red-clover swards in the first full harvest year (three-cut silage regime) (from Frame *et al.*, 1976b).

Date of sowing	DM (t ha^{-1}) Total herbage	Red clover	Clover content (%)
April 30	12.2	11.9	98
May 24	12.0	11.9	99
June 21	12.2	11.9	98
July 19	11.6	10.2	88
August 13	8.8	5.1	58
September 10	5.2	3.6	69

in several regions, e.g. Scandinavia and Atlantic Canada, since weed invasion is curbed and the grain produces a valuable cash crop. Grain yields may be unaffected by undersown red clover (Stewart *et al.*, 1980) or reduced (Kunelius *et al.*, 1992), such differences being due to differing trial conditions. The risk of poor clover establishment is greater with undersowing than with direct sowing, because of competition with the arable crop. Also, a late harvested or lodged grain crop may retard development of the undersown grass and/or clover seedlings. Conversely, lush clover growth in a cereal crop can lead to grain-harvesting difficulties. Cutting the cereal crop at a growth stage suitable for arable silage is an alternative option, since the red clover is under competitive stress for a shorter period (Laidlaw, 1979a).

Because of the small size of the seed and its limited food reserves, the optimum sowing-depth range is 10–15 mm and seed sown deeper than 25 mm will have an adverse effect on seedling population density (Sund *et al.*, 1966). A population of about 200 clover plants m^{-2} by autumn of the establishment year is a reasonable target for the establishment of a red clover-dominant sward intended for a silage cutting regime (Castle, 1984). An adequate initial plant population is essential, since plant numbers decrease with increasing age of stand (Sheldrick *et al.*, 1986; Skipp and Christensen, 1990).

Rhizobial inoculation is not usually carried out in European countries, since most soils contain *R. leguminosarum* bv. *trifolii*, although indigenous strains may not necessarily be the most effective for red clover. However, there has only been limited experimentation on the relative merits of inoculating red-clover seed under field conditions and further investigations are warranted. If red-clover use is extended to soils without a previous clover, red or white, rhizobial inoculation of the seed is essential, as discussed in the N_2-fixation section.

Red clover has proved a reliable species for direct seeding into existing grazing or conservation swards in need of renovation, although the success of establishment is directly related to the degree of destruction of the previous sward (Kunelius *et al.*, 1982; Kunelius and Campbell, 1984). It may be introduced into red clover or grass/red-clover swards, where there is need to reinforce a reduced clover contribution, or even into grass swards. Several techniques, including one-pass operations, with a variety of equipment, have been developed and reported for differing situations in many countries, for example, from New Zealand (Hay *et al.*, 1978), the USA (Koch *et al.*, 1983) Australia (Curll and Gleeson, 1987), the UK (Gracey, l98l; Haggar and Koch, 1983), Canada (Kunelius *et al.*, 1988) and Japan (Takeda, 1993). In the North American work, winter kill of red-clover plants, due to frost-heaving of the soil, is more severe with conventional sowing than with direct seeding.

Management guidelines have been outlined for successful establishment by direct drilling (Naylor *et al.*, 1983; Tiley and Frame, 1991). A major prerequisite is a low-density sward, so that competition to the introduced species is minimized. Heavy grazing, cutting for silage or the use of chemical desiccation prior to sowing are means of ensuring reduced competition. Some drills can be fitted with a chemical applicator to spray bands astride the drilled slots simultaneously with drilling the seed. Timing is important *vis-à-vis* the seasonal rainfall pattern, since adequate soil moisture is necessary for seed germination and seedling development. Postsowing management should avoid overgrazing the developing seedlings, but grazing pressure should be sufficient to defoliate the existing vegetation and suppress its competitive ability.

Limitations to the establishment of direct-drilled red clover include attack from nematodes, e.g. root knot, clover cyst and clover stem, which may be in the existing sward if clover is present (Mercer and Campbell, 1986), or competition from the existing sward, even when paraquat has been band-sprayed along the drill (Kunelius *et al.*, 1982). Optimum timing of application of chemical treatment to retard growth of the existing sward varies. In a tall-fescue sward in Illinois, red clover was adversely affected by application of mefluidide prior to sowing (Olsen *et al.*, 1981), whereas bahia grass (*Paspalum notatum*) was best controlled when treated with paraquat prior to sowing direct-drilling red clover (Kalmbacher *et al.*, 1980).

Fertilizer use

Red clover is adaptable to a wide range of soil conditions, but excessively wet or acid soils are unsuitable. Good soil drainage was identified as a key positive influence on clover persistence on Finnish farms (Pulli, 1984). Manganese and Al toxicity is a problem in acid soils (Elliott *et al.*, 1973). Ideally, soil pH should be in the range 6.0–6.5 for optimum crop development and root-nodule formation and liming acid soils to improve the pH status has also led to improved forage yields (Habovštiak, 1989). Soil analysis provides a good guide to lime and fertilizer requirements, and recommendations for different levels of soil fertility are generally available from advisory services.

Adequate P and K in the seed-bed are necessary and the phosphate fertilizer used should contain a high proportion of readily available, water-soluble P_2O_5. There has been little work reported on fertilizer-N requirements for establishment, but a 'starter' N application of 50–75 kg ha^{-1} encourages early clover development (Crowley, 1975), particularly if soil N is likely to be low, for example, after a number of successive cereal crops.

When red-clover swards are utilized mainly for silage cropping, the soil nutrients removed, particularly P and K, require replenishment; otherwise, production will be depressed. Assuming P and K concentrations of 3 and 20 g kg^{-1} DM, respectively, the offtakes from a sward with an annual production of 10 t DM ha^{-1} will be equivalent to fertilizer application of 70 kg ha^{-1} P_2O_5 and 240 kg ha^{-1} potash (K_2O). Sulphur fertilization may be required on light-textured, sandy soils low in OM.

Cattle slurry is a valuable source of available nutrients and may be substituted for part of the fertilizer requirements. An application of 50 m^3 ha^{-1} (1 : 1 dilution with water) supplies approximately 20 kg ha^{-1} P_2O_5 and 115 kg ha^{-1} K_2O, but also about 60 kg ha^{-1} N, which encourages the grass component at the expense of clover in mixed swards. Depression of red clover is reduced by

Table 4.3. Herbage production from a red-clover/tall-fescue sward under a three-cut silage regime at two annual rates of fertilizer N (from Frame, 1990).

Harvest year	N rate (kg ha^{-1} yr^{-1})	DM (t ha^{-1}) Total herbage	Red clover	Clover content (%)
1	0	12.7	8.9	70
	180	13.1	6.8	52
2	0	10.8	6.4	59
	180	12.6	2.7	21
3	0	7.9	1.5	19
	180	12.7	0.4	3

using phosphate and potash fertilizer in the spring, followed by cattle slurry after the first harvest (Gracey, 1981). Slurry application should be restricted to times when soil physical conditions are favourable for wheeled traffic, if soil rutting and compaction are to be avoided, but preferably not in winter, when substantial loss of nutrients, causing environmental pollution, occurs.

Red-clover monocultures are unlikely to respond to N, whether from fertilizer or slurry sources (Frame *et al.*, 1976a; Frankow-Lindberg, 1989) and sown clover-dominant mixtures do not require applied N if the clover is abundant and growing vigorously. Fertilizer-N application depresses clover production in a mixed sward, but nevertheless sustains or increases total herbage production by encouraging the grass component (Table 4.3; Johansen, 1984). In Sweden, it is common practice for farmers to increase N fertilization as the ley ages, often making silage from the youngest clover-rich leys and hay from swards over 2 years old. Fertilizer N maintained a satisfactory level of production in the seventh and eighth year of sown red-clover-dominant swards (McBratney, 1987). Such longevity of red-clover swards, though with low clover contents, is unusual, except in disease- and pest-free conditions, as occurred in Northern Ireland and also at an upland site in Wales (Young, 1984).

Weed control

Weed invasion is more likely to be a problem in pure-sown red clover than in mixed stands, at both the establishment phase and later in the crop's life, especially in early spring, when the sward lacks density of cover. A weed-free seedbed is an essential starting-point, but a number of selective herbicides are available for postemergence control of invasive annual and perennial broadleaved weed seedlings.

Timing of spray application is critical in relation to growth stage of the weed species, which has to be at a susceptible stage, and to the sown clover and grass, which have to be at a resistant stage of growth. Normally, clover plants should be between the first and third trifoliate-leaf stage, while grass seedlings must be at, or beyond, the two- to three-leaf stage, and manufacturers'-approved label recommendations and guidelines should be closely followed. When sowing monocultures, incorporation of EPTC into the seed-bed prior to sowing is effective against invasive grass and volunteer cereals. 'Clover-safe' herbicides for postemergence weed control in young swards and for weed invasion in established swards are usually based on 2,4-DB, MCPB and benazolin, used either singly or in combination. Topping with a mower or cutting an early silage crop are cultural measures of controlling annual broad-leaved weeds in establishing swards.

Diseases

Clover rot (*S. trifoliorum*) is a major disease limiting red-clover persistence and production in many countries, and conceivably its incidence could become more widespread if there were an expansion of red-clover use in regions where its use is currently limited. Historically, clover rot has been considered the most serious disease of red clover in Europe since the midnineteenth century (Williams, 1984). The infection appears in the ground in autumn from spores produced by the germinating resting bodies. The clover leaves become peppered with brown spots and necrotic, but over winter large black patches of rotting plant stems, leaves and roots develop and the plant population is severely reduced. The development of foliar necrosis is optimal at 15–18°C (Raynal, 1989a).

Enhanced resistance was first shown by tetraploid cultivars developed in the early 1970s, but diploid cultivars with good resistance are now also available, and resistant cultivars should be used in conjunction with a 4- to 5-year crop rotation between susceptible crops. Ability to regrow after an attack is also an important attribute (Schmidt, 1980) and genetic variation in plants exists, which warrants further selection for resistance (Matsuura *et al.*, 1985). Culturally, good autumn defoliation of the clover minimizes the amount of foliage liable to infection by the ascospores from the sclerotia if the disease is present and reduces the high humidity in the crop canopy that favours the disease.

A range of fungi, with *Fusarium* spp. featuring prominently, may cause progressive necrotic breakdown and rotting of clover crowns and roots, resulting in reduced persistence (Skipp *et al.*, 1986; Gondran and Raynal, 1989). However, they are not usually responsible for the initial damage; this is caused by agents, such as clover root borer (*Hylastinus obscurus*) (Jin *et al.*, 1992), or pathogenic bacteria, e.g. *Pseudomonas viridiflava* (Leath *et al.*, 1989). Other factors inducing root rot include overfrequent defoliation and nutrient deficiency, both of which place the clover plants under physical or physiological stress (Chi and Hanson, 1961; Rufelt, 1983). Improved persistence and production have been achieved experimentally by the application of benomyl, which is considered to reduce the development of *Fusarium* root rot (Leath *et al.*, 1973).

A number of stem and leaf diseases, with varying degrees of seriousness, can attack clover in different countries, reducing its productivity and longevity (O'Rourke, 1976; Leath, 1985). Frequently, there is no economic method of fungicide control, but progress has been achieved in cultivar-development programmes, so that resistance ratings are an important consideration when choosing seed mixtures.

Clover scorch (*K. caulivora*) can have serious effects on productivity, causing stem blackening and withering of the leaves, as if the crop had been burnt (Raynal, 1989b). It is a seed-borne disease and contaminated seed is the main vector of infection (National Institute of Agricultural Botany, 1988), but the fungus can also overwinter on infected plant material in the field. Known as northern anthracnose in the USA, it is a major disease in northern USA, while another disease with similar symptoms, southern anthracnose (*Colletotrichum trifolii*), is more common in the south. Powdery mildew (*Erysiphe trifolii*) can be severe in many countries, especially following warm, dry summer conditions, which favour its spread; yellow mottling appears on the upper surfaces of the leaves, followed by whitish grey, powdery areas. Other diseases that flare up periodically and adversely affect growth vigour and herbage quality include *Pseudopeziza* leaf spot (*Pseudopeziza trifolii*), rust (*Uromyces* spp.), spring blackstem (*Phoma trifolii*) and black patch (*Rhizoctonia leguminicola*).

Several viruses infect red clover (Barnett and Diachun, 1985; McLaughlin and Boykin, 1988), and their adverse effect on clover productivity and persistence is probably due more to multiple infection than to a single virus (Carr, 1971). However, research and, therefore, information on the precise effect of individual viruses or their control are sparse. Red-clover necrotic mosaic virus (RCNMV), transmitted by the root-infecting fungus, *Olpidium radicale*, was identified in Europe in the late 1960s and was noted in the UK in 1976 (Bowen and Plumb, 1979). The virus causes veinal sclerosis in the leaves, often followed by severe necrosis and deformation, the plants become stunted and yield is markedly reduced. Once the crop is infected, the virus can be spread by machinery or grazing. Some cultivars at that time proved highly susceptible, e.g. Hungaropoly, which was then on European recommended lists. Fortunately, there was differential resistance among cultivars. The seed-borne white-clover mosaic virus (WCMV) also reduces productivity, nodule number and nitrogenase activity (Khadhair *et al.*, 1984) and is potentially serious. The incidence of both RCNMV and WCMV was reduced in an Italian ryegrass/red-clover mixture, compared with a clover monoculture (Lewis *et al.*, 1985).

Pests

Stem eelworm (*Ditylenchus dipsaci*) is a major pest of red clover, living within the plant and causing considerable damage (National Institute of Agricultural Botany, 1988; Raynal, 1989a). At first, patches of clover appear with poor growth vigour and with stunted plants, which have abnormally enlarged leaves, stems and buds (Cook and Yeates, 1993). These patches die out and then progressively enlarge and merge, until the whole field may be affected. Its introduction to 'clean' fields can take place from particles of infected plant debris on machinery or in irrigation water, for example, or from dormant eelworm cysts attached to the introduced clover seed (Caubel, 1989), although this latter source of infection is preventable by seed fumigation with methyl bromide, carried out by the seed merchant. In the past, stem eelworm and clover rot were major factors responsible for 'clover-sick' land, on which it was difficult to sustain good clover productivity. There is no economic chemical control of stem eelworm (or clover rot), so breeding resistant cultivars has been an aim of plant breeders, and recommended lists contain a number of varieties with good dual resistance.

Nematodes of significance in North America are root-lesion nematodes (*Pratylenchus* spp.), root-knot nematodes (*Meloidogyne* spp.) and the clover-cyst nematode (*Heterodera trifolii*). Adverse effects of these pests include reduced seedling establishment, plant persistence and herbage production of red clover (Kimpinski *et al.*, 1984; Leath, 1985).

Slugs, e.g. the netted slug (*Deroceras reticulatum*), actively feed on red clover from spring to autumn and are a particular hazard during establishment, especially under wet conditions on heavy-textured clays. Control measures by pesticides, e.g. methiocarb, are available and are advisable when a grass sward is cultivated and reseeded with clover directly or where red clover is slot-seeded into an existing grass sward. Leatherjackets (*Tipula* spp.), which damage roots and shoot bases can also be chemically controlled, e.g. by chorphyrifos. These pests and any other pest damage should be identified and the cause treated, if appropriate and economic – control of weevils (*Sitona* spp.) by fenitrothian, for example.

Advisory services offer forecasts of the onset of infestation and measurement of actual populations of certain pests, such as leatherjackets; prediction of potential damage can then be made and cost-effective control measures instituted. The impact of insect herbivory has been demonstrated on plant communities containing natural red-clover populations by the use of a foliar insecticide, which increased clover survival, plant size, leaf number and length, number of seed heads and seed weight (Gange *et al.*, 1989).

Herbage production

The reported levels of annual DM production from red-clover-dominant swards have varied according to factors such as site conditions, fertilizer treatment or onset of disease. Typical European DM levels (t ha^{-1}) cited for harvest years 1–6 are 9–18, 9–15, 4–14, 8–12, 6–9 and 7–9 (Laidlaw and Frame, 1988), although data are limited for years 3–6 because of lack of clover persistence. The proportions of total annual production for specific cuts will depend upon the relative cutting dates; thus, a late first cut may contribute 50–60% or more of the annual production and a second cut 30–40%, while the late-season aftermath rarely contributes more than 10–20%.

Productivity of pure sown red-clover swards in the UK over 3 years was equivalent to that from grass swards given fertilizer N at annual rates of 250,

Table 4.4. Effect of cutting frequency on herbage production and quality from pure-sown red-clover swards (mean of six cultivars of broad red clover over 2 harvest years) (from Sheldrick et al., 1986).

	Cuts per annum	
	3	6
Annual DM (t ha^{-1})	9.2	6.3
Annual DOMD	56.7	62.0
Annual N concentrations (g kg^{-1} DM)	30.2	36.6

210 and 140 kg ha^{-1} in the successive years (Hunt et al., 1975). Progressive decline in annual DM production with age is a typical characteristic of clover swards, even in the absence of pests and diseases, due to a natural decline in plant population. The depletion of minerals by continuous silage cropping is also a possible contributory factor (Hunt et al., 1976). In Atlantic Canada, the DM production of slot-seeded red clover in the postsowing years was equivalent to that from a grass sward receiving 150 kg N ha^{-1} annually (Kunelius et al., 1988), while, in France, red-clover mixtures with cocksfoot (*Dactylis glomerata*) or annual ryegrass (*L. multiflorum*) receiving 150 kg N ha^{-1} annually yielded similarly to the grass swards given 300 kg N ha^{-1} annually (Guy, 1989).

While a strategy of increasing the number of defoliations from three cuts to four, five or six cuts, for example, improves herbage quality (digestible organic matter in the DM (DOMD), metabolizable energy (ME), CP), the gain in quality with increased frequency of defoliation is small relative to reductions of up to 33% in DM production, 28% in ME output and 15% in CP production (Table 4.4). Also, the scope for such a strategy is extremely limited in regions with short growing seasons and harsh winters (Pulli, 1988).

Red clover, sown alone or with white clover and/or lucerne, gave herbage DM yields of 14.5–16.5 t ha^{-1} in the first harvest year and 10.0–12.6 t ha^{-1} in the second, with red clover the dominant constituent in the mixed associations (Frame, 1986). Further investigational work on the potential of mixed legume associations is warranted, not least because of the high nutritive value of such mixtures and the possible synergism from individual legume characteristics.

Information about on-farm production of red clover is sparse, but reported DM levels from a spring-sown monoculture under a silage-cutting regime were 6.1 t ha^{-1} for the establishment year and 10.9 t ha^{-1} for the first full harvest year (Castle and Watson, 1974). Production of 14.7 t ha^{-1} from a red-clover/grass sward in the first harvest year was equivalent to 87% of that from a grass sward receiving fertilizer N at a rate of 310 kg ha^{-1} (Roberts et al., 1990).

Nutritive value

It is well recognized that the chemical composition of herbage species is influenced by factors such as stage of growth, season of year, plant part sampled, soil-nutrient status and soil pH. From a review (Spedding and Diekmahns, 1972), average mineral contents (g kg^{-1}) were 2.7 for P, 24.7 for K, 15.5 for Ca, 2.6 for Mg, 1.0 for Na and 2.1 for S. In later work, with different cultivars and management, P, K, Ca and Mg contents were similar to the above (Frame, 1985). Compared with grasses, red clover is usually higher in concentrations of pectin, lignin, N and the minerals Ca, Mg, iron (Fe) and cobalt (Co). It is lower in cellulose, hemicellulose, water-soluble carbohydrates (WSC), and chlorine (Cl), Mn and silicon (Si) concentrations, but similar in lipids and the minerals P, K, S, Na, zinc (Zn), copper (Cu) and molybdenum (Mo). However, data for some elements are limited and the concentration of some elements, e.g. Mo, vary widely in plants, due to different growing conditions. Mineral contents in red clover, particularly P, K, Ca, Mg, Cu and Zn, are increased by the application in early spring of the growth regulators daminozide or mefluidide (Narasimhalu and Kunelius, 1994b).

The digestibility of red-clover primary growth declines with advancing maturity in an approximately linear fashion and is typically related to the declining leaf : stem ratio. Flowering does not appear to have a direct depressive effect on digestibility. Herbage DM digestibility at the first cut is improved when treated with the growth regulator daminozide (Narasimhalu and Kunelius, 1994a). When primary growths are compared at the same stage of flowering, the digestibility of red clover is lower than that of perennial ryegrass, but the rate of decline with age and consequent stemminess are similar (Demarquilly, 1981). The reduction in digestibility is related to the increase in cell-wall constituents, compared with cell contents (Wilman et al., 1977); also the digestibility of the stems is lower than that of the leaves (Wilman and Altimimi, 1984; Åman, 1985) although another study reported that red-clover stems had slightly higher digestibility than leaves at each stage of maturity until flowering (Buxton et al., 1985). Generally, decline in digestibility in red clover is associated with an increasing lignin content and a reduction in degradability of polysaccharides, other than starch (Taylor and Quesenberry, 1996). Digestibility is higher in plants subject to moderate moisture stress than in control irrigated plants (Peterson et al., 1992), higher leaf : stem ratio, delayed maturity and higher digestibility of plant parts accounted for the higher nutritive value of stressed plants.

Although nutritive value varies widely between genotypes within ploidy levels, tetraploids generally have higher CP and WSC contents and DM digestibility but lower DM content than diploids (Mousset-Declas et al., 1993). Phenological staging schemes for predicting the nutritional quality of red clover are available for farm management and research purposes (Ohlsson and Wedin, 1989).

In studies on the quality of autumn-accumulated red clover, concentrations of N and certain minerals (P, K, S) decreased with increasing rest interval following a late summer harvest, while contents of other minerals (Ca, Mg) responded less consistently (Collins, 1983; Collins and Taylor, 1984); levels of herbage digestibility and TNC also declined with delayed harvesting, although yield declined faster than herbage quality, and the onset of subfreezing temperatures and precipitation accelerated the fall in herbage quality and yield. Increasing proportions of dead material are the probable cause of declining digestibility and TNC in autumn herbage, a significant feature being the comparatively high quality of the stems compared with the leaves, due to lower acid detergent fibre (ADF), lignin and cellulose levels. This leads to the suggestion that a high grazing pressure is needed in the autumn–early-winter period to exploit the better nutritive value of the plant tissues in the lower layers (Sheehan et al., 1985); also, in spite of falling herbage quality in late season, the inclusion of red clover in a sward slows down the decline, compared with that of a grass sward, because of its superior nutritive value relative to grass, at least until the clover completely senesces in winter.

Antiquality factors

Red clover contains oestrogens capable of reducing reproductive performance in ruminants, with formononetin being identified as the particular phytoisoflavone responsible. Formononetin content in red clover is generally lower in stems than in leaves, especially expanding leaves, and when temperatures are high (McMurray et al., 1986). Thus, the danger of oestrogenic effects is likely to be lower in summer than in spring and autumn, although this is dependent on the leaf : stem ratio when the forage is utilized. Clover cultivars differ in formononetin concentration (Sachse, 1974; Mela et al., 1993), and reselection of existing cultivars or breeding new cultivars with the objective of lower formononetin concentration has led to the release of low-formononetin cultivars.

Ewe flocks grazed on red-clover-dominant swards or fed on red-clover silage prior to and during the mating period have given lower lambing percentages than when grazing grass (Newton and Betts, 1973; Thomson, 1975). Reduced oestrus incidence and ovulation are causal factors (Shakell and Kelly, 1984). The reductions in lambing percentages are smaller after short rather than prolonged periods of feeding and with conserved compared with fresh clover. Provided red clover with a high formononetin content comprises only part of the diet of breeding ewes, fertility may not be affected. Ewes grazing high-formononetin red clover (8–12 g kg^{-1}) exhibited no fertility problems when the companion grass comprised 70–80% of the sward (Baxter et al., 1993). Red-clover hay was safe for breeding ewes as the formononetin content in the DM decreased during its making from 10 to 2 g kg^{-1} (Hay et al., 1978). Limited evidence suggests that cattle are less susceptible than sheep to adverse oestrogenic effects, probably because they metabolize plant oestrogens more rapidly. Red-clover silage fed to cattle in winter is usually part of a mixed diet and so oestrogen intake is decreased.

Bloat (tympanites) is a potential hazard when swards with a high red-clover content are grazed or large quantities of clover-rich hay are fed. However, a number of management measures or preventive treatments are available, e.g. use of the non-ionic surfactant poloxalene (Essig, 1985).

Genetic engineering may offer the best means of incorporating anti-bloating-inducing condensed tannins from other forage legumes, such as sainfoin or birdsfoot trefoil, into red clover, but such work is at an early stage (Morris and Robbins, 1997).

Conservation

Red clover in monoculture or in mixture with grasses is a major hay crop in several regions of the world, e.g. in north-east and north-west USA. There, two hay crops or a hay crop followed by a seed crop is possible from a fully established sward (Smith *et al.*, 1985); in years with adequate moisture in late season, the sward can produce additional grazing or possibly a third hay crop.

The principles of good haymaking are well documented (Nash, 1985). Stage of growth at cutting is the major determinant of hay quality, but the final hay quality is strongly influenced by the weather during field exposure, which, ideally, should be as short as possible to minimize field losses. The average fall in the digestibility of red clover caused by haymaking is greater than in grass hays, as a result of relatively greater leaf loss (Demarquilly and Jarrige, 1970).

Early mechanical conditioning and/or chemical conditioning, using desiccant agents, is advantageous in speeding up drying, compared with unconditioned hay, but only under rain-free weather (Clark *et al.*, 1989; Narasimhalu *et al.*, 1992). Early and frequent tedding also improves field drying rate, but the frequency and severity of tedding at the later stages of field drying should be reduced in order to avoid loss of leaves by shattering, the leaves being more nutritious than the stems. However, conditioned hay is more prone to loss of the soluble carbohydrates and minerals by leaching than unconditioned hay, should there be unexpected rainfall. Red clover wetted by rain twice during haymaking lost 67% of its soluble carbohydrate constituents and digestible DM fell from 76 to 63% (Collins, 1982); minerals and vitamins are also susceptible to leaching by rainfall.

Compared with field curing, cold- or warm-air blowing systems enable a reduction in field drying time and therefore less exposure to the vagaries of weather. The end result is superior acceptability and feeding value of the hay to livestock.

A fairly standard silage management for red-clover-dominant swards is a first cut at the early flowering stage and a second cut 6–8 weeks later. The subsequent autumn regrowth provides useful grazing or, less frequently, another cut. In regions with shorter growing seasons, e.g. Nordic countries or Atlantic Canada, it is more common to take two cuts of early-flowering cultivars, in order to optimize herbage yield and quality and also to allow the build-up in late season of the nutrient reserves needed for winter survival (Pulli, 1988; Narasimhalu and Kunelius, 1994a).

When red clover is cut for silage, wheel tracking may cause physical damage to the plants and compaction of the soil. These effects lead to reduced plant density and growth vigour in the tracks, especially in wet weather and

Table 4.5. Herbage production from red-clover and red-clover/perennial-ryegrass swards under a three-cut silage regime subjected to wheel tracking (3-year means) (from Frame, 1987).

Wheel tracking	DM (t ha^{-1}) Total herbage	DM (t ha^{-1}) Red clover	Clover content (%)
Nil (control)	13.2	12.0	91
Tracking	11.0	9.2	84

Table 4.6. Management guidelines for red-clover/grass swards under UK conditions (from Frame, 1990).

Seed	Choose compatible companion grasses Sow 15 kg ha^{-1} red clover plus 5 kg ha^{-1} grass Use clover seed prefumigated against stem eelworms
Establishment	Sow shallowly (10–15 mm) Aim for soil pH of 6.0 and adequate P and K status Sow direct, preferably in spring
Production	Maintain adequate soil pH, P and K status Use potash-rich slurry in manuring programmes Use only clover-safe herbicides for weed control
Utilization	Take two silage crops and an aftermath grazing Use an effective silage additive Minimize wheel traffic at all times Keep breeding ewes off red clover swards several weeks before and after mating Graze off late-autumn/early-winter growth with sheep

on sloping fields, because of wheel slip and soil smearing. Wheel tracking reduced herbage DM production in red-clover monocultures and red-clover/grass associations compared with no tracking, and frequent wheel tracking depressed production more than infrequent, as did tracking over long periods compared with short periods after cutting (Frame, 1985; Table 4.5). It may be concluded that wheel traffic should be minimized during silage or haymaking by undertaking as few operations and over as short a period as possible, consistent with good conservation techniques, and preferably using equipment with low-ground-pressure tyres. The key points for successful management of red-clover-dominant swards for silage in the UK are summarized in Table 4.6.

Traditionally, red clover was regarded as a 'difficult' crop for silage making, on account of low DM and WSC contents and a high buffering capacity, which slowed down the attainment of a satisfactorily low pH for good fermen-

Table 4.7. Analyses of wilted red-clover silage harvested in autumn from seeding-year swards at two stages of maturity (from Narasimhalu and Sanderson, 1994).

	Sward maturity stage	
	Vegetative	20% bloom
Silage DM (g kg^{-1})	227	531
DM composition (g kg^{-1} DM):		
NDF	337	445
ADF	288	369
Lignin	37	72
Total N	38.0	24.9
Silage pH	4.1	4.8

ADF, acid detergent fibre; NDF, neutral detergent fibre.

tation. Nevertheless, silage of a satisfactory quality can be made, provided the supply of WSC is not limiting, and bruising or wilting, addition of molasses and reduction of effluent loss are measures to prevent this limitation (McDonald et al., 1965). Nevertheless, although the DM loss in the effluent was greater with clover-rich herbage than with grass-only herbage, the total loss of DM from the former was less, because of reduced loss from respiration and fermentation (Randby, 1992). Contents of WSC are usually higher in second or third cuts of red-clover swards than in first cuts (Knotek et al., 1992)

A satisfactory silage fermentation is more likely to result from red-clover/grass mixtures than from clover monocultures, since the mixtures have higher DM and WSC concentrations and lower N contents. Soil contamination, which adversely affects fermentation, is less with mixtures, since the soil is more densely covered with vegetation than with clover monocultures. Satisfactory silage has been made on-farm, using modern techniques of wilting, short chopping and application of an effective silage additive, such as formic acid (Castle and Watson, 1974; Roberts et al., 1990). Table 4.7 illustrates the composition of silage made in autumn of the establishment year.

Grazing

The morphological characteristics of red clover, including its upright growth habit and solitary root stock, make it less adaptable to grazing than to conservation. Nevertheless, it is used widely in multispecies mixtures for grazing in Scandinavia, for example. Late-flowering cultivars have a greater tolerance of grazing and are more persistent than earlier-flowering cultivars, because of a more prolific initiation of growth buds from the plant crowns. Notwithstanding, persistence in grazed swards is still limited to 2 or 3 years, so that the main function of sowing a few kilograms of clover seed in a general-purpose seed mixture for long-term use is to improve production and quality in the early life of the sward.

Red-clover content declines when swards are continuously stocked with sheep, the decline accelerating with increasing grazing pressure. This applies both when the clover is a minor constituent (Brougham, 1960) and when it is the major component (Laidlaw, 1979b) of the sown mixture, or when frequently grazed by cattle (Cosgrove and Brougham, 1985) or by sheep (Hickey and Harris, 1989). Clover contribution is also reduced under rotational sheep grazing (Wilman and Asiedu, 1983). The low herbage density of red-clover-dominant swards allows selective defoliation by sheep of the clover laminae throughout the depth of the canopy, an effect that is greatest when the sward is being grazed down (Laidlaw, 1983). Winter grazing can cause considerable damage to clover if there is overgrazing of the sward and trampling damage to the plant crowns (Hay, 1985). The use of a remote-photography technique to determine the grazed-species preferences of different types of livestock has revealed a strong preference for red clover by farmed red deer (Hunt and Hay, 1990).

Sheep-treading research on a range of pasture species places red clover low on the scale of tolerance to hoof treading by sheep, since its growth may be adversely affected, both directly from physical damage and indirectly as a result of soil compaction (Edmond, 1964).

Red clover is affected relatively more than grass when a mixed sward is severely grazed to 2.5 cm than when laxly grazed to 7.5 cm above ground level (Brougham, 1959). Thus, some form of lax rotational grazing seems a better management than continuous stocking, but there has been a lack of grazing experiments aimed at devising the best grazing regimes based on morphological characteristics, such as the seasonal pattern of development of the growth buds. This lack is understandable, since red clover is usually a minor or short-term dominant constituent of grazing mixtures, but, none the less, such experimentation would be valuable for red-clover-dominant swards used for cutting and grazing.

Animal intake and performance

Intake studies with fresh or ensiled clover herbage, using sheep, beef cattle or dairy cows, demonstrate that the animals eat more red clover than grass when compared at the same digestibility and, in some cases, when the digestibility of the clover herbage is lower. On average, DM intake is 14% higher with sheep and 19% higher with beef cattle (Copeman and Younie, 1982). A 21% increase in red-clover silage intake over grass silage has been recorded with dairy cows, although the clover silage was slightly lower in digestibility (Thomas *et al.*, 1985). However, for various reasons, including the form of red clover fed, the potential of the stock and the crop management, e.g. harvesting the clover at a low digestibility, the nutritional benefits of red clover compared with grass may not always be achieved in practice.

Dairy cows consumed 13% more DM from grass/red-clover silage, compared with grass silage of similar digestibility, although intakes of the two types of silage did not differ when the digestibility of the grass/clover silage was lower (Heikkila *et al.*, 1992). In Canadian work (Narasimhalu and

Sanderson, 1994), the voluntary intakes by sheep of seeding-year red clover, harvested unwilted or wilted for silage in the autumn, approached the upper limit of the intake range of 20–75 g DM (kg live weight $^{0.75}$ day^{-1}) reported from several silage trials with sheep (Demarquilly and Weiss, 1971). In a study with goats, herbage intake increased with increasing proportions of red clover in a cocksfoot/red-clover mixture (Lee et al., 1987).

Improved performance of different animal species and classes when fed red clover, compared with grass, is commonly reported (Copeman and Younie, 1982). This is the result of higher voluntary-intake characteristics (Thomas et al., 1981), largely as a result of lower cell-wall content, higher CP and mineral concentrations and, perhaps, greater efficiency of utilization of ME. Thomson (1977) reported that the efficiency of utilization of red clover ME was higher than that of pelleted perennial ryegrass. Examples of improved performance, relative to grass, include grazing lambs (Gibb and Treacher, 1976), silage-fed fattening beef cattle (Day et al., 1978), beef calves (Thomas et al., 1981), dairy heifers (Fisher and Otter, 1991), dairy cows (Thomas et al., 1985; Randby, 1992; Table 4.8) and red deer (Barry et al., 1993). In Chile, milk production per cow and per hectare was higher from grazed red clover than from grazed lucerne or birdsfoot trefoil swards (Pedraza et al., 1988).

In New Zealand, the high quality and strong summer growth of red clover sustain high lamb growth rates (Jagusch et al., 1981). These characteristics also suit the seasonal nutritional needs of lactating red deer, and a system based on red clover for summer production and oversown Westerwolds ryegrass (*L. multiflorum*) for winter gave fawn weaning weights 10% higher than those from traditional perennial-ryegrass/white-clover pastures (Hunt, 1993). Fawns gained 410–433 g day^{-1} liveweight on grazed red clover, compared with 322 g day^{-1} on perennial ryegrass.

For well-made red-clover silage and grass silage of similar digestibility, clover silage is predicted to give 25–30% higher liveweight gain for beef cattle and 10–13% higher milk yield from dairy cows; on swards grazed by lambs, 20% more liveweight gain is expected (Copeman and Younie, 1982).

Table 4.8. Performance of dairy cows fed grass/red-clover or grass silages of similar digestibility (from Heikkila et al., 1992).

	Grass/red clover*	Grass*
Milk yields (kg day^{-1})	29.0	27.4
Solids yield (g day^{-1})		
Fat	1310	1288
Protein	950	890
Lactose	1453	1361
Liveweight change (g day^{-1})	−6	−91

* Grass = a mixture of timothy and meadow fescue.

References

Åkerberg, E., Bingefors, S., Josefson, A. and Ellestrom, S. (1963) Induced polyploids as fodder crops. In: *Recent Plant Breeding Research, Svalof 1946–61*. Almquist and Wiksell, Stockholm, pp. 125–149.

Åman, P. (1985) Chemical composition and *in vitro* degradability of major chemical constituents in botanical fractions of red clover harvested at different stages of maturity. *Journal of the Science of Food and Agriculture* 36, 775–780.

Ames-Gottfred, N.P. and Christie, B.R. (1989) Competition among strains of *Rhizobium leguminosarum* bivar. *trifolii* and use of a diallele analysis in assessing competition. *Applied and Environmental Microbiology* 55, 1599–1604.

Arines, J., Vilarino, A. and Sainz, M. (1990) Effect of vesicular-arbuscular mycorrhizal fungi on Mn uptake by red clover. *Agriculture Ecosystems and Environment* 29, 1–4.

Arnemo, B. and Steen, E. (1972) *Seed Rate Experiments with Red Clover and Timothy in North Sweden* (English summary). Reports of the Agricultural College of Sweden, Series A, No. 161, Stockholm.

Baligar, V.C., Wright, R.J., Kinraide, T.B., Foy, C.D. and Elgin, J.H. (1987) Aluminium effects on growth, mineral uptake and efficiency ratios in red clover cultivars. *Agronomy Journal* 79, 1038–1044.

Barnett, O.W. and Diachun, S. (1985) Virus diseases of clovers. In: Taylor, N.L. (ed.) *Clover Science and Technology*. ASA/CSSA/SSSA, Madison, Wisconsin, pp. 235–268.

Barry, T.N., Niezen, J.H., Semiadi, G., Hodgson, J., Wilson, P.R. and Ataja, A.M. (1993) Development of specialist forage systems for deer production. In: Baker, M.J. (ed.) *Proceedings of the XVII International Grassland Congress, New Zealand and Australia*, Vol. II. New Zealand Grassland Association *et al.*, Palmerston North, pp. 1501–1502.

Baxter, G.S., Stevens, D.R. and Harris, A.J. (1993) Integration of high formononetin red clover in sheep grazing systems. In: Baker, M.J. (ed.)*Proceedings of the XVII International Grassland Congress, New Zealand and Australia*, Vol. II. New Zealand Grassland Association *et al.*, Palmerston North, pp. 1381–1382.

Belzile, L. (1987) Effect of companion timothy on winter survival of red clover. *Canadian Journal of Plant Science* 67, 1101–1103.

Belzile, L. (1994) Potential yield of red clover seed in the seeding year. *Canadian Journal of Plant Science*. 74, 807–809.

Boller, B.C. (1996) Suitability of red clover and Italian ryegrass varieties for binary species mixtures. In: *Abstracts of the 20th Meeting of the Fodder Crops and Amenity Grasses Section*, EUCARPIA, Poland, pp. 19–20.

Boller, B.C. and Nösberger, J. (1987) Symbiotically fixed nitrogen from field grown white and red clover mixed with ryegrasses at low levels of ^{15}N fertilisation. *Plant and Soil* 104, 219–226.

Bowen, R. and Plumb, R.T. (1979) The occurrence and effects of red clover necrotic mosaic virus on red clover (*Trifolium pratense*). *Annals of Applied Biology* 9l, 227–236.

Bowley, S.R., Taylor, N.L. and Dougherty, C.T. (1984) Physiology and morphology of red clover. *Advances in Agronomy* 37, 317–347.

Bowley, S.R., Taylor, N.L. and Dougherty, C.T. (1987) Photoperiodic response and heritability of the pre-flowering interval of 2 red clover (*Trifolium pratense*) populations. *Annals of Applied Biology* 111, 455–461.

Brougham, R.W. (1959) The effects of frequency and intensity of grazing on the productivity of a pasture of short rotation ryegrass and red and white clover. *New Zealand Journal of Agricultural Research* 2, 1232–1248.

Brougham, R.W. (1960) The effects of frequent hard grazings at different times of the year on the productivity and species yields of a grass–clover pasture. *New Zealand Journal of Agricultural Research* 3, 125–136.

Butler, G.W., Greenwood, R.M. and Soper, K. (1959) Effects of shading and defoliation on the turnover of root and nodule tissue of plants of *Trifolium repens*, *Trifolium pratense* and *Lotus uliginosus*. *New Zealand Journal of Agricultural Research* 2, 415–426.

Buxton, D.R., Hornstein, J.S., Wedin, W.F. and Marten, G.C. (1985) Forage quality in stratified canopies of alfalfa, birdsfoot trefoil and red clover. *Crop Science* 25, 273–279.

Campbell, T.A., Neurenberg, N.J. and Foy, C.D. (1990) Differential responses of red clover germplasms to aluminium stress. *Journal of Plant Nutrition* 13, 1463–1474.

Carr, A.J.H. (1971) Virus diseases of forage legumes. In: Western, J.H. (ed.) *Diseases of Crop Plants*. Macmillan, London, pp. 254–307.

Castle, M.E. (1984) Red clover: syndicate report. In: Thomson, D.J. (ed.) *Forage Legumes*. Occasional Symposium No. 16, British Grassland Society, Hurley, pp. 222–224.

Castle, M.E. and Watson, J. (1974) Red clover silage for milk production. *Journal of the British Grassland Society* 29, 101–108.

Caubel, G. (1989) Les nématodes. In: Raynal, G., Gondran, J., Bournoville, R. and Courtillot, M. (eds) *Ennemis et Maladies des Prairies*. INRA, Paris, pp. 196–200.

Chi, C.C. and Hanson, E.W. (1961) Nutrition in relation to the development of wilts and root rots incited by *Fusarium* in red clover. *Phytopathology* 51, 704–711.

Chloupek, O. (1976) The size of the root system of tetraploid red clover and its relation to the chemical composition of the herbage produced. *Journal of the British Grassland Society* 31, 23–27.

Choo, T.M. and Suzuki, M. (1983) Effect of plant age on cold hardiness in red clover. *Forage Notes, Canada* 27, 26–27.

Christie, B.R. and Choo, T.M. (1990) Effects of harvest time and Alar-85 on seed yield of red clover. *Canadian Journal of Plant Science* 70, 869–871.

Christie, B.R. and Choo, T.M. (1991) Morphological characteristics associated with winter survival of five growth types of tetraploid red clover. *Euphytica* 54, 275–278.

Christie, B.R., Clark, E.A. and Fulkerson, R.S. (1992) Comparative plowdown value of red clover strains. *Canadian Journal of Plant Science* 72, 1207–1213.

Chugunova, N.G. and Romanova, A.K. (1988) Effect of cultivation temperature on growth, photosynthesis and oxygenase activity of ribulose bisphosphate carboxyl carboxylase oxygenase in red clover. *Soviet Plant Physiology* 35, 816–821. (*Herbage Abstracts* 59, abstract 1684.)

Clark, E.A., Crump, S.V. and Kondra, Z.P. (1989) Mechanical and chemical conditioning of hays. *Canadian Journal of Plant Science* 69, 133–141.

Clifford, P.T. (1973) Increasing bumble bee densities in red clover seed production areas. *New Zealand Journal of Experimental Agriculture* 1, 377–379.

Collins, M. (1982) The influence of wetting on the composition of alfalfa, red clover and birdsfoot trefoil hay. *Agronomy Journal* 74, 1041–1044.

Collins, M. (1983) Changes in composition of alfalfa, red clover and birdsfoot trefoil during autumn. *Agronomy Journal*, 75, 287–291.

Collins, M. (1993) Yields and nitrogen contents of maize receiving fertilizer nitrogen or after alfalfa and red clover. In: Baker, M.J. (ed.) *Proceedings of the XVII International*

Grassland Congress, New Zealand and Australia, Vol III. New Zealand Grassland Association *et al.*, Palmerston North, pp. 2200–2202

Collins, M. and Taylor, T.H. (1984) Quality changes of late summer and autumn produced alfalfa and red clover. *Agronomy Journal* 76, 409–415.

Connolly, V. (1970) A comparison of six red clover varieties in association with three grass species. *Irish Journal of Agricultural Research* 9, 203–213.

Cook, R. and Yeates, G.W. (1993) Nematode pests of grassland and forage crops. In: Evans, K., Trudgill, D.L. and Webster, J.M. (eds) *Plant Parasitic Nematodes in Temperate Agriculture*. CAB International, Wallingford, pp. 305–350.

Copeman, G.J.F. and Younie, D. (1982) Feed quality and utilisation of red clover swards. In: Murray, R.B. (ed.) *Legumes in Grassland. Proceedings of the Fifth Study Conference of the Scottish Agricultural Colleges*. The Scottish Agricultural Colleges, Aberdeen, Auchincruive, Edinburgh, pp. 53–58.

Cosgrove, G.P. and Brougham, R.W. (1985) Grazing management influences on seasonality and performance of ryegrass and red clover in a mixture. *Proceedings of the New Zealand Grassland Association* 46, 71–76.

Crowley, J.G. (1975) Red clover – a new look at an old crop. *Farm and Food Research, The Agricultural Institute* 6, 38–40.

Curll, M.L. and Gleeson, A.C. (1987) The introduction of red or white clover into a perennial grass sward. *Grass and Forage Science* 42, 397–403.

Day, N., Harkess, R.D. and Harrison, D.M. (1978) A note on red clover silage for cattle finishing. *Animal Production* 26, 97–100.

Demarquilly, C. (ed.) (1981) *Prévision de la Valeur Nutritive des Aliments des Ruminants*. INRA, Centre de Recherches Zootechniques et Vétérinaires de Theix, Beaumont.

Demarquilly, C. and Jarrige, R. (1970) The effect of method of forage conservation on digestibility and voluntary intake. In: *Proceedings of the XI International Grassland Congress, Surfers' Paradise, Australia*, pp. 733–737.

Demarquilly, C. and Weiss, P. (1971) Liaisons entre les quantités de matière sèche de fourrages verts par les moutons et celles ingérées par les bovins. *Annales Zootechniques* 20, 119–134.

Dennis, B.A. (1980) Breeding for improved seed production in autotetraploid red clover. In: Hebblethwaite, P.D. (ed.) *Seed Production. Proceedings of the 28th Easter School, University of Nottingham*. Butterworth, London, pp. 229–240.

Dennis, B.A. and Haas, H. (1967) Pollination and seed-setting in diploid and tetraploid red clover (*Trifolium pratense* L.) under Danish conditions in relation to the number and type of pollinating insects. (Cited in Dennis, 1980.)

Dougherty, C.T. (1972) Water stress in Turoa red clover under Aotea wheat. *New Zealand Journal of Agricultural Research* 15, 706–711.

Edmond, D.B. (1964) Some effects of sheep treading on the growth of ten pasture species. *New Zealand Journal of Agricultural Research* 7, 1–16.

Elliott, C.R., Hoyt, P.B., Nyborg, M. and Siemens, B. (1973) Sensitivity of several species of grasses and legumes to soil acidity. *Canadian Journal of Plant Science* 53, 113–117.

Essig, H.W. (1985) Quality and antiquality components. In: Taylor, N.L. (ed.) *Clover Science and Technology*. ASA/CSSA/SSSA, Madison, Wisconsin, pp. 309–324.

Evans, A.M. (1976) Clover (*Trifolium* spp.). Leguminosae – Papilionatae. In: Simmonds, N.W. (ed.), *Evolution of Crop Plants*. Longmans, London, pp. 175–179.

Evans, P.S. (1973) The effect of repeated defoliation to three levels on root growth of five pasture species. *New Zealand Journal of Agricultural Research* 16, 31–34.

Farnham, D.E. and George, J.R. (1993) Dinitrogen fixation and nitrogen transfer among red clover cultivars. *Canadian Journal of Plant Science* 73, 1047–1054.

Fejer, S.O. (1960) Response of some New Zealand pasture species to vernalisation. *New Zealand Journal of Agricultural Research* 3, 656–662.

Fergus, E.N. and Hollowell, E.H. (1960) Red clover. *Advances in Agronomy* 12, 365–436.

Fernandez, P. and Warembourg, F.R. (1987) Distribution and utilisation of assimilated carbon in red clover during the first year of vegetation. *Plant and Soil* 97, 131–143.

Fisher, G.E.J. and Offer, N.W. (1991) *In vivo* nutritional evaluation of silage made from extensive pastures containing wild flowers and herbs. *Animal Production* 52, 574.

Fitter, A.H. (1988) Water relations of red clover (*Trifolium pratense* L.) as affected by VA mycorrhizal infection and phosphorus supply before and during drought. *Journal of Experimental Botany* 39, 595–603.

Frame, J. (1976) The potential of tetraploid red clover and its role in the United Kingdom. *Journal of the British Grassland Society* 31, 139–152.

Frame, J. (1985) The effect of tractor wheeling on red clover swards. *Research and Development in Agriculture* 2, 77–85.

Frame, J. (1986) The production and quality potential of four legumes sown alone and combined in various associations. *Crop Research (Horticultural Research)* 25, 103–122.

Frame, J. (1987) The effect of tractor wheeling on the productivity of red clover and red clover/ryegrass swards. *Research and Development in Agriculture* 4, 55–60.

Frame, J. (1990) The role of red clover in the United Kingdom pastures. *Outlook on Agriculture* 19, 49–55.

Frame, J. and Harkess, R.D. (1987) The productivity of four forage legumes sown alone and with each of five companion grasses. *Grass and Forage Science* 42, 213–223.

Frame, J., Harkess, R.D. and Hunt, I.V. (1972) The effect of a ryegrass companion grass and the variety of red clover on the productivity of red-clover swards. *Journal of the British Grassland Society* 27, 241–249.

Frame, J., Harkess, R.D. and Hunt, I.V. (1976a) The effect of variety and fertilizer nitrogen level on red clover production. *Journal of the British Grassland Society* 31, 111–115.

Frame, J., Harkess, R.D. and Hunt, I.V. (1976b) The influence of date of sowing and seed rate on the production of pure-sown red clover. *Journal of the British Grassland Society*, 31, 117–122.

Frame, J., Harkess, R.D. and Hunt, I.V. (1985) Effect of seed rate of red clover and of timothy or tall fescue on herbage production. *Grass and Forage Science* 40, 459–465.

Frame, J., Fisher, G.E.J. and Tiley, G.E.D. (1994) Wildflowers in grassland systems. In: Haggar, R.J. and Peel, S. (eds) *Grassland Management and Nature Conservation*. Occasional Symposium No 28. British Grassland Society, Reading, pp. 104–114.

Francis, C.M. and Quinlivan, B.J. (1974) Selection for formononetin content in red clover. In: Iglovikov, V.G. and Movsisyants, A.P. (eds) *Proceedings of the XII International Grassland Congress, Moscow, Soviet Union*, Vol. 3. Universe Publishing House, Moscow, pp. 754–758.

Frankow-Lindberg, B.E. (1989) The effect of nitrogen and clover proportion on yield of red clover–grass mixtures. In: *Proceedings of the XVI International Grassland Congress, Nice, France*, Vol. I. Association Française pour la Production Fourragère, Versailles, pp. 173–174.

Gange, A.C., Brown, V.K., Evans, I.M. and Storr, A.L. (1989) Variation on the impact of insect herbivory on *Trifolium pratense* through early plant succession. *Journal of Ecology* 77, 537–551.

Gerke, J., Meyer, U. and Romer, W. (1995) Phosphate, Fe and Mn uptake of N_2-fixing red clover and ryegrass from an oxisol as affected by P and model humic substances application. 1. Plant parameters and soil solution composition. *Zeitschrift für Pflanzenernährung und Bodenkunde*, 158, 261–268.

Gibb, M.J. and Treacher, T.T. (1976) The effect of herbage allowance on herbage intake and performance of lambs grazing perennial ryegrass and red clover. *Journal of Agricultural Science, Cambridge* 86, 355–365.

Gondran, J. and Raynal, G. (1989) Maladies des parties souterraines des Fusarioses. In: Raynal, G., Gondran, J., Bournoville, R. and Courtillot, M. (eds) *Ennemis et Maladies des Prairies*. INRA, Paris, p. 111.

Gracey, H.J. (1981) Cattle slurry as a source of nutrients for red clover: herbage production and clover contribution. *Grass and Forage Science* 36, 291–295.

Guy, P. (1989) Essais multilocaux d'associations trifle violet graminée. *Fourrages* 117, 29–47.

Habovštiak, J. (1989) The effect of graded liming doses on meadow clover (*Trifolium pratense* L.) yields in alpine regions of Slovakia. *Pol'nohospodárstvo*, 35, 519–527.

Haggar, R.J. and Koch, D.W. (1983) Slot-seeding investigations. 3. The productivity of slot-seeded red clover compared with all-grass swards receiving nitrogen. *Grass and Forage Science* 38, 45–53.

Hall, M.H. and Eckert, J.W. (1992) Seeding year harvest: a management of red clover. *Journal of Production Agriculture* 5, 52–57.

Halling, M.A. (1988) Growth of timothy and red clover in relation to weather and time of autumn cutting 2. Storage of carbohydrates in autumn and spring. *Swedish Journal of Agricultural Research* 18, 161–170.

Hardie, K. and Leyton, L. (1981) The influence of vesicular-arbuscular mycorrhiza on growth and water relations of red clover l. In phosphate deficient soil. *New Phytologist* 89, 599–608.

Hay, R.J.M. (1985) Variety × management interactions with red clover. Unpublished PhD thesis, Lincoln College, New Zealand.

Hay, R.J.M. and Ryan, D.L. (1989) A review of l0 years' research with red clovers under grazing in Southland. *Proceedings of the New Zealand Grassland Association* 50, 181–187.

Hay, R.J.M., Kelly, R.W. and Ryan, D.L. (1978) Some aspects of the performance of Grasslands Pawera red clover in Southland. *Proceedings of the New Zealand Grassland Association* 38, 246–252.

Hegarty, T.W. and Ross, H.A. (1979) Use of growth regulators to remove the differential sensitivity to moisture stress of seed germination and seedling growth in red clover (*Trifolium pratense* L.). *Annals of Botany* 43, 657–660.

Heichel, G.H. and Henjum, K.I. (l991) Dinitrogen fixation, nitrogen transfer and productivity of forage legume–grass communities. *Crop Science* 31, 202–208.

Heichel, G.H., Vance, C.P., Barnes, D.K. and Henjum, K.I. (1985) Dinitrogen fixation and N and dry matter distribution during 4 year stands of birdsfoot-trefoil and red clover. *Crop Science* 25, 101–105.

Heikkila, T., Toivonen, V. and Mela, T. (1992) Comparison of red clover–grass silage with grass silage for milk production. In: *Proceedings of the 14th General Meeting of the European Grassland Federation, Lahti, Finland*. Finnish Grassland Association, pp. 388–391.

Hesketh, J.D. and Moss, D.N. (1963) Variation in the response of photosynthesis to light. *Crop Science* 3, 107–110.

Hickey, M.J. and Harris, A.J. (1989) Comparative performance of three early-flowering red clovers under sheep grazing in Southland, New Zealand. *New Zealand Journal of Agricultural Research* 32, 327–332.

Hunt, I.V., Frame, J. and Harkess, R.D. (1975) Potential productivity of red clover varieties in south-west Scotland. *Journal of the British Grassland Society* 30, 209–216.

Hunt, I.V., Frame, J. and Harkess, R.D. (1976) Removal of mineral nutrients by red clover varieties. *Journal of the British Grassland Society* 31, 171–179.

Hunt, W.F. (1993) Maximising red deer venison production through high quality pasture. In: Baker, M.J. (ed.) *Proceedings of the XVII International Grassland Congress, New Zealand and Australia*, Vol II. New Zealand Grassland Association et al., Palmerston North, pp. 1497–1498.

Hunt, W.F. and Hay, R.J.M. (1989) Alternative pasture species for deer production in the Waikato. In: *Proceedings of the Ruakura Deer Industry Conference*, Hamilton, pp. 31–33.

Hunt, W.F. and Hay, R.J.M. (1990) A photographic technique for assessing the pasture species preferences of grazing animals. *Proceedings of the New Zealand Grassland Association* 51, 191–196.

Jagusch, K.T., Duganzich, D.M., Winn, G.W. and Rattray, P.V. (1981) The effect of the year and pasture allowance on the growth of lambs fed different pasture species. *Proceedings of the New Zealand Society of Animal Production* 41, 117–118.

Jin, X.X., Morton, J. and Butler, L. (1992) Interactions between *Fusarium avenaceum* and *Hylastinus obscurus* (Coleoptera Scolytidae) and their influence on root decline in red clover. *Journal of Economic Entomology* 85, 1340–1346.

Joggi, D., Hofer, U. and Nösberger, J. (1983) Leaf area index, canopy structure and photosynthesis of red clover (*Trifolium pratense* L.). *Plant Cell and Environment* 6, 611–616.

Johansen, B.R. (1984) Influence of nitrogen on yield and botanical composition in monocultures and mixtures of red clover and three grass species. In: Riley, H. and Skjelvåg, A.O. (eds) *Proceedings of the 10th General Meeting of the European Grassland Federation, Ås, Norway*. The Norwegian State Agricultural Research Stations, Ås, pp.186–190.

Jones, T.W.A. (1974) The effect of leaf number on the sensitivity of red clover seedlings to photoperiodic induction. *Journal of the British Grassland Society* 29, 25–28.

Jonsson, H.A. (1985) Red clover (*Trifolium pratense*) Sara. *Agri Hortique Genetica* 42, 43–51.

Kalmbacher, R.S., Misleiry, P. and Martin P.G. (1980) Sod-seeding Bahia grass in winter with three temperate legumes. *Agronomy Journal* 72, 114–118.

Kelling, K.A. and Matocha, J.E. (1990) Plant analysis as an aid in fertilizing forage crops. In: Kelling, K.A., Matocha, J.E. and Westerman, R.L. (eds) *Soil Testing and Plant Analysis*, 3rd edn. Book Series No. 3, SSSA. Madison, Wisconsin, pp. 603–643.

Kelly, A.F. (1988) *Seed Production of Agricultural Crops*. Longman Scientific and Technical, Harlow.

Kemp, P.D. and Blair, G.J. (1991) Phosphorus efficiency in pasture species. 6. A comparison of Italian ryegrass, *Phalaris*, red clover and white clover over time. *Australian Journal of Agricultural Research* 42, 541–558.

Kendall, W.A., Stroube, W.H. and Taylor, N.L. (1962) Growth and persistence of several varieties of red clover at various temperature and moisture levels. *Agronomy Journal* 54, 345–347.

Khadhair, A.H., Sinha, R.C. and Peterson, J.F. (1984) Effect of white clover mosaic virus infection on various processes relevant to symbiotic N_2-fixation in red clover. *Canadian Journal of Botany* 62, 38–43.

Kilpatrick, R.A., Judd, R.W., Dunn, G.M. and Rich, A.E. (1966) Agents affecting survival of red and white clovers exposed to low temperatures in a freezing chamber. *Crop Science* 6, 499–501.

Kimpinski, J., Kunelius, H.T. and Willis, C.B. (1984) Plant parasitic nematodes in temperate forage grass and legume species in Prince Edward Island. *Canadian Journal of Plant Science* 6, 499–501.

Knotek, S., Žilaková, J. and Zimková, M. (1992) Differences in forage conservation of semi-natural grasslands, sown grasses, and grass/clover mixtures and legumes. In: *Proceedings of the 14th General Meeting of the European Grassland Federation, Lahti, Finland.* Finnish Grassland Assocation, pp. 144–148.

Koch, D.W., Mueller-Warrant, G.W. and Mitchell, G.R. (1983) Sod-seeding of forages. 1. Alternative to conventional establishment. *University of New Hampshire Agricultural Experiment Station Bulletin* 525, 1–29.

Kunelius, H.T. and Campbell, A.J. (1984) Performance of sod-seeded temperate legumes in grass dominant swards. *Canadian Journal of Plant Science* 64, 643–650.

Kunelius, H.T., Carter, M.R., Kimpinski, J. and Sanderson, J.B. (1988) Effect of seeding method on alfalfa and red clover establishment and growth, soil physical condition and nematode populations. *Soil and Tillage Research* 12, 163–175.

Kunelius, H.T., Harris, W., Henderson, J.D. and Baker, E.J. (1982) Comparison of tillage methods on red clover and ryegrass establishment and production under grazing in the establishment year. *New Zealand Journal of Agricultural Research* 10, 253–263.

Kunelius, H.T., Johnston, H.W. and MacLeod, J.A. (1992) Effect of undersowing barley with Italian ryegrass or red clover on yield, crop composition and root biomass. *Agriculture, Ecosystems and Environment* 38, 127–137.

Laidlaw, A.S. (1979a) Barley and westerwolds ryegrass as cover crops for red clover. *Record of Agricultural Research, Department of Agriculture for Northern Ireland* 23, 33–36.

Laidlaw, A.S. (1979b) Effects of grazing by lambs in autumn on a red clover–perennial ryegrass sward. *Grass and Forage Science* 34, 191–196.

Laidlaw, A.S. (1983) Grazing by sheep and the distribution of species through the canopy of a red clover–perennial ryegrass sward. *Grass and Forage Science* 38, 317–321.

Laidlaw, A.S. and Frame, J. (1988) Maximising the use of the legume in grassland systems. In: *Proceedings of the 12th General Meeting of the European Grassland Federation, Dublin, Ireland.* Irish Grassland Association, Belclare, pp. 199–203.

Laidlaw, A.S. and McBratney, J.M. (1980) The effect of companion perennial ryegrass on red clover productivity when timing of the first cut is varied. *Grass and Forage Science* 35, 257–265.

Lampkin, N. (1990) *Organic Farming.* Farming Press Books, Ipswich.

Leath, K.T. (1985) General diseases. In: Taylor, N.L. (ed.) *Clover Science and Technology.* ASA/CSSA/SSSA, Madison, Wisconsin, pp. 205–233.

Leath, K.T., Zeiders, K.E. and Byers, R.A. (1973) Increased yield and persistence of red clover after a soil drench application of benomyl. *Agronomy Journal* 65, 1008–1009.

Leath, K.T., Lukezic, F.L., Penningpacker, B.W., Kendal, W.A., Levine, R.G. and Hill, R.R. (1989) Interaction of *Fusarium avenaceum* and *Pseudomonas viridiflava* in root rot of red clover. *Phytopathology* 79, 436–440.

Lee, I.D., Myung, J., Song, W.S. and Chun, Y.K. (1987) Study on the use of orchardgrass–red clover mixtures on intake, digestibility and preference by Korean native goats. *Journal of the Korean Grassland Society* 7, 31–36.

Lewis, G.C., Heard, A.J., Gutteridge, R.A., Plumb, R.T. and Gibson, R.W. (1985) The effects of mixing Italian ryegrass (*Lolium multiflorum*) with perennial ryegrass (*L. perenne*) and red clover (*Trifolium pratense*) on the incidence of viruses. *Annals of Applied Biology* 106, 483–488.

Lindstrom, K. and Millyniemi, H. (1987) Sensitivity of red clover rhizobia to soil acidity factors in pure culture and in symbiosis. *Plant and Soil* 98, 353–362.

Lipsanen, P. and Lindstrom, K. (1986) Adaptation of red clover rhizobia to low temperatures. *Plant and Soil* 92, 55–62.

Maag, H.P. and Nösberger, J. (1980) Photosynthetic rate, chlorophyll content and dry matter production of *Trifolium pratense* L. as influenced by nitrogen nutrition. *Angewandte Botanik* 54, 187–194.

McBratney, J.M. (1981) Productivity of red clover grown alone and with companion grasses over a four-year period. *Grass and Forage Science* 36, 267–279.

McBratney, J.M. (1984) Productivity of red clover grown alone and with companion grasses: further studies. *Grass and Forage Science* 39, 167–175.

McBratney, J.M. (1987) Effect of fertiliser nitrogen on six-year-old red clover/perennial ryegrass swards. *Grass and Forage Science* 42, 147–152.

McDonald, P., Stirling, A.C., Henderson, A.R. and Whittenbury, R. (1965) Fermentation studies on red clover. *Journal of the Science of Food and Agriculture* 16, 549–557.

McKersie, B.D. (1997) Improving forage production systems using biotechnology. In: McKersie, B.D. and Brown, D.C.W. (eds) *Biotechnology and the Improvement of Forage Legumes*. CAB International, Wallingford, pp. 3–21.

McLaughlin, M.R. and Boykin, D.L. (1988) Virus diseases of seven species of forage legumes in the south eastern USA. *Plant Diseases* 72, 539–542.

MacLeod, L.B. and Jackson, L.P. (1965) Effect of concentration of the aluminium ion on root development and establishment of legume seedlings. *Canadian Journal of Soil Science* 45, 221–234.

McMurray, C.H., Laidlaw, A.S. and McElroy, M. (1986) The effect of plant development and environment on formononetin concentration in red clover (*Trifolium pratense* L.). *Journal of the Science of Food and Agriculture* 37, 333–340.

Mallarino, A.P., Wedin, W.F., Perdomo, C.H., Goyenola, R.S. and West, C.P. (1990a) Legume species and proportion effects on symbiotic dinitrogen fixation in legume–grass mixtures. *Agronomy Journal* 82, 785–789.

Mallarino, A.P., Wedin, W.F., Perdomo, C.H., Goyenola, R.S. and West, C.P. (1990b) Nitrogen transfer from white clover, red clover and birdsfoot trefoil to associated grasses. *Agronomy Journal* 82, 790–795.

Martensson, A.M. (1990) Competitiveness of inoculant strains of *Rhizobium leguminosarum* bivar. *trifolii* in red clover using repeated inoculation and increased inoculum levels. *Canadian Journal of Microbiology* 36, 136–139.

Matsuura, M., Matsumoto, N., Sawa, A., Grau, M. and Weda, S. (1985) Resistance of red clover cultivars to clover rot in pure and mixed stands with reference to breeding work. In: *Proceedings of the XV International Grassland Congress, Kyoto, Japan*.

The Science Council of Japan and the Japanese Society of Grassland Science, Tochigi-kem,pp. 288–289.

Mela, T. (1969) The effects of N dimethylamino-succinamic acid (B-995) on the seed cultivation characteristics of late flowering red clover. *Acta Agraria Fennica* 115, 1–114.

Mela, T., Laakso, I., Mäenpää, T. and Ihamäki, H. (1993) Variation in the isoflavone and sterocarpan contents of red clover. In: Baker, M.J. (ed.) *Proceedings of the XVII International Grassland Congress, New Zealand and Australia*, Vol. II. New Zealand Grassland Association *et al.*, Palmerston North, pp. 1385–1386.

Mercer, C.F. and Campbell, B.D. (1986) Early invasion of overdrilled red clover seedlings by nematodes. *New Zealand Journal of Agricultural Research* 29, 495–499.

Montpetit, J.M. and Coulman, B.E. (1991a) Relationship between spring vigor and the presence of adventitious roots in established stands of red clover (*Trifolium pratense* L.). *Canadian Journal of Plant Science* 71, 749–754.

Montpetit, J.M. and Coulman, B.E. (1991b) Response to divergent selection for adventitious growth in red clover (*Trifolium pratense* L.). *Euphytica* 58, 119–127.

Morris, P. and Robbins, M.P. (1997) Manipulating condensed tannins in forage legumes. In: McKersie, B.D. and Brown, D.C.W. (eds) *Biotechnology and the Improvement of Forage Legumes*. CAB International, Wallingford, pp. 147–173.

Mouat, M.C.H. and Anderson, L.B. (1974) Effect of ploidy change on root cation-exchange capacity and mineral composition in red clover. *New Zealand Journal of Agricultural Research* 17, 55–58.

Mousset-Declas, C., Faurie, F. and Tisserand, J.L. (1993) Is there variability for quality in red clover? In: Baker, M.J. (ed.) *Proceedings of the XVII International Grassland Congress, New Zealand and Australia*, Vol. I New Zealand Grassland Association *et al.*, Palmerston North, pp. 442–443.

Myers, J.R., Grosser, J.W., Taylor, N.L. and Collins, G.B. (1989) Genotype dependent whole plant regeneration from protoplasts of red clover (*Trifolium pratense* L.). *Plant Cell, Tissue and Organ Culture* 19, 113–127.

Narasimhalu, P. and Kunelius, H.T. (1994a) Yield and composition of red clover treated with growth regulators in spring. *Grass and Forage Science* 49, 138–145.

Narasimhalu, P. and Kunelius, H.T. (1994b) Mineral composition of red clover treated with growth regulators in early spring. *Grass and Forage Science* 49, 146–151.

Narasimhalu, P. and Sanderson, J.B. (1994) Composition and utilization in sheep of unwilted and wilted silages prepared from seeding year red clover (*Trifolium pratense* L.) cut at two maturity stages. *Canadian Journal of Plant Science* 74, 87–91.

Narasimhalu, P., Kunelius, H.T. and McCrae, K.B. (1992) Chemical and mechanical conditioning for field drying of *Trifolium pratense* L. *Canadian Journal of Plant Science* 72, 1193–1198.

Nash, M.J. (1985) *Crop Conservation and Storage in Cool Temperate Climates*. Pergamon Press, Oxford.

National Institute of Agricultural Botany (1988) *Diseases of Grass and Herbage Legumes*. Leaflet, NIAB, Cambridge, 37 pp.

Naylor, R.E.L., Marshall, A.H. and Matthews, S. (1983) Seed establishment in directly drilled sowings. *Herbage Abstracts* 53, 73–91.

Nelson, S.O., Ballard, A.T., Stetson, L.E. and Wolf, W.W. (1976) Radiofrequency electrical treatment effects in dormancy and longevity of seed. *Journal of Seed Technology* 1, 31–43.

Newton, J.E. and Betts, J.E. (1973) The effect of red clover (*Trifolium pratense* var. Redhead), white clover (*Trifolium repens* var. S 100) or perennial ryegrass (*Lolium perenne* var. S 23) on the reproductive performance of sheep. *Journal of Agricultural Science, Cambridge* 80, 323–327.

Nösberger, J. and Joggi, D. (1981) Canopy structure and photosynthesis in red clover. In: Wright, C.E. (ed.) *Plant Physiology and Herbage Production*. Occasional Symposium No. 13, British Grassland Society, Hurley, pp. 37–40.

Ohlsson, C. and Wedin, W.F. (1989) Phenological staging schemes for predicting red clover quality. *Crop Science* 29, 416–420.

Oliva, R.N., Steiner, J.J. and Young, W.C. III (1994a) Red clover seed production. 1. Crop water requirements and irrigation timing. *Crop Science* 34, 178–184.

Oliva, R.N., Steiner, J.J. and Young, W.C. III (1994b) Red clover seed production. 2. Plant water status on yield and yield components. *Crop Science* 34, 184–192.

Olsen, F.J., Jones, J.H. and Patterson, J.H. (1981) Sod-seeding forage legumes in a tall fescue sward. *Agronomy Journal* 73, 1032–1036.

O'Rourke, C.J. (1976) *Diseases of Grasses and Forage Legumes in Ireland*. Agricultural Institute, Dublin.

Palmer-Jones, T., Forster, I.W. and Clinch, P.G. (1966) Observations on the pollination of Montgomery red clover (*Trifolium pratense* L.). *New Zealand Journal of Agricultural Research* 9, 738–747.

Pedraza, G.C., Tames, A.I. and Olguin, H.H. (1988) Milk production from direct grazing of birdsfoot trefoil, lucerne and red clover. *Agricultura Tecnica* 48, 97–101.

Peterson, P.R., Sheaffer, C.C. and Hall, M.H. (1992) Drought effects on perennial forage legume yield and quality. *Agronomy Journal* 84, 774–779.

Phillips, G.C., Grosser, J.W., Berger, S., Taylor, N.L. and Collins, G.B. (1992) Interspecific hybridisation between red clover and *T. alpestre* using *in vitro* embryo rescue. *Crop Science* 32, 1113–1115.

Planquaert, P. and Raphalen, J.L. (1973) La luzerne et le trèfle violet en Bretagne. *Bulletin Technique d'Information ITCF, Paris* 281, 519–524.

Pulli, S. (1984) Adaptation and persistence of red clover on Finnish farms. In: Riley, H. and Skjelvåg, A.O. (eds) *Proceedings of the 10th General Meeting of the European Grassland Federation, Ås, Norway*. The Norwegian State Agricultural Research Stations, Ås, pp. 297–301.

Pulli, S. (1988) Adaptation of red clover to the long day environment. *Journal of Agricultural Science in Finland* 60, 201–214.

Puri, K.P. and Laidlaw, A.S. (1983) The effect of time of harvest on seed production of three red clover cultivars. *Grass and Forage Science* 39, 221–228.

Puri, K.P. and Laidlaw, A.S. (1984) The effect of temperature on components of seed yield and on hardseededness in three cultivars of red clover (*Trifolium pratense* L.). *Journal of Applied Seed Production* 2, 18–23.

Randby, A.T. (1992) Grass–clover silage for dairy cows. In: *Proceedings of the 14th General Meeting of the European Grassland Federation, Lahti, Finland*. Finnish Grassland Association, pp. 272–275.

Raynal, G. (1989a) La sclérotiniose des trèfles et luzernes à *Sclerotinia trifoliorum* Eriks. II. Variabilité du parasite, résistance des plantes en conditions contrôlées. *Agronomie* 1, 573–578.

Raynal, G. (1989b) Maladies les légumineuses: sclérotiniose. In: Raynal, G, Gondran, J., Bournoville, R. and Courtillot, M. (eds) *Ennemis et Maladies des Prairies*. INRA, Paris, pp. 109–110.

Rice, W.A., Penney, D.C. and Nyborg, M. (1977) Effects of soil acidity and *Rhizobium* numbers, nodulation and nitrogen fixation by alfalfa and red clover. *Canadian Journal of Soil Science* 57, 197–203.

Rincker, C.M. and Rampton, H.H. (1985) Seed production. In: Taylor, N.L. (ed.) *Clover Science and Technology*. ASA/CSSA/SSSA, Madison, Wisconsin, pp. 417–443.

Roberts, D.J., Lawson, A. and Fisher, G.E.J. (1990) A comparison of a red clover/grass sward with a grass sward plus fertilizer nitrogen under a three-cut silage regime. In: *British Grassland Society Second Research Conference*, Session III, Paper 5 British Grassland Society, Reading.

Rufelt, S. (1983) Root rot – an unavoidable disease? A discussion of factors involved in the root rot of forage legumes. *Vaxtskyddsnotiser* 6, 123–127.

Ruiz, I. (1973) Cutting management effects on growth, morphology and physiology of red clover. *Dissertation Abstracts International* 33B, 4621–4622.

Rumball, W. (1983) Red clover. In: Wratt, G.S. and Smith, H.C. (eds) *Plant Breeding in New Zealand*. Butterworths/DSIR, Wellington, pp. 237–241.

Ryle, G.J.A., Powell, C.E. and Gordon, A.J. (1981) Patterns of ^{14}C-labelled assimilate partitioning in red and white clover during vegetative growth. *Annals of Botany* 47, 505–514.

Sachse, J. (1974) Die Bestimmung östrogener Isaflavone und Cumöstrol in Klee (*Trifolium pratense* L. and *Trifolium repens* L.). *Journal of Chromatography* 96, 123–136.

Sainz, M.J. and Arines, J. (1988a) Effects of native vesicular-arbuscular mycorrhiza fungi and phosphate fertiliser on red clover growth in acid soils. *Journal of Agricultural Science, Cambridge* 111, 67–73.

Sainz, M.J. and Arines, J. (1988b) P absorbed from soil by mycorrhizal red clover plants as affected by soluble P fertilisation. *Soil Biology and Biochemistry* 20, 61–67.

Schmidt, D. (1980) Etudes sur la résistance du trèfle violet à *Sclerotinia trifoliorum* Eriks. *Revue Suisse Agriculture* 12, 197–206.

Shakell, G.H. and Kelly, R.W. (1984) Residual effects of short-term grazing of red clover cultivar Grasslands Pawera pastures on the reproductive performance of ewes. *New Zealand Journal of Experimental Agriculture* 12, 113–118.

Sheehan, W., Fontenot, J.P. and Blaser, R.E. (1985) In-vitro dry matter digestibiliity and chemical composition of autumn-accumulated tall fescue, orchard grass and red clover. *Grass and Forage Science* 40, 317–322.

Sheldrick, R.D., Lavender, R.H. and Tewson, V.J. (1986) The effects of frequency of defoliation, date of first cut and heading date of a perennial ryegrass companion on the yield, quality and persistence of diploid and tetraploid red clover. *Grass and Forage Science* 41, 137–149.

Skipp, R.A. and Christensen, M.J. (1990) Selection for persistence in red clover: influence of root disease and stem nematode. *New Zealand Journal of Agricultural Research* 33, 319–333.

Skipp, R.A., Christensen, M.J. and Biao, N.Z. (1986) Invasion of red clover (*Trifolium pratense*) roots by soil borne fungi. *New Zealand Journal of Agricultural Research* 29, 305–313.

Smith, D. and Smith, R.R. (1977) Responses of red clover to increasing rates of top dressed potassium fertilizer. *Agronomy Journal* 69, 45–48.

Smith, R.R., Taylor, N.L. and Bowley, S.R. (1985) Red clover. In: Taylor, N.L. (ed.) *Clover Science and Technology*. ASA/CSSA/SSSA, Madison, Wisconsin, pp. 457–470.

Smith, R.S. and Bishop, D.J. (1993) Astred – a stoloniferous red clover. In: Baker, M.J. (ed.)*Proceedings of the XVII International Grassland Congress, New Zealand and Australia*, Vol. 1. New Zealand Grassland Association *et al.*, Palmerston North, pp. 421–423.

Smurygin, M.A. and Brazhnikova, T.S. (1989) Carbohydrate metabolism and hardiness of clover. *Soviet Agricultural Sciences* 3, 16–20.

Spedding, C.R.W. and Diekmahns, E.C. (eds) (1972) *Grasses and Legumes in British Agriculture*. Bulletin 49, Commonwealth Bureau of Pastures and Field Crops, Commonwealth Agricultural Bureaux, Farnham Royal.

Stevens, D.R., Casey, M.J., Lucas, R.J., Baxter, G.S. and Miller, K.B. (1992) Angora goat production from different legumes mixed with ryegrass. *Proceedings of the New Zealand Society of Animal Production* 52, 97–99.

Stewart, R.H., Lynch, K.W. and White, E.M. (1980) The effect of growing clover cultivars in association with barley cultivars upon grain yield of the barley crop in the year of sowing and in the subsequent year. *Journal of Agricultural Science, Cambridge* 95, 715–720.

Sund, J.M., Barrington, G.P. and Scholl, J.M. (1966) Methods and depth of sowing forage grasses and legumes. In: Hill, A.G.G. (ed.) *Proceedings of the X International Grassland Congress, Helsinki, Finland*. Finnish Grassland Association, Helsinki.

Takeda, Y. (1993) Improvement of grass-dominant swards by sod-seeding in eastern Hokkaido, Japan. In: Baker, M.J. (ed.) *Proceedings of the XVII International Grassland Congress, New Zealand and Australia*, Vol. I. New Zealand Grassland Association et al., Palmerston North, pp. 669–670.

Taylor, N.L. and Berger, S. (1989) Polyploids from 2x and 4x and 4x–2x crosses in red clover. *Crop Science* 29, 233–235.

Taylor, N.L. and Quesenberry, K.H. (1996) *Red Clover Science*. Kluwer Academic Publishers, Dordrecht.

Taylor, N.L. and Smith, R.R. (1979) Red clover breeding and genetics. *Advances in Agronomy* 31, 125–154.

Taylor, N.L. and Smith, R.R. (1995) Red clover. In: Barnes, R.F., Miller, D.A. and Nelson, C.J. (eds) *Forages*. Vol. 1, *An Introduction to Grassland Agriculture*, 5th edn. Iowa State University Press, Ames, Iowa, pp. 217–226.

Ten Holte, L. and Van Keulen, H. (1988) Effects of white and red clover as a green manure crop on growth, yield and nitrogen response of sugar beet and potatoes. *Plant and Soil Sciences* 37, 16–24.

Thomas, C., Gibbs, B.G. and Tayler, J.C. (1981) Beef production from silage. 2. The performance of beef cattle given silages of either perennial ryegrass or red clover. *Animal Production* 32, 149–153.

Thomas, C., Aston, K. and Daley, S.R. (1985) Milk production from silage. 3. A comparison of red clover with grass silage. *Animal Production* 41, 23–31.

Thomson, D.J. (1975) The effect of feeding red clover conserved by drying or ensiling on the reproductive performance of ewes. *Journal of the British Grassland Society* 30, 149–152.

Thomson, D.J. (1977) The role of legumes in improving the quality of forage diets. In: *Proceedings of an International Meeting on Animal Production from Temperate Grassland, Dublin, Ireland*. Irish Grassland and Animal Production Association/The Agricultural Institute, Dublin, pp. 131–135.

Tiley, G.E.D. and Frame, J. (1991) Improvement of upland permanent pastures and lowland swards by surface sowing methods. In: *Proceedings of a Conference of the European Grassland Federation, Graz, Austria*. Federal Research Institute for Agriculture in Alpine Regions, Gumpenstein (BAL), pp. 89–94.

Umaerus, M. (1963) The influence of photoperiod treatment on the overwintering of red clover. *Zeitschrift für Pflanzenzüchtung* 50, 167–193.

Williams, R.D. (ed.) (1984) *Crop Production Handbook – Grass and Clover Swards*. British Crop Protection Council; Croydon.

Wilman, D. and Altimimi, M.A.K. (1984) The *in vitro* digestibility and chemical composition of plant parts in white clover, red clover and lucerne during primary growth. *Journal of the Science of Food and Agriculture* 35, 135–138.

Wilman, D. and Asiedu, F.H.K. (1983) Growth, nutritive value and selection by sheep of sainfoin, red clover, lucerne and hybrid ryegrass. *Journal of Agricultural Science, Cambridge* 100, 115–126.

Wilman, D., Koocheki, A., Lwoga, A.B. and Samaan, S.F. (1977) Digestion *in vitro* of Italian and perennial ryegrasses, red clover, white clover and lucerne. *Journal of the British Grassland Society* 72, 13–24.

Wright, R.J., Carter, M.C., Kinraide, T.B. and Bennett, O.L. (1984) Phosphorus requirements for the early growth of red clover, trefoil and flat pea. *Communications in Soil Science and Plant Analysis* 15, 49–63.

Young, N.R. (1984) Red clover for conservation on upland farms. In: Thomson, D.J. (ed.) *Forage Legumes*. Occasional Symposium No. 16, British Grassland Society, Hurley, pp. 132–135.

Subterranean Clover 5

Introduction

Subterranean clover (*Trifolium subterraneum* L.), hereafter referred to as subclover, is a winter annual legume of Mediterranean origin, which has been developed for pastoral use and soil improvement, especially in Australia, where it was pivotal in rotation with cereal cropping, and, to a lesser extent, in north-western USA, southern Europe, southern Latin America and New Zealand. It is adapted to regions with hot, dry summers and moist winters, with mild temperatures, ranging from 6 to 14°C, and with annual rainfall levels of 350–700 mm. It is outstanding among the annual forage legumes for its tolerance to grazing, particularly the prostrate types (Rossiter, 1974; Williams *et al.*, 1980).

There are many distinct types within the species, all of which follow the same annual growth pattern of germinating rapidly in the moist autumn, growing during winter and spring, where there is sufficient rainfall, flowering and seeding prolifically in late winter/early spring, and then surviving dry summers as dormant seed buried in the upper soil layers following the natural death of the plants. In effect, its life cycle is designed to escape summer drought. The plant's name comes from its characteristics of burying or embedding its small seed heads in the soil, where the seed-containing burrs protect the next generation over the summer period.

Subclover occurs naturally as far north as the English Midlands, although it has no economic importance in the UK. Its eastern limits are the regions west of the Caspian Sea, but its natural homelands are France, Spain, the Balkan countries, North Africa, Asia Minor and southern Russia. It spread in Australia during the past two centuries and from there to other New World countries, probably as impurities in seed shipments and as a contaminant of livestock feeds in the early days. Settlers introduced subclover to Australia during two periods, from 1829 to 1842, and again during the 1860s (Gladstones, 1966), probably from the Atlantic coastal regions of Western Europe. Selective introduction of improved cultivars, and trade, have since ensured its spread to other suitable pastoral areas.

Commercialization of subclover began in the 1890s, when Mr A.L. Howard promoted the 'Mediterranean' clover he found near Mount Barker, South Australia. It rapidly gained popularity during the 1920s, once its basic fertilizer requirements were known, and its use expanded as minor- or trace-element needs were identified. Improved ecotypes and cultivars made the species adaptable to different environments.

During the mid-1980s, Australia was producing 5400 t of certified subclover seed annually, more than a third of the country's total production of herbage seed. At that stage, Australia had 18 cultivars of subclover registered and was exporting nearly 500 t to the USA, Latin America, South Africa and Japan (Reed, 1987). It has also been reintroduced to the Mediterranean countries, where it originated (Piano and Francis, 1993).

Subclover is now the most important legume in Australia, especially in the state of Western Australia, where the area sown to subclover peaked at around 7 million ha during the 1970s, although the area in the 1980s was nearer 6 million ha (Collins *et al.*, 1984). The development of a wide range of cultivars enabled subclover to be used through much of the state's dry wheat belt to the higher rainfall areas in the south-west. More than 15 million ha in Australia had been colonized with this legume by the mid-1980s (McGuire, 1985), but one estimate of the potential area for its use in Australian agriculture is 40 million ha (Morley, 1961).

Subclover was first introduced into the USA in the early 1920s, but the species was not promoted to any extent until the late 1930s. Approximately 300,000 ha have been sown for pastures in areas of California, where the rainfall varies between 400 and 2000 mm annually, and further north, in Oregon, more than 200,000 ha have been sown on steep hill soils, while its use has spread to several south-eastern states (McGuire, 1985).

The species has been used for erosion control, by hydroseeding roadside banks and disturbed land cuttings, and as a green manure or weed-smothering cover in horticultural and orchard situations (Caporali *et al.*, 1993). Its tolerance to shade and its molecular nitrogen (N_2)-fixing ability make it suitable as an understorey in sylvipastoral systems, providing both protein-rich forage for livestock and nitrogen (N) for tree growth (Bellon, 1995).

Historically, subclover has been regarded as a complex species, comprising three subspecies of *T. subterraneum*, namely *subterraneum*, *brachycalycinum* and *yanninicum*. However, these subspecies were raised to species status, since they were distinguishable by morphological features and isolated genetically by strong sterility barriers (Katznelson, 1974). The first two species are by far the most common and widely distributed, whereas *Trifolium yanninicum* occurs naturally in water-logged soils in the Balkan region of the Mediterranean Basin. *Trifolium brachycalycinum* is the least hardy of the three species. Cocks *et al.* (1982) refer to the three types as subterranean species and describe the following characteristic differences:

- *Trifolium subterraneum*: high spring herbage yield; low-growing and small-leaved in winter; relatively early-flowering; most seed produced above ground; high hard-seed content.
- *Trifolium brachycalycinum*: low spring herbage yield; tall-growing in winter; relatively late-flowering; low hard-seed content.
- *Trifolium yanninicum*: low spring herbage yield but high yield during winter; large-leaved; tall-growing in winter but shorter in spring; relatively early-flowering; large seeds and seedlings.

Subclover is treated hereafter as one general legume, but any particular differences among the three species will be noted.

The Plant

Growth habit

Seeds germinate and produce two large fleshy cotyledons and a strong radicle. Subclover is a tap-rooted plant with a prostrate growth habit, but is not adapted to deep sandy soils, since it does not develop a deep enough root system to survive dry periods (Hoveland and Evers, 1995). Horizontal stems and stolons, which grow along the soil surface, are produced as the plant develops, but they do not root at the nodes. Some recent cultivars tend to be more upright in growth habit and less ground-hugging than their predecessors and the leaves have longer petioles.

Leaves

Subclover has trifoliate leaves, which are generally heart-shaped (Fig. 5.1), but the three species of subclover have some major morphological differences. *Trifolium subterraneum* is a variably hairy plant, whereas *T. brachycalycinum* has glabrous foliage and the third species, *T. yanninicum*, has glabrous foliage, except for the upper leaf surfaces (McGuire, 1985).

There is a very wide range of characteristic leaf marks on the upper surface of the leaflets, and these help to identify the vast range of different types and cultivars (Collins *et al.*, 1984). These marks are either white, red or brown, can be solid-coloured, or with just a flecking of colour and are a good diagnostic aid.

Subclover's stipules are also a diagnostic feature. They are small and partly free from the stem, and vary in hairiness and striping colour. There are four categories of stipule pigmentation, ranging from pale green through to red flush, which covers at least half of the stipule.

Flowers

Flower-heads of subclover are different from those usually seen in the other forage clovers, in that they are small and much less significant. They consist of only three to six florets per head, each just over 1 cm long. The corolla varies

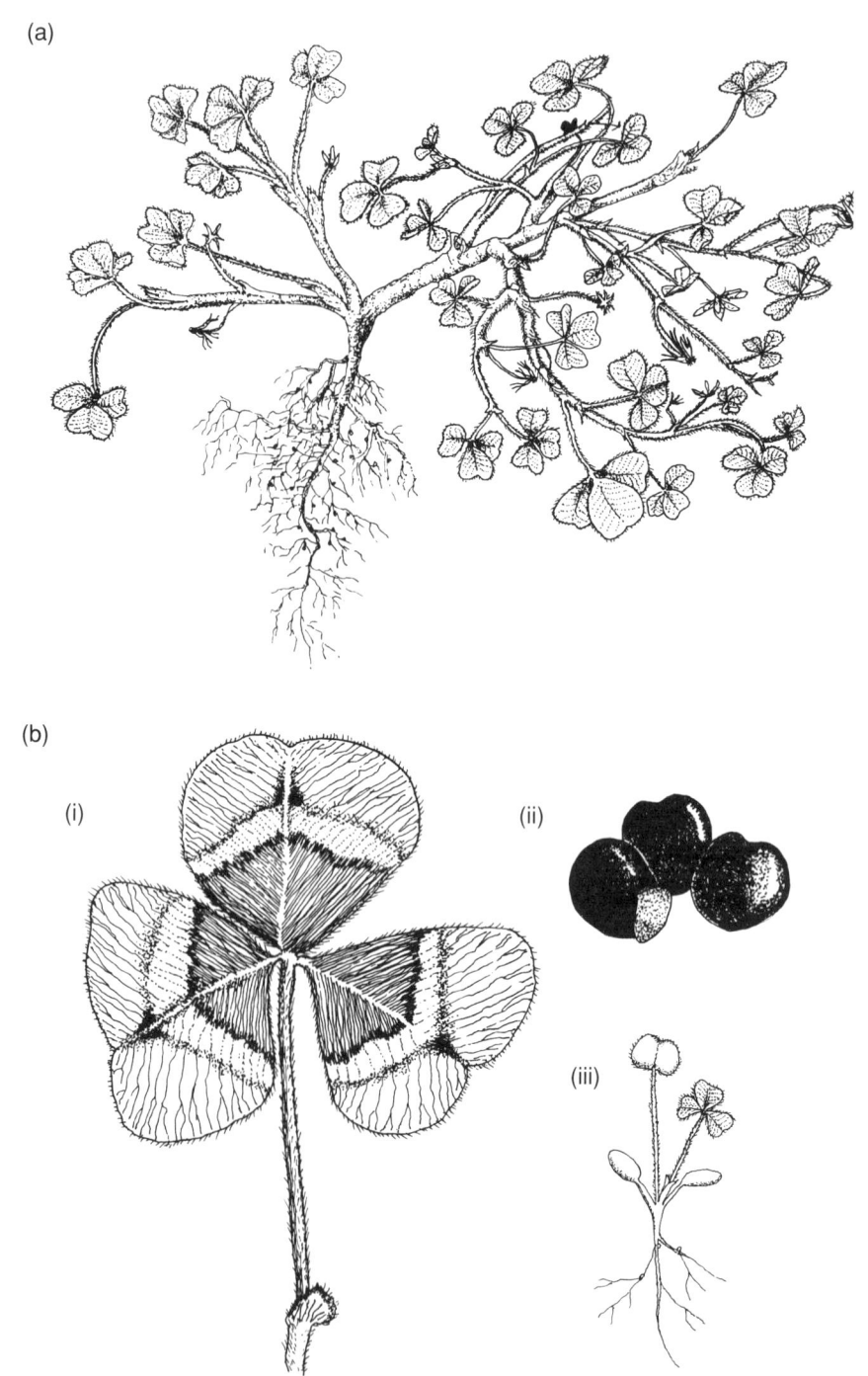

Fig. 5.1. (a) Subterranean clover plant. (b) (i) Leaf, (ii) seeds and (iii) seedling. (c) (i) Inflorescence and (ii) burr.

Fig. 5.1 continued

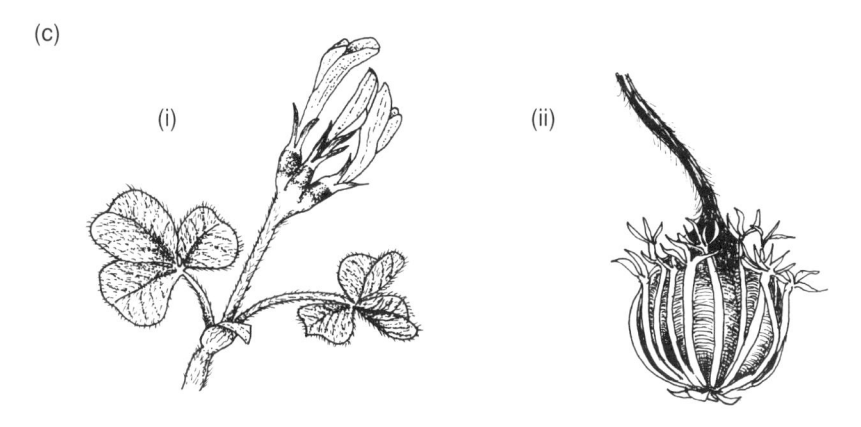

(c)

(i)   (ii)

in colour from white to cream, or with a pale tinge of pink, depending on the cultivar. The calyx also varies, with one or two narrow red rings around it, or none at all.

Subclover also differs from the red and white clovers in that its flowers are strongly self-fertile and, although honey-bees are often seen visiting flowerheads, this is usually after self-fertilization (McGuire, 1985).

Following fertilization, the florets reflex and turn brown and the peduncle turns them towards the soil surface. Several sterile florets, situated at the tip of the head, develop spiny growths, and the very characteristic subclover burr develops to protect the seeds. The burrs of subclover are then partly or wholly buried in the soil where the seed mature and await the first rains for germination.

Seed and seed burrs

Subclover seeds are among the largest of the common forage legumes, with 1000-seed weights of 6.7–8.0 g. The seeds are oval, have a pronounced hilum and are deep purple or black in colour in *T. subterraneum* and *T. brachycalycinum* whereas the species *T. yanninicum* has creamy-amber-coloured seeds. The only cultivar of *T. brachycalycinum*, Clare, commercially available in Australia, has relatively large, unusually flat seeds.

Being an annual plant species, subclover's seeds are vital to its survival and subsequent performance. Buried burrs are larger than those left on the soil surface, contain more and larger seeds, which have superior viability, and tend to be more hard-seeded. The buried seeds usually establish more successfully than those on the soil surface, which are very prone to desiccation during germination.

When selecting a subclover cultivar for a particular situation, the first consideration is its flowering season, which has to be before the onset of dry

Table 5.1. Mean seed yields (kg ha^{-1}) of several Australian-bred subterranean-clover cultivars, not defoliated, at two dryland sites (from Scott, 1969).

Cultivar	Flowering	Site 1	Site 2
Clare	Early	845	421
Geraldton	Early	739	615
Mt Barker	Late	239	1408
Tallarook	Late	70	1145
Woogenellup	Late	499	831
Yarloop	Early	96	559

weather, so that enough seeds are produced to ensure subsequent establishment the following autumn. The threshold seed yield for success for survival and performance is considered to be about 600 kg ha^{-1} (Rossiter, 1966). A period of sustained annual droughts in southern Australia restricted dry-matter (DM) production to below 3 t ha^{-1} year^{-1} and reduced subsequent seed banks, particularly of soft-seeded cultivars (Blumenthal and Ison, 1993). In south-eastern New Zealand, there were large differences in the ability of cultivars to produce seed yields above the level considered critical for satisfactory survival and performance (Table 5.1) although only 16 mm of rain fell in the seed-producing months at site 1, compared with almost 200 mm at site 2.

The characteristic subclover burr develops during seed set. The florets reflex and the old flower-head is pushed towards the soil surface. Spiny growths develop at the tip to protect the large seeds during the process of burial in the soil surface, and the ability to bury the burrs varies among cultivars. Some later-maturing types are rather poor at burr burial, but burial also depends on soil texture, since even the weakest cultivars are able to penetrate soft, sandy soils. More burrs tend to be buried during the early part of the plant's seed-producing period, when soils are still moist, than in the later stages, when the soil surface is drying off and hardening. Where soil surfaces set hard early in spring, even the best burr-burying cultivars may be prevented from achieving this vital process (Collins *et al.*, 1984). This results in rapid thinning of subclover pastures, although use of cultivars that are particularly good at setting seed on the soil surface will alleviate this problem. Notwithstanding, burr burial always gives better subclover establishment and performance than seed from burrs that are on the soil surface or shallowly buried.

Where diurnal soil temperatures fluctuate in summer over a long period, softening of seed-coat impermeability takes place and there is premature germination, with the likelihood of desiccation, and so cultivars with seed resistance to the softening process are beneficial (Taylor, 1981). Higher levels of soft seed result when the seeds ripen under warm, moist soil conditions (Smith, 1988; Fairbrother and Lowe, 1995). Hard-seededness is associated with late maturity under some conditions (Evans and Hall, 1995) and also varies among subclover cultivars (Blumenthal and Ison, 1994). Small seed size may be advantageous in some environments, since they need less water for germination, because of a high ratio of surface to volume, and, during seed ripening, they mature more rapidly than large seeds (Pecetti and Piano, 1994).

Nitrogen fixation
When a pasture is being renewed by sowing using subclover alone or in mixture with grass, inoculation with an effective *Rhizobium* strain recommended for local use is advisable, unless there has been a long history of satisfactory subclover growth. Using the correct strain of *Rhizobium leguminosarum* bv. *trifolii* has a marked influence on establishment and performance, and some strains are specific to certain subclover cultivars. Indigenous strains will nodulate subclover seedlings, but may cause subsequent plant failure because they are ineffective (Holland, 1970; Hagedorn, 1978). In soils with an indigenous strain, the success, as measured by nodule occupancy of the inoculant strain, was inversely related to the size of the indigenous population (Ireland and Vincent, 1968).

Various well-documented methods for inoculating legume seeds are available, and contact with fertilizer should be avoided. Lime applied to acid soils increases nodulation and the proportion of legume N derived from N_2 fixation (Unkovitch *et al.*, 1996). In addition to the direct effect of increasing pH, lime application reduces aluminium ion (Al^{3+}) availability, which also improves nodulation and N_2 fixation per nodule. Use of lime-coated, rhizobially inoculated seed overcomes acidic soil conditions at establishment and increases *Rhizobium* numbers. Freshly inoculated seed, with adequate numbers of rhizobia, should always be sown as soon as possible after inoculation, and preferably drilled into moist soil shallowly at 12–20 mm depth. Broadcasting or oversowing inoculated seeds is best done when weather is damp, to ensure that rhizobia survive long enough to invade the radicles of the emerging seedlings.

Estimates of N_2 fixation by subclover in Australia include ranges of 50–125 kg N ha^{-1} (Bolger *et al.*, 1995) and 103–188 kg N ha^{-1} (Sanford *et al.*, 1993), the highest value in the latter range being in grazed subclover stands. In the south-eastern USA, N_2 fixation ranged from 104 to 206 kg ha^{-1} in pure-sown stands of subclover (Brink, 1990), while, in Bermuda-grass (*Cynodon dactylon*) or Bahia-grass (*Paspalum notatum*) swards oversown with subclover, DM yields were equivalent to those from the pure-sown grass swards given fertilizer N at annual rates of 160–254 kg ha^{-1} (Evers, 1985). As with other forage legumes, the proportion of fixed N assimilated by the legume is higher when grown with grass than in monoculture, since most of the available soil mineral N is taken up by the grass (Unkovitch *et al.*, 1996). Because of reduced N_2 fixation and/or low uptake of N after defoliation, N from root-protein degradation and from residual vegetation is mobilized to provide the N required for the growth of new leaves (Phillips *et al.*, 1983; Culvenor and Simpson, 1991).

Breeding

Almost all breeding of subclover has been undertaken in Australia during this century and no cultivars from any other country are registered in the current Organization for Economic and Cooperation Development (OECD) list. Considerable natural variation has developed within subclover under Australian conditions and a range of new ecotypes, adaptable to different local environments, has emerged, which exist alongside the sown bred cultivars. Early work consisted mainly of selecting improved types with the desirable stage of maturity to meet local farming needs, and now a national breeding programme is seeking to improve major agronomic features, e.g. seedling vigour, DM production, seed production and resistance to disease and pests. This programme has involved all possible methods of plant improvement, including hybridization between existing ecotypes or cultivars, collection of material within Australia and from other countries, particularly in the Mediterranean basin, and developing improved forms from induced mutations. Subclover is strongly self-fertile and hybridization is achieved through emasculation by corolla removal, with pollination being done 24–48 h afterwards. Improvements in traits such as resistance to disease, pests and viruses and in herbage quality are likely to emerge as genetic engineering is increasingly utilized in the development of new cultivars.

The paucity of subclover cultivars registered in other countries probably stems from the large number of Australian cultivars commercially available that perform satisfactorily under a wide range of conditions, although better cold tolerance is required in some Mediterranean areas. Breeding programmes in Italy, Spain and France have developed several cultivars, but they have not yet been released commercially (Piano and Talamucci, 1996).

A comprehensive collection of naturalized subclovers was made in New Zealand, with the aim of characterizing the types and determining any possible genetic shift from their Australian origins, but nearly all collected proved to be either Mount Barker or Tallarook types and there was little evidence to suggest that natural selection had favoured one type over another (Suckling *et al.*, 1983). Cultivar evaluation at eight hill-pasture sites showed that the ranking order of cultivar performances differed under different local climatic conditions (Chapman *et al.*, 1986). More recently, lines have been identified which are potential parents in a breeding programme for subclover that will be suitable for hill-land areas in New Zealand. These genotypes are late heading, have a prostrate growth habit and a long growing season. The best lines have the potential to regenerate more than 200 plants m^{-2}, produce at least 1000 kg DM ha^{-1} in spring in combination with grass and have a very low formononetin content (Dodds *et al.*, 1995).

Some cultivars exhibit resistance to fungal diseases, such as clover scorch (*Kabatiella caulivora*), known as northern anthracnose in the USA, or viruses, such as bean yellow-mosaic virus or subclover mottle virus, but introduction of new races may result in resistance breakdown, e.g. to clover scorch

(Barbetti, 1995) and so breeding has to keep up with the emergence of new virulent strains of disease agents.

Over 30 Australian subclovers have been registered since Mount Barker was released in 1907. Based on a Western Australian state classification, their maturities vary from early season to late midseason over a period of 10 weeks, their oestrogen levels from very low to very high, with most recent releases being low or very low, and their hard-seed level on a scale from 1 (little or no hard seed) to 10 (very high, i.e. at least 40% hard seed). *Trifolium subterraneum* is the most widely grown species and currently there are 24 cultivars of *T. subterraneum*, five of *T. yannicum* and three of *T. brachycalycinum* registered. The range of cultivars available with different characteristics means that subclover can be grown successfully over a wide range of edaphic and climatic conditions.

Variation among cultivars in their tolerance of some herbicides for the control of broad-leaved weeds suggests that herbicide tolerance could also be included in subclover breeding programmes (Dear *et al.*, 1995).

Seed production

The major seed-producing country is Australia and, because of the burial characteristics of the seed, special combine-harvesters, with a suction action, have been developed to harvest the seed. Seed yields up to 2000 kg ha^{-1} are achievable, but 500–1000 kg ha^{-1} are more common (McGuire, 1985). Early hard grazing to create an open canopy benefits subsequent seed production, due to increased inflorescence number and to increased seed-burr burial (Collins, 1978; Steiner and Grabe, 1986). Seed yield increases with herbicide treatment to control weeds, especially if applied sufficiently early to avoid harming the developing clover inflorescences, as both yield and germination of seed are adversely affected by inflorescence damage (Dear *et al.*, 1995). Bromoxynil or MCPA applied 6 weeks before flowering has successfully increased seed production by more than 50%, due mainly to an increase in seed number.

Agronomy and Management

Establishment

Where there is extreme summer drought, subclover is usually sown alone at 10–15 kg ha^{-1}, but in situations where summer drought is not so extreme, subclover is used in mixture with grasses in order to supplement pasture quality during the cool, moist months. Suitable companion grasses include annual or perennial species, depending on the circumstances; for example, annual ryegrass (*Lolium rigidum*) in Western Australia, warm-season Bermuda grass in south-eastern USA or perennial ryegrass (*Lolium perenne*) in north-west USA. Red or white clovers are also sometimes added to the seed mixture to improve production at different times of the season.

Germination rate of subclover is high at day/night temperatures of 25/15°C and reduces as temperatures decrease (Evers, 1991). At soil temperatures above 15°C, germination rate is high, and the large seed size ensures that seed vigour is also high, but excessively high soil temperatures can induce high-temperature dormancy, while low soil temperatures can markedly reduce germination rate of some cultivars (Hampton *et al.*, 1987). Germination may also be inhibited by residues of the previous stand, e.g. wheat or Harding grass (*Phalaris aquatica*), on the surface of the soil and removal or incorporation of these residues into the soil reduces this allelopathic effect (Leigh *et al.*, 1995).

When sown in monoculture, incorporation of the herbicide, EPTC, into the seed-bed prior to sowing controls most germinating grass weeds and volunteer cereals. For subclover/grass mixtures, 'clover-safe' single herbicides, such as 2,4-DB, or multicomponent formulations control a range of broad-leaved weeds. Early grazing at 6 weeks after emergence almost doubled subclover yield, compared with first grazing at 11 weeks, because of the control of weeds (Gillespie *et al.*, 1983). The seedlings may be attacked by pathogenic fungi, insect pests or slugs, depending upon local conditions, and the problem of seedling depredation can be a major factor in failure of subclover establishment and persistence (Charlton, 1978). Once established, lenient grazing is necessary in the first season in order to ensure that the first year's seed set is high and subsequent seedling density adequate, firstly, to prevent weed ingress and, secondly, to provide the basis for good productive performance. Oversowing or minimum-cultivation techniques have been used to introduce subclover into warm-season grass pastures (Evers, 1985), into degraded natural pastures, and in the reclamation of shrub-encroached areas (Bullitta *et al.*, 1989). In Portugal, sowing subclover-based seeds mixtures on to scarified grassland can result in higher subclover seedling density than conventional cultivation, especially if lime is also applied (Moreira *et al.*, 1994).

Subclover is well known for its ability to 'false strike', i.e. seeds germinate after early autumn rain and then die off, due to drought, but the subsequent establishment of other seeds, including hard seeds, which are buried in the soil, enables the species to re-establish, provided the soil seed bank is adequate. Summer rainfall can also cause premature germination.

Buried seed populations in the top 5 cm of soil in New Zealand pastures ranged from 58 kg ha^{-1} in intensively stocked lowland cattle pastures to almost 700 kg ha^{-1} in similar pastures grazed by sheep and in pastures on sandy/shingly soils prone to summer drought (Suckling and Charlton, 1978); in summer-dry hill pastures, the seed load was above 600 kg ha^{-1}, but it was only 7–22 kg ha^{-1} under medium summer rainfall and less than 8 kg ha^{-1} with high summer rainfall.

Fertilizer use

Subclover grows best when soil-fertility levels are relatively high, especially with high phosphorus (P) and sulphur (S); nevertheless, it is valued for its ability to grow in less fertile, acid soils. *Trifolium brachycalycinum* is better adapted than the other two subspecies to heavy-textured, moderately alkaline soils and *T. yanninicum* to waterlogged soils (Katznelson, 1974). *Trifolium subterraneum* is susceptible to iron (Fe)-deficiency chlorosis on alkaline soils (Gildersleeve and Ocumpaugh, 1989). In all countries where the species is sown, application of superphosphate, which contains P and S, has been the key to improved performance, the most suitable application time being autumn/early winter (Charlton, 1983). In Chile, beef cattle and lamb liveweight gains were increased four to five times, from 40 to 180–200 kg ha^{-1}, by applying fertilizer P to subclover swards at 15–20 kg P ha^{-1} annually (Klee *et al.*, 1985). Some cultivars of subclover have performed adequately at soil pH levels of around 5.0 and with Olsen phosphate levels as low as 3 (Chapman *et al.*, 1986). Nevertheless, liming infertile, acid soils is important for successful establishment and persistence (Moreira *et al.*, 1994).

Molybdenum may also be required in small quantities, e.g. sodium (Na) molybdate at 1 kg ha^{-1}, to ensure good N_2 fixation and plant growth, particularly on acidic soils, and potassium (K) may be needed after a hay or silage cut has been removed or if the land has been previously cropped. Subclover was not particularly sensitive to high rates of copper (Cu) applied in sewage sludge, but excessive amounts of zinc (Zn) were harmful (Domingues *et al.*, 1995), as was high Al^{3+}, usually associated with very acid soils (Burnet *et al.*, 1994). Some subclover cultivars are relatively tolerant of high concentrations of salt (NaCl) in the soil, tolerance being associated with the ability to restrict uptake of salt into their shoots and to maintain a high K : Na ratio (Shannon and Noble, 1995).

Diseases and pests

Fewer diseases and pests affect subclover than most other major forage legumes. Economic chemical control is rarely possible and so breeding for resistance is the main way forward. Sward performance in Australia can be badly affected on occasion by clover scorch. Also, root rot, from a complex of *Fusarium*, *Pythium* and *Rhizoctonia*, results in discoloration of the foliage and reduced yield (Rossiter, 1978). Root rot caused by *Phytophthora clandestina* has become a serious problem in both irrigated and dryland pastures in Australia (Taylor and Greenhalgh, 1987). Diseases are not a major problem in the USA (Hoveland and Evers, 1995).

A number of virus diseases, such as subclover mottle and stunt viruses, which are transmitted by aphids, can affect plant growth adversely (Helms *et al.*, 1993). Yields of DM have been nearly halved and seed production reduced by 75%, due to the alfalfa mosaic virus and bean yellow-mosaic virus (Jones, 1992). Resistance to some aphid-transmitted diseases is associated with early flowering, with aphids being less able to infect mature plants than those with succulent vegetative organs (Ferris and Jones, 1996). Damage from the blue-green aphid (*Acyrthosiphon konda*) has also been reported (Gillespie and Sandow, 1981).

Herbage production

Potential biological yield under optimum management conditions has been calculated to be as high as 35 t DM ha^{-1} under southern Australian conditions, but actual yields over a 5-month period have reached over 16 t DM ha^{-1}, equivalent to a conversion coefficient of radiant energy of 4.5% (Cocks *et al.*, 1980). However, yields in practice are more likely to be around a quarter to a third of the above potential and sometimes even less, the actual values depending upon the specific growing conditions, but especially rainfall during the growing season; for example, there was a linear relationship between water use and DM yield up to 440 mm of growing-season rainfall, with DM yields ranging from 3 to 12 t ha^{-1} (Bolger *et al.*, 1993). In mixed associations, the yield of subclover is determined by the competitive abilities of the mixture components and how well the dynamic association is managed to avoid over-dominance of the grass. In 'permanent' pastures, this will be encouraged by N from the legume. An ideal management is to use associations where the components have complementary rather than competing growth rhythms, thus enabling reliable seed setting and annual regeneration of the subclover (Piano and Talamucci, 1996).

Herbage quality

In common with other legume species, subclover is rich in crude protein, compared with grasses, but the protein concentration declines steadily with advancing plant maturity, as does herbage digestibility. When grazing pastures of subclover/Bermuda grass with steers at three levels of forage availability, the quality of the herbages was similar, since leaf laminae and petioles were the primary components of subclover at the three levels (Fairbrother and Lowe, 1995). Clover seeds are rich in protein, and grass-seed burrs can contribute to the nutritional needs of grazing stock, albeit at the expense of the seed bank.

Grazed subclover herbage in irrigated swards had consistently high digestibility and N content, i.e. greater than 75% and 32 g kg^{-1}, respectively, and low neutral detergent fibre (NDF), acid detergent fibre (ADF) and lignin content (Fig. 5.2). Unlike most other temperate forage legumes, leaf laminae of subclover are of lower digestibility than petioles and stems (Stockdale, 1992a; Mulholland *et al.*, 1996). Effective rumen-degradable protein in the leaf lamina can be so low that microbial protein synthesis in the rumen is limited, adversely affecting animal production (Mulholland *et al.*, 1996).

The phyto-oestrogens formononetin, genistein, biochanin A and daidzein are present in foliage of subclover, sometimes in high concentrations. Formononetin is the oestrogen that was largely responsible for serious sheep infertility on Australian farms (Marshall *et al.*, 1971), particularly when cultivars such as Yarloop and Dinninup, which have very high levels of oestrogens, were used, but the problem was minimal with cattle. The disorder is less likely to arise in mixed swards, and so the use of cultivars with high formononetin content persists where other traits make them valuable. Several cultivars, e.g.

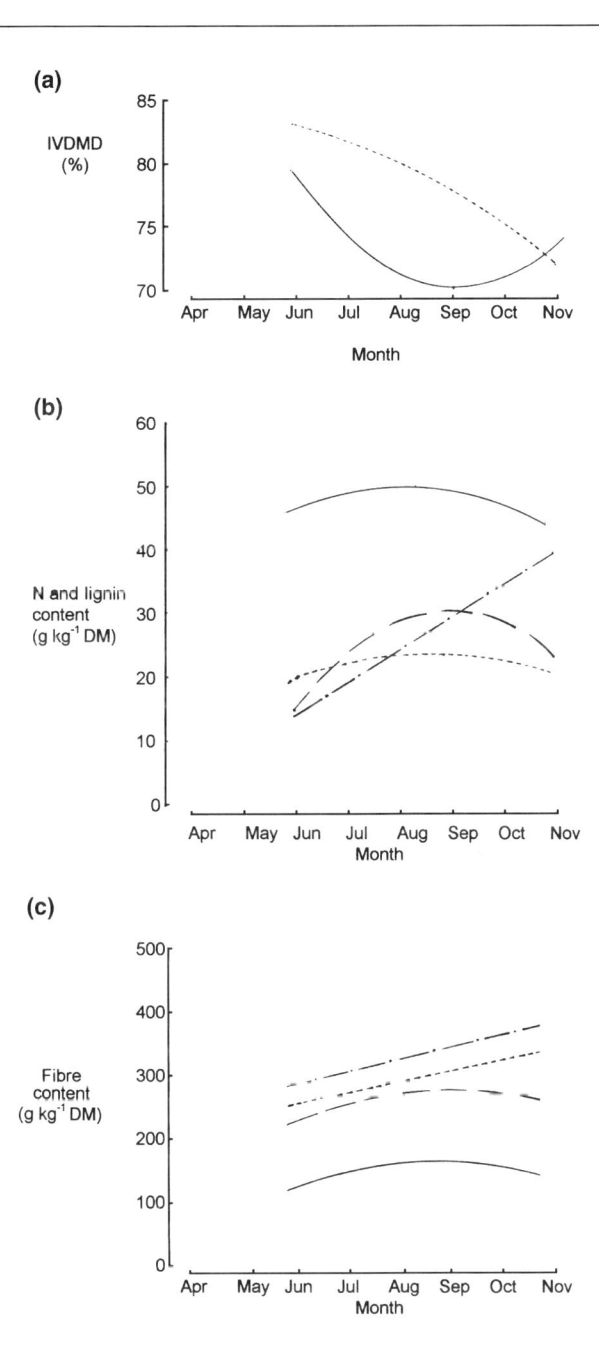

Fig. 5.2. Nutritive value of leaf laminae and stems (including petioles) of irrigated subterranean clover from winter to early summer in Victoria, Australia. (a) *In vitro* DM digestibility (IVDMD) (laminae ——, stems -----). (b) Nitrogen (laminae ——, stems -----) and lignin (laminae – –, stems –·–). (c) ADF (laminae —— stems -----) and NDF (laminae – –, stems –·–) (from Stockdale, 1992a).

Seaton Park, have a naturally occurring low-formononetin mutation (Rossiter *et al.*, 1985) and commercial cultivars with low formononetin levels have been released (Stern *et al.*, 1983).

Bloat in grazing livestock may occur on lush, rapidly growing subclover swards, but management can minimize the risk and specific preventive measures are available, e.g. the use of the surfactant poloxalene (Essig, 1985).

Grazing

An annual seed crop is essential for subclover persistence in pasture, so it is important that the sward's potential to produce an adequate seed crop is not militated against by overgrazing. Thus, a compromise must be effected between utilization efficiency and seed production to optimize sustained pasture performance. Notwithstanding, there should always be sufficient grazing pressure to keep any sown companion grasses or weed species in check; otherwise, the seed yields of subclover may be reduced. Prostrate cultivars are more suited to continuous stocking by sheep than upright cultivars, although the latter are well suited to rotational grazing.

Due to the lower nutritive value of leaf laminae than that of petioles and stems, grazing systems that maximize the proportion of laminae in the diet can reduce animal production. Therefore, individual animal performance may be lower at low than at intermediate stocking rates, for example, as the low utilization of the former will result in a high content of laminae in the diet (Stockdale, 1992b).

Treading damage severely reduced herbage yields of subclover (Table 5.2), but, provided damage is minimized during grazing, continuous stocking throughout the growing season to maintain pasture height at 6–8cm has given optimum yields of both herbage and seed (Carter, 1987), although, in south-eastern USA, 3–6 cm is advocated as optimum (Fairbrother *et al.*, 1992).

Conservation

Stands should be closed from grazing by early spring if conservation is planned. Subclover grows tall quickly under these conditions, reaching over 40 cm, but it then tends to subside suddenly by several centimetres and should be cut for hay by this stage, though before this if being cut for high

Table 5.2. Effect of stocking rate on establishment and yield of subterranean clover in mixture with annual ryegrass (*Lolium rigidum*) (from Carter and Silavingam, 1977).

Sheep equivalent ha^{-1}	Subterranean clover Plants m^{-2}	Subterranean clover kg DM ha^{-1}	Mixture Plants m^{-2}	Mixture kg DM ha^{-1}
0	560	3150	1110	4340
25	370	1790	1130	3450
50	300	1550	890	2820

quality silage. Where hay or silage production is part of the farming system, the use of upright types of subclover is an advantage (Quinlivan and Francis, 1976). Hay yields average 5 t DM ha^{-1}, but where annual rainfall exceeds 750 mm and soil fertility is good, yields may reach 8 t DM ha^{-1}.

Animal performance

Intake characteristics of subclover are generally good, with dairy cows capable of grazing up to 22 kg DM day^{-1} and producing 28 kg milk day^{-1} from unsupplemented subclover pasture (Stockdale, 1992b). Daily liveweight gains of 150 g day^{-1} have been recorded for lambs grazing subclover swards, when petioles accounted for a high proportion of the diet (Mulholland *et al.*, 1996).

References

Barbetti, M.J. (1995) Breakdown in resistance of subterranean clovers to clover scorch disease (*Kabatiella caulivora*). *Australian Journal of Agricultural Research* 46, 645–653.

Bellon, S. (1995) Stratégies sylvopastorales en région méditerranéenne. *Options Méditerranéennes l2*, (Systèmes Sylvopastoraux), 195–198.

Blumenthal, M. and Ison, R. (1993) Use of water balance models to examine the role of climate in annual legume decline in southern Australia. In: Baker, M.J. (ed.) *Proceedings of the XVII International Grassland Congress, New Zealand and Australia*, Vol. I. New Zealand Grassland Association *et al.*, Palmerston North, pp. 61–62.

Blumenthal, M.J. and Ison, R.L. (1994) Plant population dynamics in subterranean clover and murex medic swards. 1. Size and composition of seed bank. *Australian Journal of Agricultural Research* 45, 913–928.

Bolger, T.P., Turner, N.C. and Leach, B.J. (1993) Water use and productivity of annual legume-based pasture systems in the south-west of Western Australia. In: Baker, M.J. (ed.) *Proceedings of the XVII International Grassland Congress, New Zealand and Australia*, Vol. I. New Zealand Grassland Association *et al.*, Palmerston North, pp. 274–275.

Bolger, T.P., Pate, J.S., Unkovitch, M.J. and Turner, N.C. (1995) Estimates of seasonal nitrogen fixation of annual subterrranean clover-based pastures using the ^{15}N natural abundance technique. *Plant and Soil* 175, 57–66.

Brink, G.E. (1990) Seasonal dry matter, nitrogen and dinitrogen fixation patterns of crimson and subterranean clovers. *Crop Science* 30, 1115–1118.

Bullitta, P., Bullitta, S. and Roggero, P.P. (1989) Agronomic methods to increase pastureland production in Mediterranean marginal areas. In: *Proceedings of the XVI International Grassland Congress, Nice, France*, Vol. 2. Association Française pour la Production Fourragère, Versailles, pp. 1591–1592.

Burnet, V.C., Coventry, D.R., Hirth, J.R. and Greenhaulgh, E.C. (1994) Subterranean clover decline in permanent pastures in N.E. Victoria. *Plant and Soil* 164, 231–241.

Caporali, F., Campiglia, E. and Paolini, R. (1993) Prospects for more sustainable cropping systems in central Italy based on subterranean clover as a cover crop. In: Baker, M.J. (ed.) *Proceedings of the XVII International Grassland Congress, New Zealand and Australia*, Vol. III. New Zealand Grassland Association *et al.*, Palmerston North, pp. 2197–2198.

Carter, E.D. (1987) Establishment and natural regeneration of annual pastures. In: Wheeler, J.L., Pearson, C.J. and Robards, G.E. (eds) *Temperate Pastures*. Australian Wool Corporation/CSIRO, Melbourne, pp. 35–51.

Carter, E.D. and Silavingam, T. (1977) Some effects of treading by sheep on pastures of the Mediterranean climate zone of South Australia. In: Wojahn, E. and Thöns, H. (eds)*Proceedings of the XIII International Grassland Congress, Leipzig, East Germany*. Akademie-Verlag, Berlin, pp. 641–644.

Chapman, D.F., Sheath, G.W., Macfarlane, M.J., Rumball, P.J., Cooper, B.M., Crouchley, G., Hoglund, J.H. and Widdup, K.G. (1986) Performance of subterranean and white clover varieties in dry hill country. *Proceedings of the New Zealand Grassland Association* 47, 53–62.

Charlton, J.F.L. (1978) Slugs as a possible cause of establishment failure in pasture legumes oversown in boxes. *New Zealand Journal of Experimental Agriculture* 6, 313–317.

Charlton, J.F.L. (1983) Lotus and other legumes. In: Wratt, G.S. and Smith, H.C. (eds) *Plant Breeding in New Zealand*. Butterworths/DSIR, Wellington, pp. 253–262.

Cocks, P.S., Mathison, J.M. and Crawford, E.J. (1980) From wild plants to pasture cultivars: annual medics and subterranean clover in southern Australia. In: Summerfield, R.J. and Bunting, A.H. (eds*) Advances in Legume Science*. Kew Royal Botanic Gardens, London, pp. 569–596.

Cocks, P.S., Craig, A.D. and Kenyon, R.V. (1982) Evolution of subterranean clover in South Australia. II. Change in genetic composition of a mixed population after 19 years grazing on a commercial farm. *Australian Journal of Agricultural Research* 33, 679–695.

Collins, W.J. (1978) The effect of defoliation on inflorescence production, seed yield and hard seededness in swards of subterranean clover. *Australian Journal of Agricultural Research* 29, 789–841.

Collins, W.J., Francis, C.M. and Quinlivan, B.J. (1984) *Registered Cultivars of Subterranean Clover – Their Origin, Identification and Potential Use in Western Australia*. Bulletin 4083, Western Australia Department of Agriculture, Perth.

Culvenor, R.A. and Simpson, R.J. (1991) Mobilization of nitrogen in swards of *Trifolium subterraneum* L. during regrowth after defoliation. *New Phytologist* 117, 81–90.

Dear, B.S., Sandral, G.A. and Coombes, N.E. (1995) Differential tolerance of *Trifolium subterraneum* (subterranean clover) cultivars to broadleaf herbicides. 1. Herbage yield. *Australian Journal of Experimental Agriculture* 35, 467–474.

Dear, B.S., Sandral, G.A., Pratley, J.E. and Coombes, N.E. (1996) Effect of time of application of MCPA and Bromoxynil in relation to flowering in subterranean clover seed yield and quality. *Australian Journal of Experimental Agriculture* 36, 177–184.

Dodds, M.B., Sheath, G.W. and Richardson, S. (1995) Development of subterranean clover (*Trifolium subterraneum* L.) genotypes for New Zealand pastures. 1. Whatawhata persistence evaluation. *New Zealand Journal of Agricultural Research* 38, 33–47.

Domingues, H., Pires, F.P., Monteiro, O. and Sequeira, E.M. (1995) Growth of *Trifolium subterraneum* in schistic soils treated with sewage sludge, having high Cu content. *Arid Soil Research and Rehabilitation* 9, 327–333.

Essig, H.W. (1985) Quality and antiquality components. In: Taylor, N.L. (ed.) *Clover Science and Technology*. ASA/CSSA/SSSA, Madison, Wisconsin, pp. 309–324.

Evans, P.M.and Hall, E.J. (1995) Seed softening patterns from single seed crops of subterranean clover (*Trifolium subterraneum* L.) in a cool temperate environment. *Australian Journal of Experimental Agriculture* 35, 1117–1121.

Evers, G.W. (1985) Forage and nitrogen contributions of arrowleaf and subterranean clovers overseeded on bermudagrass and bahiagrass. *Agronomy Journal* 77, 960–963.

Evers, G.W. (1991) Germination response of subterranean, berseem and rose clovers to alternating temperatures. *Agronomy Journal* 83, 1000–1004.

Fairbrother, T.E. and Lowe, D.E. (1995) Temperature and soil-water effects on dormancy and mortality of subclover seed. *Agronomy Journal* 87, 252–257.

Fairbrother, T.E., Brink, G.E. and Ivy, R.L. (1992) Effect of forage availability on steer performance of a subterranean clover–bermudagrass forage system. *Journal of Production Agriculture* 5, 28–33.

Ferris, D.G. and Jones, R.A.C. (1996) Natural resistance to bean yellow mosaic potyvirus in subterranean clover. *Australian Journal of Agricultural Research* 47, 605–624.

Gildersleeve, R.R. and Ocumpaugh, W.R. (1989) Greenhouse evaluation of subterranean clover species for susceptibility to iron-deficiency chlorosis. *Crop Science* 29, 949–951.

Gillespie, D.J. and Sandow, J.D. (1981) Selection for bluegreen aphid resistance in subterranean clover. In: Smith, J.A. and Hayes, V.W. (eds) *Proceedings of the XIV International Grassland Congress, Lexington, USA*. Westview Press, Boulder, Colorado, pp. 105–108.

Gillespie, D.J., Ewing, M.A. and Nicholas, D.A. (1983) Subterranean clover establishment techniques. *Journal of Agriculture of the Western Australia Department of Agriculture, 4th Series* 24, 16–20.

Gladstones, J.S. (1966) Naturalised subterranean clover (*Trifolium subterraneum* L.) in Western Australia: the strains, their distribution and possible origins. *Australian Journal of Botany* 14, 329–354.

Hagedorn, C. (1978) Effectiveness of *Rhizobium trifolii* populations associated with *Trifolium subterraneum* L. in southwest Oregon soils. *Soil Science Society of America Journal* 42, 447–451.

Hampton, J.G., Charlton, J.F.L., Bell, D.D. and Scott, D.J. (1987) Temperature effects on the germination of herbage legumes in New Zealand. *Proceedings of the New Zealand Grassland Association* 48, 177–183.

Helms, K., Müller, W.J. and Waterhouse, P.M. (1993) National survey of viruses in pastures of subterranean clover. 1. Incidence of four viruses assessed by ELISA. *Australian Journal of Agricultural Research* 44, 1837–1862.

Holland, A.A. (1970) Competition between soil and seed borne *Rhizobium trifolii* in nodulation of introduced *Trifolium subterraneum* L. *Plant and Soil* 32, 293–302.

Hoveland, C.S. and Evers, G.W. (1995) Arrowleaf, crimson and other annual clovers. In: Barnes, R.F., Miller, D.A. and Nelson, C.J. (eds) *Forages*, 5th edn, Vol. 1, *An Introduction to Grassland Agriculture*. Iowa State University Press, Ames, Iowa, pp. 249–260.

Ireland, J.A. and Vincent, J.M. (1968) A quantitative study of competition for nodule formation. Transactions of the 9th International Congress, Sydney, Australia. *Soil Science* 2, 85–93.

Jones, R.A.C. (1992) Further studies on losses in productivity caused by infection of annual pasture legumes with three viruses. *Australian Journal of Agricultural Research* 43, 1229–1241.

Katznelson, J. (1974) Biological flora of Israel. 5. The subterranean clovers of *Trifolium subterraneum* sect. *Calycomorphum* Katzn. *Trifolium subterraneum* L. (s.l.). *Israeli Journal of Botany* 23, 69–108.

Klee, G., Crempien, Ch., Acuña, H. and Fernandez, M. (1985) Elaboración de un sistema de producción ovina para la precordillera de Bio-Bio. *Agricultura Técnica (Chile)* 45, 1–7.

Leigh, J.H., Halsall, D.M. and Holgate, M.D. (1995) The role of allelopathy in legume decline in pastures 1. Effects of pasture and crop residues on germination and survival of subterranean clover in the field and nursery. *Australian Journal of Agricultural Research* 46, 179–188.

McGuire, W.S. (1985) Subterranean clover. In: Taylor, N.L. (ed.) *Clover Science and Technology*. ASA/CSS/SSSA, Madison, Wisconsin, pp. 515–534.

Marshall, T., Fells, H.H., Neil, H.G. and Rossiter, R.C. (1971) Pasture legume varieties and ewe fertility. *Journal of Agriculture of Western Australia* 12, 110–112.

Moreira, N., Trindade, H., Coutinho, J. and Almeida, J.F. (1994) Effects of liming and cultivation on the establishment and persistence of rainfed Mediterranean pastures. *Experimental Agriculture* 30, 453–459.

Morley, F.H.W. (1961) Subterranean clover. *Advances in Agronomy* 13, 57–123.

Mulholland, J.G., Nandra, K.S., Scott, G.B., Jones, A.W. and Coombes, N.E. (1996) Nutritive value of subterranean clover in a temperate environment. *Australian Journal of Experimental Agriculture* 36, 803–814.

Pecetti, L. and Piano, E. (1994) Observations on the rapidity of seed and burr growth in subterranean clover. *Journal of Genetics and Breeding* 48, 225–228.

Phillips, D.A., Center, D.M. and Jones, M.B. (1983) Nitrogen turnover and assimilation during regrowth in *Trifolium subterraneum* L. and *Bromus mollis* L. *Plant Physiology* 71, 472–476.

Piano, E. and Francis, C.M. (1993) The annual species of *Medicago* in the Mediterranean region: ecogeography and related aspects of plant introduction and breeding. In: Rotili, P. and Zannone, L. (eds) *The Future of Lucerne Biotechnology, Breeding and Variety Constitution. Proceedings of the X International Conference EUCARPIA* Medicago *Species Group, Lodi, Italy*. Istituto Sperimentale per le Colture Foraggere, Lodi, pp. 373–385.

Piano, E. and Talamucci, P. (1996) Anual self-regenerating legumes in Mediterranean areas. In: Parente, G., Frame, J. and Orsi, S. (eds) *Grassland and Land Use Systems. Proceedings of the 16th General Meeting of the European Grassland Federation, Grado, Italy*, published as *Grassland Science in Europe*, Vol. 1. ERSA, Gorizia, pp. 895–909.

Quinlivan, B.J. and Francis, C.M. (1976) *Subterranean Clover in Western Australia*. Bulletin 3995, Western Australia Department of Agriculture, Perth.

Reed, K.F.M. (1987) Agronomic objectives for pasture plant improvement. In: Wheeler, C.J., Pearson, C.J. and Robards, G.E. (eds) *Temperate Pastures –Their Production, Use and Management*. Australian Wool Corporation/CSIRO, Canberra, pp. 265–271.

Rossiter, R.C. (1966) The success or failure of strains of *Trifolium subterraneum* L. in a Mediterranean environment. *Australian Journal of Agricultural Research* 17, 425–446.

Rossiter, R.C. (1974) The relative success of strains of *Trifolium subterraneum* L. in binary mixtures under field conditions. *Australian Journal of Agricultural Research* 25, 757–766.

Rossiter, R.C. (1978) The ecology of subterranean clover-based pastures. In: Wilson, J.R. (ed.) *Plant Relations in Pastures*. CSIRO, Melbourne, pp. 325–339.

Rossiter, R.C., Collins, W.J. and Haynes, Y. (1985) Genetic variability in Seaton Park subterranean clover. *Australian Journal of Agricultural Research* 36, 43–50.

Sanford, P., Pate, J.S., Unkovitch, M.J., and Thompson, A.N. (1995) Nitrogen fixation in grazed and ungrazed subterranean clover pasture in south western Australia assessed by the ^{15}N natural abundance technique. *Australian Journal of Agricultural Research* 46, 1427–1443.

Scott, W.R. (1969) An agronomic evaluation of subterranean clover cultivars. MAgrSc thesis, Lincoln College, University of Canterbury, New Zealand.

Shannon, M.C. and Noble, C.L. (1995) Variation in salt tolerance and ion accumulation among subterranean clover cultivars. *Crop Science* 35, 798–804.

Smith, G.R. (1988) Screening subterranean clover for persistent hard seed. *Crop Science* 28, 998–1000.

Steiner, J.J. and Grabe, D.F. (1986) Sheep grazing effects on subterranean clover development and seed production in western Oregon. *Crop Science* 26, 367–372.

Stern, W.R., Gladstones, J.S., Francis, C.M., Collins, W.J., Chatel, D.L., Nicolas, D.A., Gillespie, D.J., Wolfe, E.C., Southwood, O.R., Beale, P.E. and Curnon, B.C. (1983) Subterranean clover improvement: an Australian program. In: Smith, J.A. and Hayes, V.W. (eds) *Proceedings of the XIV International Grassland Congress, Lexington, USA*. Westview Press, Boulder, Colorado, pp. 116–119.

Stockdale, C.R. (1992a) The nutritive value of subterranean clover herbage grown under irrigation in Northern Victoria. *Australian Journal of Agricultural Research* 43, 1265–1280.

Stockdale, C.R. (1992b) The productivity of dairy cows fed irrigated subterranean clover herbage. *Australian Journal of Agricultural Research* 43, 1281–1295.

Suckling, F.E.T. and Charlton, J.F.L. (1978) A review of the significance of buried legume seeds with particular reference to New Zealand agriculture. *New Zealand Journal of Experimental Agriculture* 6, 211–215.

Suckling, F.E.T., Forde, M.B. and Williams, W.M. (1983) Naturalised subterranean clover in New Zealand. *New Zealand Journal of Agricultural Research* 26, 35–43.

Taylor, G.B. (1981) Effect of constant temperature treatments followed by fluctuating temperatures on the softening of hard seeds of *Trifolium subterraneum* L. *Australian Journal of Plant Physiology* 8, 547–558.

Taylor, P.A. and Greenhalgh, F.C. (1987) Significance, causes and control of root rot of subterranean clover. In: Wheeler, J.L., Pearson, C.J. and Robards, G.E. (eds) *Temperate Pastures: Their Production, Use and Management*. AWC/CSIRO, Melbourne, pp. 349–351.

Unkovitch, M.J., Sanford, P. and Pate, J.S. (1996) Nodulation and nitrogen fixation by subterranean clover in acid soils as influenced by lime application, toxic aluminium, soil mineral N and competition from annual grass. *Soil Biology and Biochemistry* 28, 639–648.

Williams, W.M., Charlton, J.F.L. and Caradus, J.R. (1980) Comparative performance of annual legume species at three sites in the southern North Island. *New Zealand Journal of Experimental Agriculture* 8, 185–190.

Birdsfoot Trefoil and Greater Lotus 6

Introduction

Lotus species comprise both perennials and annuals and are often referred to as pioneer legumes, since they are suitable for developing pastures on acid, infertile soils under extensive-farming situations in cool, moist regions of the world. Early references to the use of *Lotus* species date from the mid-eighteenth century in the UK, when pasture seed was first used.

The number of *Lotus* species distributed throughout the world is estimated at 100–120 (Blumenthal *et al.*, 1993) to 176 (Grant, 1995). The greatest number of species is found in the Mediterranean region, indicating this area as their probable centre of origin, although at least 60 are endemic in North America along the west coast, but especially in coastal California (Allen and Allen, 1981), and so it is possible that there was coevolution in the two regions.

Two perennial species are in general use for grassland farming on a wide range of soils, at altitudes ranging from sea level to 3000 m above sea level (a.s.l.), and this chapter concentrates on these two:

- *Lotus corniculatus* L., commonly called birdsfoot trefoil, although there are many other local names, 20 in English alone, mainly based on characteristic features of the plant at flowering or seeding;
- *Lotus uliginosus* Schkuhr. (syn. *Lotus pedunculatus* Cav.), commonly known as marsh birdsfoot trefoil in the UK, big trefoil in the USA and lotus or greater lotus in Australasia. It is referred to as greater lotus in the following text.

A third species of much less importance is used in more local situations:

- *Lotus tenuis* Waldst and Kit. ex Willd, commonly called narrow-leaf birdsfoot trefoil.

Birdsfoot trefoil was not naturalized in North America until the early 1900s, but in Europe, Africa and Asia it was one of the most common and widespread native plants (Seaney and Henson, 1970). This species grows in Europe from northern Russia (72°N) through to West Africa. *Inter alia*, it has also been recorded in South Africa, India and Japan. It was introduced into some areas of Australasia during the early 1900s but has not naturalized there.

Birdsfoot trefoil is now widely developed as a sown species, *c.* 1.2 Mha, in north-eastern North America where its growth requirements are suited to acid, infertile soils and low-input management systems (Beuselinck and Grant, 1995), although its range of adaptation is extending into south-eastern USA, with the release of cultivars containing Mediterranean germplasm (Hoveland *et al.*, 1990). It has also been used successfully in pastures throughout eastern Europe, southern Latin America and parts of Asia. It is the most drought-tolerant of the three above-mentioned *Lotus* species, and even more so than lucerne (Petersen *et al.*, 1992). Birdsfoot trefoil persists and yields better under poor soil-drainage conditions than lucerne and white or red clovers, is more tolerant of flooding than lucerne (Barta, 1986) and is highly tolerant of saline soils (Schachtman and Kelman, 1991). Prostrate selections proved more tolerant of poorly drained soils than erect-growing cultivars. It also thrives in sand-dune associations at sea level and is used in mixture with grass for dune stabilization, and yet it also naturalizes well at high altitudes, such as in parts of the Swiss Alps, where it tolerates severe winters. In southern Latin America, it is estimated that about 2000 tonnes of seed are used each year, which approximates to almost 250,000 ha sown annually (Asuaga, 1994).

Use of greater lotus is documented from the mid-1800s in Britain, France and Germany and it subsequently adapted well when sown in north-western USA. It is less winter-hardy than birdsfoot trefoil and this limits the areas where it can be grown. In New Zealand, it was hand-sown by farmers as a pioneer legume to develop scrub-covered land, since it enticed stock to graze and trample undergrowth, leading to improved pastures. It is now resident there in mixed pastures throughout moist regions and is also sown as a forest understorey in pine (*Pinus radiata*) plantations, where the forage is grazed, and its molecular nitrogen (N_2) fixation boosts tree growth (Gadgil *et al.*, 1986). It grows successfully in cattle-grazed pastures under the high-rainfall conditions of the south-eastern Australian coast.

Birdsfoot trefoil and greater lotus are well known in the honey production industry as species that produce high-quality honey.

Narrow-leaf birdsfoot trefoil is adapted to poorly drained soils and is sown in parts of central Europe and north-eastern and western states of the USA, especially on saline and alkaline soils.

The Plant

Growth habit

In birdsfoot trefoil, all forms of growth habit are found, from erect to prostrate and all intermediate stages. Branches develop in the leaf axils and, together with secondary branching, produce multibranched shoots. The base of the shoot system forms a crown and numerous stems are then produced from this source (Fig. 6.1). The surface stems are thicker than the stems of other *Lotus* species and sometimes behave as stolons in a manner similar to buried stolons of white clover. Spring growth of birdsfoot trefoil is observed to be from the crown, but, after cutting, buds are formed at nodes on the upper end of cut stems.

Greater lotus is regarded as prostrate, because of its stoloniferous habit and spreading rhizomes, but it is a climbing plant in ungrazed situations, striving to pass through tall natural vegetation to reach better light conditions and sometimes attaining 4–5 m in height (Fig. 6.2).

Narrow-leaf birdsfoot trefoil is usually prostrate in habit since the stems are weak and tend to lodge.

Roots

Birdsfoot trefoil is a tap-rooted species, developing roots that penetrate the soil almost as deeply as those of lucerne, but with a greater lateral spread, which leads to its better persistence than lucerne on undrained shallow soils. However, rhizome production, initiated from axillary buds on basal portions of shoots, has been observed in Moroccan genotypes (Beuselinck *et al.*, 1996; Li and Beuselinck, 1996).

In greater lotus, rhizomes are regarded as the main perennating organs, allowing the plant to spread and develop in a similar manner to white clover by initiating numerous daughter plants, a characteristic that aids the persistence of the species. Rhizomes arise at the crown and develop adventitious roots at nodes. Greater lotus tolerates waterlogged conditions better than other *Lotus* species and a number of other forage legumes, including lucerne – the most susceptible – by root suberization, the production of adventitious roots, the thickening of submerged stems and the transportation of higher quantities of water (Shiferaw *et al.*, 1992).

Narrow-leaf birdsfoot trefoil has a short, pronounced tap root, with extensive lateral root branching, and usually has very dense root growth in the upper soil layers.

Leaves

Lotus species are characteristically pentafoliate, because their two stipules are leaflet-like and similar in size and shape to the leaflets, which can vary widely. All three perennial *Lotus* species resemble other forage legumes in their ability to close together the terminal leaflets at night, or in daytime during conditions conducive to wilting, in order to reduce transpiration. The leaves of birdsfoot trefoil are fairly similar to those of greater lotus, but tend to be paler green and

248   Chapter 6

(a)

Fig. 6.1. Birdsfoot trefoil. (a) Plant. (b) (i) Leaves on stem and (ii) inflorescences. (c) iI) Seed head with pods and (ii) seeds.

to have more slender leaflets. The tetraploid selections of greater lotus have the largest leaves, particularly when grown under optimum conditions, and can attain up to 5 cm across. Narrow-leaf birdsfoot trefoil has characteristically slender leaves, as its name suggests.

Flowers
The inflorescence of *Lotus* species is an umbel borne at the end of a long axillary peduncle, with up to eight florets in birdsfoot trefoil, five to 12 in greater lotus and rarely more than four in narrow-leaf birdsfoot trefoil. The calyx is a dentate

Fig. 6.1 continued
(b)

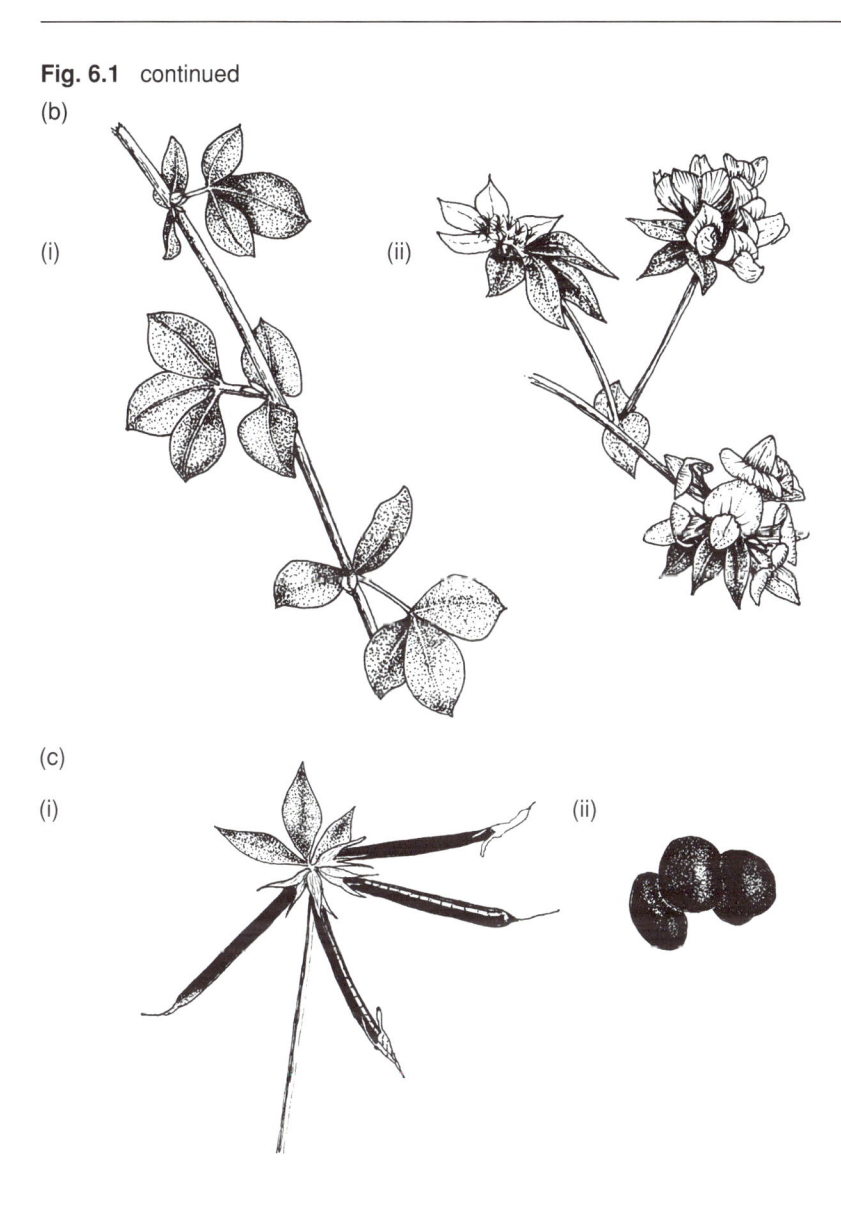

(i)   (ii)

(c)

(i)   (ii)

tube between 3 and 5 mm long. In greater lotus, the tips of the calyx turn outwards and have a star-like appearance until they are mature, this being one of the main botanical features that distinguish it from the other two perennials.

The corolla is that of a typical legume, having five petals, which are generally yellow, though often tinged with red at some stage. The yellow shade of birdsfoot trefoil is usually lighter than that of greater lotus when seen in mass, such as in flowering seed crops. There are nine joined or fused stamens and one free stamen, and the gynoecium has an ovary in the shape of a cylindrical tube, 6–8 mm long at flower opening.

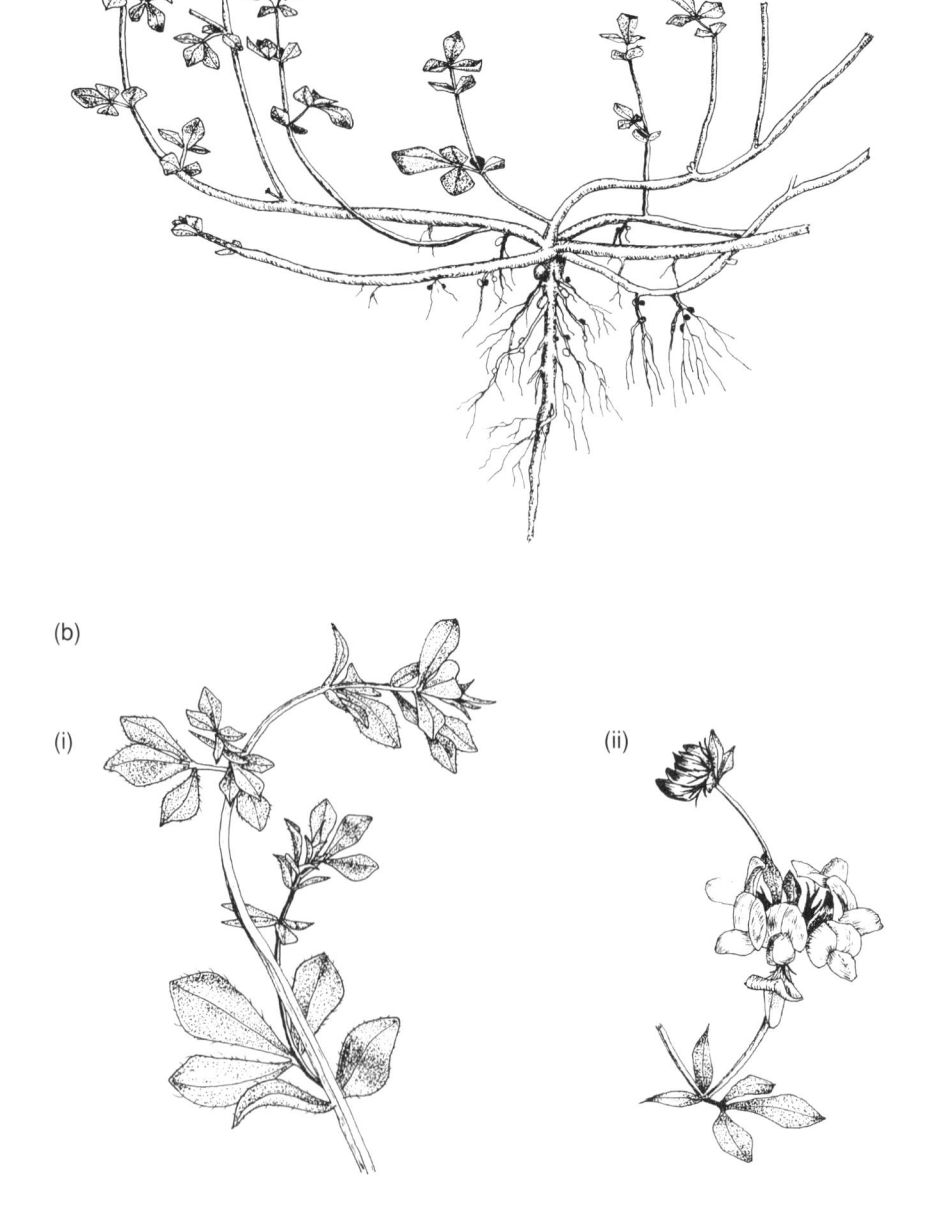

Fig. 6.2. Greater lotus. (a) Plant. (b) (i) Aerial parts and (ii) inflorescences. (c) (i) Seeds and (ii) seed head with pods.

Fig. 6.2 continued

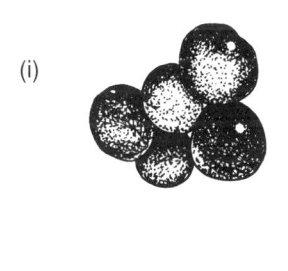

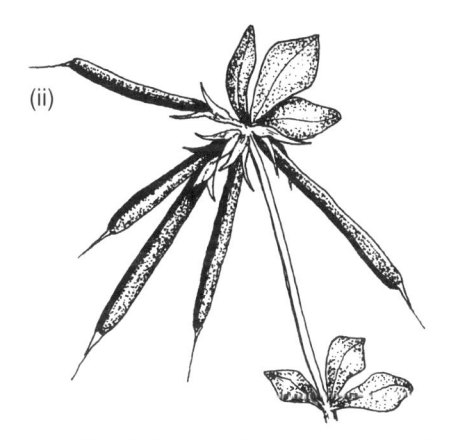

(c)

(i)   (ii)

The anthers of *Lotus* species shed pollen before the flower is open and pollination is mainly by honey-bees and other insects. The pollen starts to germinate approximately 30 min after pollination and fertilization usually occurs 1–2 days after cross-pollination. Less than half the ovules, around 20 per ovary, develop into mature seed. Birdsfoot trefoil and narrow-leaf birdsfoot trefoil are highly self-sterile, whereas greater lotus is highly self-fertile. The seed pods develop rapidly once pollination has occurred, reaching their maximum length after 3 weeks. As they ripen, the pods change colour from green to brown, becoming darker as they mature. Dehiscence of pods begins within 1–2 weeks of seed ripening, depending on climatic conditions. Rapid drying causes rapid dehiscence, so seed crops are best dried out slowly. These pod features have considerable effects on seed harvesting and largely determine seed yield.

Seed

En masse, seeds of the *Lotus* species appear chocolate-brown when mature and have a green tinge when less mature, although seed of birdsfoot trefoil are less green in bulk than those of greater lotus. Individual seed colour ranges from greenish-yellow to almost black, and the shape from oval to spherical. Seeds of birdsfoot trefoil are the largest, with a 1000-seed weight of 1.2–1.4 g, while seeds of greater lotus are the smallest, with an average 1000-seed weight of 0.5 g for diploids and 0.8 g for tetraploids; narrow-leaf birdsfoot trefoil has an average 1000-seed weight of 0.95 g.

Seed quality varies considerably and depends upon ripeness at harvesting and the crop handling processes used. *Lotus* seed samples may contain up to 50% or more hard seeds, i.e. seeds that are viable but do not absorb moisture during standard germination-test periods, due to an impermeable seed-coat. However, their germination capacity can be high. For example, over a 3-year period, more than half of the seed lots of tetraploid greater lotus cv. Grasslands Maku submitted for official testing in New Zealand had a germination of over

80% (Hampton *et al.*, 1987). Expert handling at all stages of producing and storing seed crops minimizes germination problems and a number of treatments, e.g. scarification or soaking the seeds in a gypsum solution, are available to reduce the percentage of hard seed when sowing out.

Physiology

It is well documented that *Lotus* species have weak seedling emergence and slow establishment. Seedling vigour and herbage yield of birdsfoot trefoil increased progressively and branching commenced earlier as seed size and weight increased (Carleton and Cooper, 1972; Beuselinck and McGraw, 1983). Seed-size effects depended on cotyledon area and stored energy, although the latter's role is not well understood, and rate of leaf area expansion is also important (Nelson *et al.*, 1995). Respiration rates in four birdsfoot trefoil cultivars were similar and increased with rising temperature, but germination, emergence rate, plant-elongation rate and subsequent herbage yields differed (Qualls and Cooper, 1968).

Birdsfoot trefoil fails to survive heavy shade and needs at least 50% of normal daylight during early seedling growth for satisfactory establishment, whereas greater lotus seedlings grow best at over 80% daylight (Grime and Jeffrey, 1965). Birdsfoot trefoil has shown a greater relative growth rate than lucerne under several shading regimes, attributable to its greater leaf area, but this did not outweigh the initial growth advantage of lucerne, gained from its larger seed (Cooper, 1966).

The percentage germination and germination rate (days to 75% germination) of tetraploid greater lotus cv. Grasslands Maku were reduced by low temperatures (5°C), and fluctuating temperatures of 5/10°C, compared with white clover and several other legumes (Hampton *et al.*, 1987). In a comparison of tetraploid greater lotus with birdsfoot trefoil, the germination rate of greater lotus declined with decreasing temperature, with maximum germination level reached by 2 weeks at 20°C, while at 10°C only 30% of seeds and at 5/10°C only 1% had germinated by this stage (Charlton, 1989); in addition, large seeds of greater lotus germinated faster than small seeds. Nevertheless, it has proved possible to improve the rate of germination in greater lotus by selection (Keoghan and Burgess, 1986).

Lotus species are long-day plants and need a day length of at least 16 h to flower profusely. Even at 15 h, restrictions in flowering primordia are evident, and abortive buds appear at temperatures of 14°C. Length and quality of photoperiods required are similar for birdsfoot trefoil and greater lotus.

Plant nutrition

The effect of nutrient deficiency in birdsfoot trefoil has been well reviewed (Russelle *et al.*, 1985). Generally, nutrient stress decreases branching and leaf-area index and increases specific leaf mass. Although only one-third of the dry matter (DM) accumulates by late vegetative stage, more than half the maximum amount of nutrients accumulates by that stage, especially sodium (Na), calcium (Ca), magnesium (Mg), sulphur (S) and zinc (Zn) (McGraw et al., 1986a); leaves generally have a higher mineral content than shoots as a whole, especially after commencement of flowering, and concentrations in the shoots of all minerals, except Na, decline with advancing maturity, with potassium (K) declining more strongly than phosphorus (P) or nitrogen (N). Seed yield is affected less than forage production by nutrient deficiency, and N_2 fixation is adversely affected more by deficiency of macro- than of micro-nutrients, except in the case of boron (B) (Russelle and McGraw, 1986).

Estimates of critical concentration of P for *Lotus* species vary; for example, for birdsfoot trefoil, it ranges from 2.3 g P kg^{-1} DM (Davis, 1991) to over 3.5 g kg^{-1} DM (Russelle et al., 1989). The critical concentration of P for greater lotus is greater than that for birdsfoot trefoil, i.e. 3.0 compared with 2.3 g kg^{-1}(Davis, 1991). Greater lotus has a higher efficiency in utilizing P than white clover (Hart and Collier, 1994).

The critical concentration of K for birdsfoot trefoil declines with age, the petiole having a higher critical concentration than the lamina in the youngest leaf, but K concentration increased from 10 to 35 g kg^{-1} without any increase in yield (Russelle et al., 1989). Nitrogen fixation is sensitive to K deficiency in birdsfoot trefoil, with application of K causing the fixation rate to increase per nodule in particular (Collins and Lang, 1985).

Greater lotus has a higher tolerance of manganese (Mn) than lucerne but lower than subterranean clover, i.e. a critical toxicity level of 0.76 g kg^{-1} DM compared with 2 g kg^{-1} for subterranean clover and 0.19 for lucerne (Wheeler and Dodd, 1995). It is more tolerant of soil acidity and high soil aluminium (Al) than birdsfoot trefoil (Floate et al., 1989). Greater lotus proved capable of withstanding 5 μmol Al in the soil without adverse effect, while 7 μmol resulted in a 50% reduction, whereas 2–3 μmol caused a 50% reduction in birdsfoot trefoil (Edmeades et al., 1991). In an examination of the tolerance of 87 cultivars of 34 plant species to soil Al, greater lotus was included among the most tolerant species (Wheeler et al., 1992).

Nitrogen fixation

Inoculation with effective strains of the appropriate, species-specific *Rhizobium* is required to achieve successful establishment of *Lotus* species in most situations where these legumes are introduced for the first time. Seedling establishment of birdsfoot trefoil is compromised if there is inadequate survival of the *Rhizobium* inoculum (Lowther and Patrick, 1995). There were striking differences in the plant growth of birdsfoot trefoil between seeds inoculated with effective and ineffective *Rhizobium* strains within 6 months of sowing (Charlton et al.,1981). In contrast to lucerne, root nodules senesce after defoliation and a new generation of nodules is formed during plant regrowth (Vance et al., 1982). In general, *Rhizobium* strains effective on birdsfoot trefoil may also be beneficial on narrow-leaf trefoil but are ineffective on greater lotus.

With greater lotus, better nodulation and subsequent performance on soil of low pH have resulted from seed inoculated with *Rhizobium* and coated with rock phosphate/dolomite, or else where lime was broadcast, than from seed only inoculated (Wedderburn, 1986); also, increasing the inoculum load increased the number of rhizobia surviving, which then resulted in better nodulation and improved establishment. Under acid conditions, higher N_2-fixation rates have been recorded from the use of slow-growing *Bradyrhizobium* strains than the faster-growing strains of *Rhizobium lotii* (Vance et al., 1987).

Optimum soil temperatures for N_2 fixation in birdsfoot trefoil are the same as for other legumes, i.e. between 18 and 24°C (Kunelius and Clark, 1970). Some strains of rhizobia fix up to 150% more N than less effective strains when root temperatures are low, indicating that strain selection can improve early growth in regions where soil temperatures are low for long periods in spring and autumn.

Nitrogen-fixation rates in birdsfoot trefoil ranged from 60 to 138 kg N ha^{-1} in North America (Heichel et al., 1985; Farnham and George, 1994a), and the proportion of N in the legume derived from N_2 fixation varied from 0.30 to 0.85 in monoculture, compared with 0.95 to 0.98 in grass/birdsfoot trefoil mixtures, indicating the competitive vigour of the grass in taking up available soil N. Cocksfoot derived 8–46% of its N from transfer from birdsfoot trefoil (Farnham and George, 1994b), while tall fescue received 46–65% of its N when the legume comprised 80% of the herbage in the sward (Mallarino et al., 1990).

In Northern Ireland, N_2 fixation of greater lotus on deep peat to which lime and ground rock phosphate were applied reached a maximum of 35 kg N ha^{-1} when the legume made up 22% of the total herbage (Laidlaw, 1981). However, when grown under glasshouse conditions, N_2 fixation of greater lotus was similar to that of white clover, being related to the DM produced, and the N_2 fixation was more sensitive to increase in P at low pH than that of white clover (Gibson et al. 1975).

Breeding

The haploid chromosome number of the genus *Lotus* is 6. Birdsfoot trefoil is a tetraploid, with $2n = 4x = 24$ somatic chromosomes, whereas greater lotus and narrow-leaf birdsfoot trefoil are both diploid ($2n = 2x = 12$). However, diploid birdsfoot trefoil has been identified in populations in the French southwestern Alps (Jay et al., 1991). Successful crosses have been made between birdsfoot trefoil and both diploid and tetraploid forms of greater lotus, and between greater lotus and narrow-leaf birdsfoot trefoil.

Local ecotypes of birdsfoot trefoil were used in some European countries before breeding began in Denmark and Czechoslovakia during the 1930s and some years later in North America and other European countries. Its breeding and use in several countries, such as Denmark, declined as more productive white and red clovers were developed for soils where birdsfoot trefoil was previously grown and where fertility had subsequently been improved. Despite its relatively limited use in Europe, there are currently 13 entries in the European Common Catalogue. There was considerable breeding activity in North America and in Latin America in the 1950s and 1960s, and there are still active programmes. Australasia's first cultivar, Grasslands Goldie, was released in 1991.

Depending on the region, breeding aims include improvements in germination capacity at low temperatures, seedling vigour, spring growth, annual herbage yield, winter-hardiness, seed yield, seed-pod indehiscence, herbicide tolerance (especially in seed crops) and disease resistance. As genotypes within birdsfoot trefoil vary in tolerance to acid conditions, there is potential for breeding for acid tolerance (Alison and Hoveland, 1989b). Selection for capability to withstand low-input conditions generally offers promise under UK conditions (Bullard and Crawford, 1995).

Cultivars

Two distinct types of birdsfoot trefoil are grown in North America: Empire and European. Empire types are related to a naturalized persistent ecotype discovered in New York State and, compared with cultivars of European origin, it is slower in seedling growth, finer-stemmed, more decumbent in growth habit, 10–14 days later in flowering, more indeterminate in flowering habit and slower in recovery rate after defoliation (Grant and Marten, 1985). At least 25 cultivars have been developed in North America, with Empire, Dawn, Leo, Viking, Norcen and Carroll among the most extensively used (Beuselinck and Grant, 1995). Latin American selections include El Boyero, Quinley and San Gabriel.

There have been fewer breeding programmes concerned with greater lotus than with birdsfoot trefoil. The cultivar Grasslands Maku, released in New Zealand in 1975, is a tetraploid selection derived from local and winter-active Portuguese material. A diploid cultivar, Grasslands Sunrise, based on similar germplasm, was released in 1991. American examples are cv. Marshfield, tolerant to winter flooding, and cv. Kaiser. Future breeding objectives for greater lotus include improvements in rate of establishment, possibly by improving germination at low temperatures (5–10°C), competitiveness in mixed grass/greater-lotus sowings, regrowth rate following defoliation, drought resistance and annual herbage yield.

In southern Australia, considerable effort is being made in evaluating the potential of greater-lotus germplasm from a wide range of sources for a complementary role with traditional legumes, and for breeding programmes with objectives such as better establishment, optimal content of condensed tannins (CT) and better seed production (Blumenthal *et al.*, 1993). A diploid cultivar, Sharnae, has been selected that is earlier-flowering than Grasslands Maku and better able to form a seed bank at low (< 32°S) latitudes (Wilson, 1993).

Some breeding of narrow-leaved trefoil has taken place in the USA, e.g. the cultivar Tretana from Montana.

Seed production

A drawback to the use of *Lotus* species is the high cost of seed relative to other forage legumes, because of large annual fluctuations in seed production, the weather at harvesting time being a major factor. The seed pods mature on the plants over a long period and then split and expel the seeds with explosive force in response to warming by the sun. Optimal timing of seed harvesting is therefore critical. Pod shatter can be lessened by drying the crop in the cut swaths after mowing at seed maturity or by desiccation of the standing crop (Hare and Lucas, 1984). Indeterminate flowering and limited distribution of the products of photosynthesis to reproductive growth are other factors that can limit seed yield (McGraw and Beuselinck, 1983). Areas that are most suitable for seed production are those with cool summers and long photoperiods (McGraw *et al.*, 1986a); over a range of genotypes at three contrasting sites in the USA, seed set and filling were highest at the most northerly site.

The optimum plant density for seed production is *c.* 20 ha^{-1} (McGraw *et al.*, 1986c). Average seed yields of birdsfoot trefoil in the USA range from 50 to 175 kg ha^{-1}, with maximum yields up to 600 kg ha^{-1} (McGraw *et al.*, 1986b). In New Zealand, average yields have reached 350 kg ha^{-1} in practice, although, experimentally, they have exceeded 1000 kg ha^{-1} for both birdsfoot trefoil and tetraploid greater lotus. About 200 ha of the tetraploid cultivar, Grasslands Maku, are harvested annually, with average seed yields of 250 kg ha^{-1} from commercial crops, whereas those of the diploid greater lotus, Grasslands Sunrise, average 400 kg ha^{-1}. In general, seed yields of birdsfoot trefoil are one-third higher than those of tetraploid greater lotus.

Application of chemical growth regulators to increase seed yields has given mixed success. In four experiments, there was no effect of daminozide or mepiquat on birdsfoot trefoil seed production (White *et al.*, 1987). However, application of paclobutrazol between the prebud stage and the commencement of flowering increased seed yield by 35%, mainly due to an increase in inflorescence density (Li and Hill, 1989). Cyclocel increased seed yield in greater-lotus crops by increasing pod number and seeds per pod (Tabora and Hill, 1992). Control of weeds in seed crops can also markedly increase seed yields in heavily infested crops. For example, controlling couch grass (*Elymus repens*) in birdsfoot trefoil with haloxyfop increased seed yield from 80 kg ha^{-1} to 130 kg ha^{-1} (Linscott and Vaughan, 1989).

Agronomy and Management

Establishment

A major factor in the slow adoption of *Lotus* species for sown pastures by farmers has undoubtedly been their slow establishment and poor competitive ability (McKersie *et al.*, 1981). However, in recent years, cultivars have been released with improved seedling vigour and growth characteristics. In addition, there are now better techniques and more experience, both in establishing productive

Lotus pastures with minimum delay and in maintaining their persistence. The optimum sowing depth is 10–15 mm and establishment is best in seed-beds with a fine but firm tilth, under warm moist conditions (Charlton, 1984).

For monocultures, birdsfoot trefoil is generally sown directly in spring, when soil moisture is adequate, at seed rates of 6–10 kg ha^{-1}. When sown in autumn, there has to be sufficient time for plants to develop winter-hardiness (Laskey and Wakefield, 1978). In simple mixtures with non-aggressive companion grasses, such as timothy (*Phleum pratense*), smooth-stalked meadow-grass (*Poa pratensis*), brome grasses (*Bromus* spp.) or a non-aggressive cocksfoot (*Dactylis glomerata*) cultivar, birdsfoot trefoil is included at 3–5 kg ha^{-1}, together with 9–10 kg ha^{-1} of the grass. The higher seed rates are used when oversowing natural pastures in North America (Seaney and Henson, 1970). Germination of birdsfoot trefoil seed is allegedly inhibited by extracts from tall fescue (*Festuca arundinacea*), with organic acids such as succinic or lactic likely to be implicated (Lim *et al.*, 1989). Generally, compatibility with birdsfoot trefoil is inversely proportional to yield of the grass in the mixture (Sheaffer *et al.*, 1992). Densely tillering grasses, such as perennial ryegrass (*Lolium perenne*) and bent grasses (*Agrostis* spp.), are not compatible with birdsfoot trefoil during establishment (Davies, 1969), although, if non-competitive species are sown, an opportunity is created for ingression of the more aggressive volunteer grasses, which may also hinder establishment (Sheldrick and Martyn, 1992).

North American experience indicates that more uniform swards result from drilling than from broadcasting. Undersowing birdsfoot trefoil in a cereal cover crop is possible but not widely advocated, because of interspecies competition.

In cold temperate areas, spring sowing of greater lotus is generally more successful than sowing at other times. Its inherent low seedling vigour is a drawback in autumn sowings and seedlings may fail to develop to a stage capable of survival after the onset of winter. In warmer regions, spring and autumn sowings are equally successful, provided soil moisture is adequate. In New Zealand, diploid greater lotus is sown at 1–3 and the tetraploid at 1–5 kg ha^{-1} (Charlton, 1992). The higher seed rates are used when establishment conditions are difficult or where the legume is surface-sown alone. Establishment has been boosted by a small application of N, at 25 kg ha^{-1} (Charlton and Brock, 1980). A method of 'hoof-and-tooth' surface sowing, which involves deliberately trampling a pasture by sheep or cattle in order to open up the sward surface, seeding the area and then trampling the seed into the soil by driving a flock of sheep across the sown area, has proved successful.

In Australia, dense stands of greater lotus have established within 18 months from oversowing at rates as low as 125 g ha^{-1} in subtropical situations. A cheap on-farm method used there is the sprinkling of seeds from boxes above the rear tractor wheels and the use of the wheels to press the seed into the soil.

Typically, narrow-leaf trefoil seed is sown at 5–8 kg ha^{-1}.

Fertilizer use

Although birdsfoot trefoil is adapted to acid, infertile soils and, as such, may be regarded as a pioneer legume, it responds to improved soil pH and fertility. Establishment and N_2 fixation may be adversely affected at a soil pH lower than 6.2 (Beuselinck and Grant, 1995). Herbage yields, seed yield and winter-hardiness are improved by P and K application on infertile but not on fertile soils (Russelle et al., 1991).

Weed control

Because of their non-aggressive early growth, *Lotus* swards are often invaded by broad-leaved weeds during establishment. Such invasion is minimized by good seed-bed preparation to ensure an initial weed-free environment, but, once weeds have developed, cultural control by cutting, or grazing in some cases, is effective particularly for tall-growing annual weeds. Herbicide application may be required for more decumbent weed species, but timing of application is critical, since the *Lotus* seedlings must be adequately developed; for example, a postemergence spray of 2,4-DB should not be applied until at least the second true-leaf stage. For monocultures, the herbicide EPTC, incorporated into the seed-bed prior to sowing, controls invasive grasses and volunteer cereals.

In a comparison of herbicide use, mowing and a cover crop of oats as weed-control measures in establishing birdsfoot trefoil, herbicide use was most successful, as judged by first harvest-year yield of the legume, while mowing reduced the legume-plant population the most (Scholl and Brunk, 1962). White campion (*Silene alba*) can be controlled with trifluralin or ethalfluralin (Wyse and McGraw, 1987), while, in seed crops, Canadian thistle (*Cirsium arvense*) was effectively controlled by picloram or glyphosate (Davis and Linscott, 1986) and couch grass (*Elymus repens*) by haloxyfop (Linscott and Vaughan, 1989).

The scope for using selective herbicides in birdsfoot trefoil stands will be increased in the future by selection and breeding for increased resistance to herbicides, e.g. sulphonylurea herbicides, by using *in vitro* techniques to select for resistance (Pofelis et al., 1992). Tolerance to non-selective glyphosate has also been developed in birdsfoot trefoil by recurrent selection (Boerboom et al., 1991).

Pests and diseases

Pests are not usually a major problem in birdsfoot trefoil grown for conserved forage or grazing, although, experimentally, regular application of insecticide (cyfluthrin) increased yield of birdsfoot trefoil by 6% (Mackun and Baker, 1990). Seed-crop yields can be reduced due to damage by mirids (*Lygus lineolaris*) or the alfalfa-plant bug (*Adelphocoris lineolaris*) (Wipfli et al., 1990a), but, generally, these pests are controlled by standard pesticides, e.g. malathion, dimethoate and trichlorphon (Wipfli et al., 1990b). Pesticides are ineffective against seed chalcids (*Bruchophagus platypterus*), which can cause considerable loss of seed yield by parasitizing the seed, but early-season harvesting and burning the leaves, stems and other debris after combining reduces populations (Beuselinck and Grant, 1995). Birdsfoot trefoil is also susceptible to the root-lesion nematode (*Pratylenchus penetrans*) and root-knot nematodes (*Meloidogyne* spp.). The good resistance of greater lotus to New Zealand grass grub (*Costelytra zealandica*) is noteworthy. Condensed tannins in birdsfoot trefoil and greater lotus may confer resistance to insects (Briggs, 1991).

Crown and root rots are the most important diseases affecting the productivity of birdsfoot trefoil, with the most serious effects occurring under warm, humid conditions, such as in southern USA (Beuselinck *et al.*, 1984) or southern Latin America (Chao *et al.*, 1995). A range of interacting fungi, with *Fusarium*, *Sclerotinia* and *Rhizoctonia* spp. featuring prominently, is responsible. The changing nature of the complex under different environmental conditions makes breeding for resistance difficult, but some success has been achieved; the cultivar 'Dawn' was selected for resistance to root rots and to leaf and stem diseases. A number of leaf and stem diseases, e.g. *Sclerotinia trifoliorum*, and *Stemphylium loti*, flare up from time to time and adversely affect growth vigour and yield

Use in roughland

Birdsfoot trefoil proved particularly suitable for use on marginal land and selected roughland, e.g. renovation of permanent grassland in north-east USA by surface sowing (MacDonald, 1946); it was further suggested that *Lotus* species in general would be suitable for the maintenance of large areas, not needed then or not economic to farm, in a condition which would facilitate their conversion to more intensive agricultural use, if required in the future. A number of reports from different regions of the world later confirmed the use of *Lotus* species, especially birdsfoot trefoil, for roughland-pasture renovation in a diversity of situations, ranging from lowland rough grazing to alpine pastures.

When oversowing, manipulation of management to reduce sward density is a major requirement to reduce competition from the existing sward, and this is achievable by chemical means, cutting or minimal cultivation. The use of herbicides, such as dalapon or paraquat, to kill or retard the growth of existing vegetation as a precursor to oversowing birdsfoot trefoil was an important advance in the 1960s since it improved the chances of the weakly competitive seedlings to develop fully. Subsequently, a four- to fivefold increase in herbage DM yield and an extension of the grazing season by 2–5 months from renovated compared with unimproved pasture was reported (Winch *et al.*, 1969). Sod-seeding also proved successful (Olsen *et al.*, 1981; Kunelius *et al.*, 1982), and many types of drill have been developed, with different modes of operation, such as removal of strips of dense existing sward, creation of slits, strip cultivation by minirotavation units, with or without fertilizer-application attachments, or supplementary equipment for band spraying of herbicides to kill or suppress adjacent grass (Lowther *et al.*, 1996). The management guidelines for successful sward renovation by such methods are now more clearly understood (Tiley and Frame, 1991).

Birdsfoot trefoil's role in low-input systems in the UK is being reassessed, and the results suggest that it has particular potential on marginal or free-draining calcareous soils, where it could be a suitable substitute for white or red clover, especially if subject to moisture stress (Bullard and Crawford, 1995). Previous work had indicated its potential as an alternative to white clover on dry sites of low fertility and under low grazing intensity (Charlton, 1973), but the species has never been exploited.

Greater lotus proved suitable as a pioneer legume for use when developing new pastures from native bush in New Zealand, particularly under wet conditions, where soil fertility is poor and where shade restricts other legumes. It also grows vigorously among ferns, scrub and rushes, encouraging stock grazing, which crushes the weed vegetation (Suckling, 1965). Natural reseeding and spread by rhizomes of greater lotus have proved major assets, enabling it to colonize bare ground after scrub has been thinned by stock or burned during land development.

Herbage quality

There is less information available on the chemical composition of birdsfoot trefoil in relation to major legume species, such as white clover, but its nutritive value is similar to that of lucerne (Marten and Jordan, 1979). Total herbage *in vitro* DM digestibility (IVDMD) of birdsfoot trefoil declines from the prebud stage onwards, because of the development of stems. In early summer, the rate of decline is slower than in later regrowths, i.e. falling from 68 to 66% over a 40-day period in the early season, compared with a decline from 67 to 58% later (Buxton *et al.*, 1985); stem IVDMD decreased at twice the rate of total herbage and declined towards the base at 2% per node, while cell-wall content increased during regrowth from about 30 to 45%, a change similar to that in red clover but lower than that in lucerne. Lignin content is consistently lower in birdsfoot trefoil than in other legumes, such as white and red clovers and lucerne. However, *in vivo* digestion rates of organic matter (OM), hemicellulose and cellulose are similar for lucerne, sainfoin and birdsfoot trefoil (Kraiem *et al.*, 1990). Drought increases the overall forage quality of birdsfoot trefoil, due to an increased leaf : stem ratio, delayed maturity and increased quality of each plant fraction (Petersen *et al.*, 1992).

Considering greater lotus, it loses quality in comparison with white clover when utilized out of season in areas that have reasonable growing conditions over winter but may be prone to frosts. For example, in the northern tablelands of Australia, N content and IVDMD declined by 20–25%, whereas they remained relatively constant in white clover (Schiller and Ayers, 1993). Yet 87% of beef farmers and 75% of dairy farmers who responded in a survey in eastern Australia considered greater lotus cv. Maku to be of higher feeding value than clover (Harris *et al.*, 1993).

Condensed tannins (CT), which are polymeric flavonol compounds, are perhaps the most important quality feature of *Lotus* species. Present in the foliar and stem tissues and at lower levels in birdsfoot trefoil than in greater lotus (Kelman and Tanner, 1990), the CT's well-documented property is the ability to bond with herbage proteins and so prevent the formation of the stable foams that can cause bloat in ruminants. Bloat prevention may also be aided by the slow rupture rate in the rumen of the herbage cell walls, a rate that is slower than that for clovers or lucerne. The CT also protect the proteins from degradation to ammonium in the rumen, and this allows more protein to reach the abomasum and the small intestine (John and Lancashire, 1981) and absorption of essential amino acids (Waghorn *et al.*, 1987), the types and combination of which present in birdsfoot trefoil are ideal for producing high-quality animal products (Waghorn *et al.*, 1990). Amino acid absorption in sheep is greater in birdsfoot trefoil than in greater lotus (Waghorn and Shelton, 1992). The nutritional effects of CT depend on the concentration in the herbage, and levels of 20–40 g kg^{-1} DM are thought to be beneficial (Barry, 1989), while, expressed as catechin equivalent, levels of 25–85 g kg^{-1} DM have been suggested (Miller and Ehlke, 1994).

Animal performance

Increased wool production and slightly increased liveweight gain in sheep were attributed to the CT in birdsfoot trefoil (Wang *et al.*, 1994) and, in a comparison of birdsfoot trefoil and lucerne grazed by ewes and lambs, voluntary intake, carcass weight and wool production were higher for birdsfoot trefoil pastures, on account of increased protein-utilization efficiency as a consequence of CT in its herbage (Douglas *et al.*, 1995).

In the USA, birdsfoot trefoil pastures with accompanying grass, usually tall fescue, produced liveweight gains from beef cattle of 0.7–1.2 kg day^{-1} and lambs gained 135 g day^{-1} on pure-sown birdsfoot trefoil (Alison and Hoveland, 1989a). Over a 2-year period, dairy heifers grazing birdsfoot trefoil had superior daily liveweight gains and annual gains ha^{-1} to those grazing lucerne, although lucerne had the higher carrying capacity (Marten *et al.*, 1987). When stocked on a birdsfoot trefoil/tall fescue pasture at an allowance of 7.4% of body weight, lambs gained 200 g day^{-1} (Beuselinck *et al.*, 1992).

In comparisons of sheep performance from pure-sown swards, greater lotus gave 43% more liveweight gain from young sheep than perennial ryegrass, although this superiority was less than from lucerne at 70% and white clover at 86% (Ulyatt, 1981). Lambs grazing greater lotus in late season produced leaner meat, which was more acceptable to the consumer market, than those grazing white clover (Purchas and Keogh, 1984). Compared with birdsfoot trefoil, the higher CT concentration in greater lotus resulted in an improved cysteine supply and utilization at whole-body level in sheep, although wool production was not increased (Lee *et al.*, 1992), but a very high concentration of CT in the herbage adversely affects voluntary food intake and rumen digestion (Barry and Duncan, 1984).

There is evidence that grazed *Lotus* species may minimize effects of the intestinal worm burden, so leading to improved performance. In addition, CT seem to reduce 'dagginess' (faeces-impregnated wool near the anus) and consequent 'fly strike' in sheep, when blowfly maggots (*Lucilia serricata* and *L. cuprina*) eat into the flesh and cause considerable animal suffering. The latter is a particularly serious livestock problem on Australasian farms (Robertson *et al.*, 1995). This beneficial effect was also confirmed in New Zealand with birdsfoot trefoil, greater lotus and sulla (*Hedysarum coronarium*). Drenched ewe lambs grazing these legumes grew faster than those on perennial ryegrass/white clover/plantain (*Plantago lanceolata*), the highest liveweight gains of undrenched lambs being obtained when fed tetraploid greater lotus and sulla.

Conservation

The more erect types of birdsfoot trefoil are best suited for hay or silage harvests, and two to three cuts are possible in a season, depending upon its length, which is longer in warm than in cool temperate regions. After a grazing in early season with a low utilization efficiency, a cut for hay or silage may then be taken following a rest period, the cut being in effect a 'topping' measure to remove flowering stems, in order to ensure a high nutritive value of the sward for subsequent grazing. Provided procedures are followed for good haymaking, namely, careful mechanical conditioning, short field-exposure time and gentle handling at all stages in the process, the legume will retain high proportions of the nutritious leaf fraction.

Highest forage and protein yields in birdsfoot trefoil are attained when cut at the full-bloom stage (Gervais, 1988) and yields are optimal at 6-week cutting intervals (Hoveland and Fales, 1985). Long cutting intervals during summer, while probably reducing plant density, increase the possibility of self-seeding (Taylor *et al.*, 1973), and a lack of self-seeding under continuous grazing explains the poor persistence of the stand in some environments.

Spring growth in birdsfoot trefoil has been shown to be directly proportional to the concentration of total non-structural carbohydrates (TNC) in roots in the 2 previous years (Alison and Hoveland, 1989c) and increasing cutting interval and cutting height result in increased concentration of TNC. Swards may be thinned out by plant loss, due to poor recovery before winter, when the final harvest is taken late in the autumn. The cutting height should be above 10 cm; otherwise, there is insufficient foliage left for subsequent photosynthesis and too few axillary buds to enable vigorous regrowth. Regrowth after cutting is dependent on a sufficiency of photosynthetic tissue, since the root carbohydrate decreases during spring growth and thereafter remains low, although, unlike lucerne, birdsfoot trefoil does not exhibit the wide fluctuations in root carbohydrate (Grant and Marten, 1985).

There is some advantage in using mixed swards for cutting, since a companion grass improves sward density, thus assisting in the prevention of weed ingress and compensating for the inevitable decline in birdsfoot trefoil plant population with time (Marten and Jordan, 1979). The grass may prevent excessive lodging of the legume and the crop is easier to handle mechanically, particularly during hay curing. However, associated grass may reduce the quality of conserved forage relative to a sward of pure-sown birdsfoot trefoil and may also become too competitive.

Grazing

Because of its high acceptability to livestock, birdsfoot trefoil pastures are best managed under some form of controlled grazing, such as a rotational grazing system (Van Keuren and Davis, 1968; Van Keuren *et al.*, 1969). The need for rotational grazing, with its periods of rest, rather than continuous stocking, is greater under warm than cool growing conditions, since root carbohydrate reserves are lower in warmer climates (Nelson and Smith, 1968, 1969). Spring stockpiling for summer utilization has also proved feasible, although forage quality declines with increasing length of the stockpiling period, and so the period used depends on the class of livestock to be fed (Collins, 1982).

Erect-growing cultivars are less persistent than prostrate cultivars under severe grazing. One advantage of the prostrate types is that more leaves are left on the stems following grazing, thus assisting more vigorous regrowth and aiding plant persistence, although these types are less productive than the erect-growing types. However, low-growing stems may escape even close grazing and this, allied to the characteristic of intermittent flowering during summer, can result in seed set and shedding of ripe seed, with subsequent regeneration and benefit to the plant population (Templeton *et al.*, 1967). The acceptability to grazing livestock of the companion grass in mixed swards can also have a bearing on birdsfoot trefoil persistence, since a grass species with low acceptability may result in the birdsfoot trefoil being preferentially grazed. Nevertheless, it is best to avoid grazing closer than 5–10 cm above the soil surface. Good persistence of birdsfoot trefoil stands has been associated with development of a buried seed bank and subsequent stand regeneration, in northern regions of the USA (Taylor *et al.*, 1973).

Persistence of greater lotus is usually associated with rhizome growth and spread. During late summer and autumn, growth is primarily in underground organs under New Zealand conditions, whereas in Scotland both aerial and rhizome growth rates peak in autumn (Wedderburn and Gwynne, 1981). However, in south-eastern Queensland at 29°S rhizome growth is greatest during spring, possibly being influenced by soil-moisture levels (Vos and Jones, 1986). Defoliation as early as midsummer reduced rhizome volume in autumn (Sheath, 1980a, b), while defoliation during early autumn almost eliminated rhizome growth (Wedderburn and Lowther, 1985). When the stand density of greater lotus in a pasture is sparse, it therefore pays to avoid grazing during late summer and autumn, since this enables a better plant density to develop.

In a buried seed-bank survey of greater lotus in south-eastern Australia, the seed content was positively associated with latitude but negatively associated with the summer mean maximum temperatures (Blumenthal and

Harris, 1993). The seed bank assisted persistence only at high latitudes (32° S), when buried-seed rates were recorded as high as 6000 seeds m^{-2}. Breakdown of seed dormancy was much slower than in seed of some white-clover cultivars, and this severely limited the ability of greater lotus to develop seedling populations to increase sward density (Kelman and Blumenthal, 1992). In New Zealand, buried-seed rates of greater lotus were found to be highest (up to 5720 seeds m^{-2}) on wet soils in hill pastures receiving high summer-rainfall levels (Suckling and Charlton, 1978).

References

Alison, M.W. and Hoveland, C.S. (1989a) Root and herbage growth response of birdsfoot trefoil entries to subsoil activity. *Agronomy Journal* 81, 677–680.
Alison, M.W. and Hoveland, C.S. (1989b) Birdsfoot trefoil management. l. Root growth and carbohydrate storage. *Agronomy Journal* 81, 739–745.
Alison, M.W. and Hoveland, C.S. (1989c) Birdsfoot trefoil management. 2. Yield, quality and stand evaluation. *Agronomy Journal* 81, 745–749.
Allen, O.N. and Allen, E.K. (1981) *The Leguminosae*. University of Wisconsin Press, Madison, Wisconsin, pp. 401–405.
Asuaga, A. (1994) Use and production of *Lotus corniculatus* in Uruguay. In: Beuselink, P.R. and Roberts, C.A. (eds) *Proceedings of the First International Lotus Symposium, St Louis, USA*. University of Missouri-Columbia Press, pp. 134–141.
Barry, T.N. (1989) Condensed tannins: their role in ruminant protein and carbohydrate digestion and possible effects upon the rumen ecology. In: Nolan, J.V., Leng, R.A. and Deneyer, D.I. (eds) *The Role of Protozoa and Fungi in Ruminant Digestion*. Pernambul Books, Armidale, Australia, pp. 153–169.
Barry, T.N. and Duncan, S.J. (1984) The role of condensed tannins in the nutritional value of *Lotus pedunculatus* for sheep. l. Voluntary intake. *British Journal of Nutrition* 51, 484–491.
Barta, A.L. (1986) Metabolic response of *Medicago sativa* L. and *Lotus corniculatus* L. roots to anoxia. *Plant Cell and Environment* 9, 127–131.
Beuselinck, P.R. and Grant, W.F. (1995) Birdsfoot trefoil. In: Barnes, R.F, Miller, D.A. and Nelson, C.J. (eds) *Forages*, 5th edn, Vol. l, *An Introduction to Grassland Agriculture*. Iowa State University Press, Ames, Iowa, pp. 237–248.
Beuselinck, P.R. and McGraw, R.L. (1983) Seedling vigour of three *Lotus* species. *Crop Science* 23, 390–391.
Beuselinck, P.R., Peters, E.J. and McGraw, R.L. (1984) Cultivar and management effects on stand persistence of birdsfoot trefoil. *Agronomy Journal* 76, 490–492.
Beuselinck, P.R., Sleper, D.A., Bughrara, S.S. and Roberts, C.A. (1992) Effects of harvest frequency on mono and mixed cultures of tall fescue and birdsfoot trefoil on yield and quality. *Agronomy Journal* 84, 133–137.
Beuselinck, P.R., Li, B. and Steiner, J.J. (1996) Rhizomatous *Lotus corniculatus* L.: I. Taxonomic and cytological study. *Crop Science* 36, 179–185.
Blumenthal, M.J. and Harris, C.A. (1993) Maku lotus soil seed banks in farmers' fields. In: Oram P.A. (ed.)*Proceedings of the 7th Australian Agronomy Conference*, Vol. 7. Australian Society of Agronomy, Parkville Victoria, p. 414.
Blumenthal, M.J., Kelman, W.M., Lolicato, S., Hare, M.D. and Bowman, A.M. (1993)

Agronomy and improvement of *Lotus*: a review. In: Michalk, D.L., Craig, A.D. and Collins, W.J. (eds) *Alternative Pasture Legumes 1993*. Technical Report 219, Department of Primary Industries, South Australia, pp. 74–85.

Boerboom, C.M., Ehlke, N.J., Wyse, D.L. and Somers, D.A. (1991) Recurrent selection for glyphosate tolerance in birdsfoot trefoil. *Crop Science* 31, 1124–1129.

Briggs, M.A. (1991) Influence of herbivory and nutrient availability on biomass, reproduction and chemical defences of *Lotus corniculatus* L. *Functional Ecology* 5, 780–786.

Bullard, M.J. and Crawford, T.J. (1995) Productivity of *Lotus corniculatus* L. (bird's-foot trefoil) in the UK when grown under low-input conditions as spaced plants, monoculture or mixed swards. *Grass and Forage Science* 50, 439–446.

Buxton, D.R., Hornstein, J.S., Wedin, W.F. and Marten, G.C. (1985) Forage quality in stratified canopies of alfalfa, birdsfoot trefoil and red clover. *Crop Science* 25, 273–279.

Carleton, A.E. and Cooper, C.S. (1972) Seed size effects upon seedling vigour of three forage legumes. *Crop Science* 12, 661–665.

Chao, L., De Battista, J. and Santinaque, F. (1995) Incidence of birdsfoot trefoil crown and root rot in west Uruguay and Entre Rios (Argentina). In: Beuselink, P.R. and Roberts, C.A. (eds) *Proceedings of the First International Lotus Symposium, St Louis, USA*. University of Missouri-Columbia Press, pp. 206–209.

Charlton, J.F.L. (1973) The potential value of birdsfoot trefoils (*Lotus* spp.) for the improvement of natural pastures in Scotland. 1. Common birdsfoot trefoil (*Lotus corniculatus* L.). *Journal of the British Grassland Society* 28, 91–96.

Charlton, J.F.L. (1984) Establishment and persistence of new herbage species and cultivars. *Proceedings of the New Zealand Institute of Agricultural Science Convention*, Hamilton, New Zealand 18, 130–135.

Charlton, J.F.L. (1989) Temperature effects on germination of Grasslands Maku lotus and three other experimental lotus selections. *Proceedings of the New Zealand Grassland Association* 50, 197–201.

Charlton, J.F.L. (1992) Some basic concepts of pasture seed mixtures for New Zealand farms. *Proceedings of the New Zealand Grassland Association* 53, 37–40.

Charlton, J.F.L. and Brock, J.L. (1980) Establishment of *Lotus pedunculatus* and *Trifolium repens* in newly developed hill country. *New Zealand Journal of Experimental Agriculture* 8, 243–248.

Charlton, J.F.L., Greenwood, R.M. and Clark, K.W. (1981) Comparison of the effectiveness of *Rhizobium* strains during establishment of *Lotus corniculatus* in hill country. *New Zealand Journal of Experimental Agriculture* 9, 173–177.

Collins, M. (1982) Yield and quality of birdsfoot trefoil stockpiled for summer utilization. *Agronomy Journal* 74, 1036–1041.

Collins, M. and Lang, D.J. (1985) Shoot removal and potassium fertilisation effects on growth, nodulation and dinitrogen fixation of red clover and birdsfoot trefoil. *Field Crops Research* 10, 251–256.

Cooper, C.S. (1966) *The Establishment and Production of Birdsfoot Trefoil-Grass Compared to Alfalfa-Grass Mixtures under Several Cultural Practices*. Montana Agriculture Experimental Station Bulletin No. 603, Bozeman, Montana.

Davies, W.E. (1969) The potential of *Lotus* species for hill country in Wales. *Journal of the British Grassland Society* 24, 264–270.

Davis, C. and Linscott, D.L. (1986) Tolerance of birdsfoot trefoil to 2,4-D. *Weed Science* 34, 373–376.

Davis, M.R. (1991) The comparative phosphorus requirements of some temperate perennial legumes. *Plant and Soil* 133, 17–30.

Douglas, G.B., Wang, Y., Waghorn, G.C., Barry, T.N., Purchas, R.W., Forte, A.G. and Wilson, G.F. (1995) Liveweight gain and wool production of sheep grazing *Lotus corniculatus* and lucerne (*Medicago sativa*). *New Zealand Journal of Agricultural Research* 38, 95–104.

Edmeades, D.C., Blamey, F.P.C., Asher, C.J. and Edwards, D.G. (1991) Effects of pH and aluminium on the growth of temperate pasture species. I. Temperate grasses and legumes supplied with inorganic nitrogen. *Australian Journal of Agricultural Research* 42, 559–569.

Farnham, D.E. and George, J.R. (1994a) Harvest management effects on productivity, dinitrogen fixation, and nitrogen transfer in birdsfoot trefoil–orchard grass communities. *Crop Science* 34, 1650–1653.

Farnham, D.E. and George, J.R. (1994b) Dinitrogen fixation and nitrogen transfer in birdsfoot trefoil–orchard grass communities. *Agronomy Journal* 86, 690–696.

Floate, M.J.S., Enright, P.D. and Woodrow, K.E. (1989) Effects of lime and altitude on the performance of pastures based on six alternative legumes for acid tussock grasslands. *Proceedings of the New Zealand Grassland Association* 50, 213–218.

Gadgil, R.L., Charlton, J.F.L., Sandford, A.M. and Allen, P.J. (1986) Relative growth and persistence of planted legumes in a mid-rotation radiata pine plantation. *Forestry Ecology and Management* 14, 113–124.

Gervais, P. (1988) Influence of growth stage on yield, chemical composition and nutrient reserves of birdsfoot trefoil. *Canadian Journal of Plant Science* 68, 755–762.

Gibson, D.I., Hayes, P. and Laidlaw, A.S. (1975) The influence of phosphate and lime on the growth and N fixation of *Lotus uliginosus* and *Trifolium repens* under greenhouse conditions. *Journal of the British Grassland Society* 30, 295–301.

Grant, W.F. (1995) A chromosome atlas and interspecific-intergeneric index for *Lotus* and *Tetragonolobus* (Fabaceae). *Canadian Journal of Botany* 73, 1787–1809.

Grant, W.F. and Marten, G.C. (1985) Birdsfoot trefoil. In: Heath, M.E., Barnes, R.F. and Metcalfe, P.S. (eds) *Forages: the Science of Grassland Agriculture*, 4th edn. Iowa State University Press, Ames, Iowa, pp. 98–108.

Grime, J.C.and Jeffrey, G.W. (1965) Seedling establishment in vertical gradients of sunlight. *Journal of Ecology* 53, 621–642.

Hampton, J.G., Charlton, J.F.L., Bell, D.D. and Scott, D.J. (1987) Temperature effects on the germination of herbage legumes in New Zealand. *Proceedings of the New Zealand Grassland Association* 48, 177–183.

Hare, M.D. and Lucas, R.J. (1984) Grasslands Maku lotus (*Lotus pedunculatus* Cav.) seed production. I. Development of Maku lotus seed and determination of time of harvest for maximum seed yields. *Journal of Applied Seed Production* 2, 58–64.

Harris, C.A., Blumenthal, M.J. and Scott, J.M. (1993) Survey of use and management of *Lotus pedunculatus* cv. Grasslands Maku in eastern Australia. *Australian Journal of Experimental Agriculture* 33, 41–47.

Hart, A.L. and Collier, W.A. (1994) The effect of phosphorus and form of nitrogen on leaf cell size and nutrient content in *Trifolium repens* and *Lotus uliginosus*. *Grass and Forage Science* 49, 96–104.

Heichel, G.H., Vance, C.P., Barnes, D.K. and Henjum, K.I. (1985) Dinitrogen fixation and N and DM distribution during four-year stands of birdsfoot trefoil and red clover. *Crop Science* 25, 101–105.

Hoveland, C.S. and Fales, S.L. (1985) Mediterranean germplasm trefoils in the southeastern USA, Piedmont. In: *Proceedings of the XV International Grassland Congress, Kyoto, Japan.* The Science Council of Japan and the Japanese Society of Grassland Science, Tochigi-Kem, pp. 126–128.

Hoveland, C.S., Alison, M.W. Jr, Hill, N.S., Lowrey, R.S. Jr, Fales, S.L., Durham, R.G., Dobson, J.W. Jr, Worley, E.E., Worley, P.C., Calvert, V.H. II and Newsome, J.F. (1990) *Birdsfoot Trefoil Research in Georgia.* Georgia Agricultural Experimental Station Research Bulletin 396, Athens, Georgia.

Jay, M.J., Reynaud, S.B. and Cartier, D. (1991) Evolution and differentiation of *Lotus corniculatus/Lotus alpinus* populations from French south western Alps. III. *Evolutionary Trends in Plants* 5, 157–160.

John, A. and Lancashire, J.A. (1981) Aspects of the feeding and nutritive value of *Lotus* species. *Proceedings of the New Zealand Grassland Association* 42, 152–159.

Kelman, W.M. and Blumenthal, M.J. (1992) Lotus in south-eastern Australia: aspects of forage quality and persistence. In: Ragless, D. (ed.) *Proceedings of the 6th Australian Agronomy Conference.* Australian Society of Agronomy, Parkville, Victoria, pp. 460–463.

Kelman, W.M. and Tanner, G.J. (1990) Foliar condensed tannins in *Lotus* species growing on limed and unlimed soils in south-eastern Australia. *Proceedings of the New Zealand Grassland Association* 52, 51–54.

Keoghan, J.M. and Burgess, R.E. (1986) The search for an improved *Lotus pedunculatus* for high country pastoral systems. *Proceedings of the New Zealand Grassland Association* 48, 125–130.

Kraiem, K. Garrett, J.E., Meiske, J.C. Goodrich, R.D. and Marten, G.C. (1990) Influence of method of forage preservation on fibre and protein digestion in cattle given lucerne, birdsfoot trefoil and sainfoin. *Animal Production* 50, 221–230.

Kunelius, H.T. and Clark, K.W. (1970) Influence of root temperature on the early growth and symbiotic nitrogen fixation of nodulated *Lotus corniculatus* plants. *Canadian Journal of Plant Science* 50, 569–575.

Kunelius, H.T., Campbell, A.J., McCrea, K.B. and Ivany, J.A. (1982) Effects of vegetation suppression and drilling techniques on the establishment and growth of sod seeded alfalfa and birdsfoot trefoil in grass dominant swards. *Canadian Journal of Plant Science* 62, 667–675.

Laidlaw, A.S. (1981) Establishment, persistence and nitrogen fixation of white clover and marsh trefoil on blanket peat. *Grass and Forage Science* 36, 227–230.

Laskey, B.C. and Wakefield, R.C. (1978) Competitive effects of several grass species and weeds on the establishment of birdsfoot trefoil. *Agronomy Journal* 70, 146–148.

Lee, J., Harris, P.M., Sinclair, B.R. and Treloar, B.P. (1992) The effect of condensed tannin containing diets on whole body amino acid utilisation in Romney sheep: consequences for wool growth. *Proceedings of the New Zealand Society of Animal Production* 52, 243–245.

Li, B. and Beuselinck, P.R. (1996) Rhizomatous *Lotus corniculatus* L.: II. Morphology and anatomy of rhizomes. *Crop Science* 36, 407–411.

Li, Q. and Hill, M.J. (1989) Effect of the growth regulator PP 333 (Paclobutrazol) on plant growth and seed production of *Lotus corniculatus* L. *New Zealand Journal of Agricultural Research* 32, 507–514.

Lim, K.T., Matches, A.G., Nelson, C.J., Peters, E.J. and Garner, G.B. (1989) Characterisation of inhibitory substances of tall fescue and birdsfoot trefoil. *Crop Science* 29, 407–411.

Linscott, D.L. and Vaughan, R. (1989) Quackgrass (*Agropyron repens*) control in birdsfoot trefoil (*Lotus corniculatus*) seed production. *Weed Technology* 3, 102–104.

Lowther, W.L. and Patrick, H.N. (1995) *Rhizobium* strain requirements for improved nodulation of *Lotus corniculatus*. *Soil Biology and Biochemistry* 27, 721–724.

Lowther, W.L., Horrell, R.F., Fraser, W.J. Trainor, K.D. and Johnstone, P.D. (1996) Effectiveness of a strip seeder direct drill for pasture establishment. *Soil and Tillage Research* 38, 161–174.

MacDonald, H.A. (1946) *Birdsfoot trefoil* (Lotus corniculatus L.) – *Its Characteristics and Potentialities as a Forage Legume*. Memoir 261, Cornell University Agricultural Experimental Station, Ithaca, New York, 182 pp.

McGraw, R.L. and Beuselinck, P.R. (1983) Growth and yield characteristics of birdsfoot trefoil. *Agronomy Journal* 75, 443–446.

McGraw, R.L., Russelle, M.P. and Grava, J. (1986a) Accumulation and distribution of dry mass and nutrients in birdsfoot trefoil. *Agronomy Journal* 78, 124–131.

McGraw, R.L., Beuselinck, P.R. and Ingram, K.T. (1986b) Plant population density effects on seed yield of birdsfoot trefoil. *Agronomy Journal* 78, 201–205.

McGraw, R.L., Beuselinck, P.R. and Smith, R.R. (1986c) Effect of latitude on genotype and environment interactions for seed yield in birdsfoot trefoil. *Crop Science* 26, 603–605.

McKersie, B.D., Tomes, D.T. and Yamamoto, S. (1981) Effect of seed size on germination, seedling vigour, electrolyte leakage and establishment of bird's foot trefoil (*Lotus corniculatus* L.). *Canadian Journal of Plant Science* 61, 337–343.

Mackun, I.R. and Baker, B.S. (1990) Insect populations and feeding damage among birdsfoot trefoil–grass mixtures under different cutting schedules. *Journal of Economic Entomology* 83, 260–267.

Mallarino, A.P., Wedin, W.F., Perdomo, C.H., Goyenola, R.S. and West, C.P. (1990) Nitrogen transfer from white clover, red clover and birdsfoot trefoil to associated grass. *Agronomy Journal* 82, 790–795.

Marten, G.C. and Jordan, R.M. (1979) Substitution value of birdsfoot trefoil for alfalfa–grass in pasture systems. *Agronomy Journal* 71, 55–59.

Marten, G.C., Ehle, F.R. and Ristau, E.A. (1987) Performance and photosensitization of cattle related to forage quality of four legumes. *Crop Science* 27, 138–145.

Miller, P.R. and Ehlke, N.J. (1994) Condensed tannin relationship with *in vitro* forage quality analyses for birdsfoot trefoil. *Crop Science* 34 1074–1079.

Nelson, C.J. and Smith, D. (1968) Growth of birdsfoot trefoil and alfalfa. II. Morphological development and dry matter distribution. *Crop Science* 8, 21–25.

Nelson, C.J. and Smith, D. (1969) Growth of birdsfoot trefoil and alfalfa. III. Changes in carbohydrate reserves and growth analysis under field conditions. *Crop Science* 8, 25–29.

Nelson, C.J., Hur, S.N. and Beuselinck, P.R. (1995) Physiology of seedling vigour of birdsfoot trefoil. In: Beuselink, P.R. and Roberts, C.A. (eds) *Proceedings of the First International Lotus Symposium, St Louis, USA*. University of Missouri-Columbia Press, pp. 68–73.

Olsen, F.J., Jones, J.H. and Patterson, J.J. (1981) Sod-seeding forage legumes in a tall fescue sward. *Agronomy Journal* 73, 1032–1036.

Petersen, P.R., Sheaffer, C.C. and Hall, M.H. (1992) Drought effects on perennial forage legume yields and quality. *Agronomy Journal*, 84 774–779.

Pofelis, S., Le, H., Grant, W.F. (1992) The development of sulfonylurea herbicide resistant birdsfoot trefoil (*Lotus corniculatus*) plants from *in vitro* selection. *Theoretical and Applied Genetics* 83, 480–488.

Purchas, R.W. and Keogh, R.G. (1984) Fatness of lambs grazed on Grasslands Maku lotus and Grasslands Huia white clover. *Proceedings of the New Zealand Society of Animal Production* 44, 219–221.

Qualls, M. and Cooper, C.S. (1968) Germination, growth and respiration rates of birdsfoot trefoil at three temperatures during the early non-photosynthetic stage of development. *Crop Science* 8, 758–760.

Robertson, H.A., Niezen, J.H., Waghorn, G.C., Charleston, W.A.G. and Jinlong, M. (1995) The effect of six herbages on liveweight gain, wool growth and faecal egg count of parasitised ewe lambs. *Proceedings of the New Zealand Society of Animal Production* 55, 199–201.

Russelle, M.P. and McGraw, R.L. (1986) Nutrient status in birdsfoot trefoil. *Canadian Journal of Plant Science* 66, 933–944.

Russelle, M.P., McGraw, R.L., Grava, J. and Beuselinck, P.R. (1985) Elemental composition of birdsfoot trefoil. *Communications in Soil Science and Plant Analysis* 16, 987–1013.

Russelle, M.P., McGraw, R.L. and Leep, R.H. (1991) Birdsfoot trefoil response to phosphorus and potassium. *Journal of Production Agriculture* 4, 114–120.

Russelle, M.P., Meyers, L.L. and McGraw, R.L. (1989) Birdsfoot trefoil seedling response to soil phosphorus and potassium availability indices. *Soil Science Society of America Journal* 53, 828–836.

Schachtman, D.P. and Kelman, W.M. (1991) Potential of *Lotus* germplasm for the development of salt, aluminium and manganese tolerant pasture plants. *Australian Journal of Agricultural Research* 42, 139–149.

Schiller, K.N. and Ayers, J.F. (1993) The effects of winter conditions on the nutritive value of *Lotus pedunculatus* cv. Grasslands Maku and *Trifolium repens* cv. Haifa. *Tropical Grasslands* 27, 43–47.

Scholl, J.M. and Brunk, R.E. (1962) Birdsfoot trefoil stand establishment as influenced by control of vegetative competition. *Agronomy Journal* 54, 142–144.

Seaney, R.R. and Henson, P.R. (1970) Birdsfoot trefoil. *Advances in Agronomy* 22, 119–157.

Sheaffer, C.C., Marten, G.C., Jordan, R.M. and Rislan, E.A. (1992) Forage potential of kura clover and birdsfoot trefoil when grazed by sheep. *Agronomy Journal* 84, 176–180.

Sheath, G.W. (1980a) Effects of season and defoliation on the growth habit of *Lotus pedunculatus* Cav., cv. 'Grasslands Maku'. *New Zealand Journal of Agricultural Research* 23, 191–200.

Sheath, G.W. (1980b) Production and regrowth characteristics of *Lotus pedunculatus* Cav., cv. 'Grasslands Maku'. *New Zealand Journal of Agricultural Research* 23, 201–209.

Sheldrick, R.D. and Martyn, T.M. (1992) Lotus species and varieties for acid, low phosphate soils. In: *Proceedings of the Third Research Conference, British Grassland Society, Northern Ireland.* British Grassland Society, Reading, p. 41.

Shiferaw, W., Shelton, H.M. and So, H.B. (1992) Tolerance of some subtropical pasture legumes to waterlogging. *Tropical Grasslands* 26, 187–195.

Suckling, F.E.T. (1965) *Hill Pasture Improvement.* Newton King Group and DSIR, Wanganui.

Suckling, F.E.T. and Charlton, J.F.L. (1978) A review of the significance of buried legume seeds with particular reference to New Zealand agriculture. *New Zealand Journal of Experimental Agriculture* 6, 211–215.

Tabora, R.S. and Hill, M.J. (1992) Effects of cyclocel on growth and seed yield of *Lotus uliginosus* Schk. cv. Grasslands Maku. *New Zealand Journal of Agricultural Research* 35, 259–268.

Taylor, T.H., Templeton, W.C. and Wyles, J.W. (1973) Management effects on persistence and productivity of birdsfoot trefoil (*Lotus corniculatus*). *Agronomy Journal* 65, 646–648.

Templeton, W.L., Jr, Bucks, C.F. and Wattenborge, D.W. (1967) Persistence of birdsfoot trefoil under pasture conditions. *Agronomy Journal* 59, 385–386.

Tiley, G.E.D. and Frame, J. (1991) Improvement of upland permanent pasture and lowland swards by surface seeding methods. In: *Grassland Renovation and Weed Control in Europe, Proceedings of the European Grassland Federation Conference, Graz, Austria*. Federal Research Institute for Agriculture in Alpine Regions, Gumpenstein (BAL), pp. 89–94.

Ulyatt, M.J. (1981) The feeding value of temperate pastures. In: Morley, F.M.W. (ed.) *Grazing Animals*. Elsevier, Amsterdam, pp. 89–94.

Vance, C.P., Johnson, L.E.B., Stade, S. and Groat, R.G. (1982) Birdsfoot trefoil *(Lotus corniculatus)* root nodules: morphogenesis and the effect of forage harvest on structure and function. *Canadian Journal of Botany* 60, 505–518.

Vance, C.P., Reibach, P.H. and Pankhurst, C.E. (1987) Symbiotic properties of *Lotus pedunculatus* root nodules induced by *Rhizobium lotii* and *Bradyrhizobium* spp. *Physiologia Plantarum* 69, 435–442.

Van Keuren, R.W., and Davis, R.R. (1968) Persistence of birdsfoot trefoil, *Lotus corniculatus* L., as influenced by plant growth habit and grazing management. *Agronomy Journal* 60, 92–95.

Van Keuren, R.W., Davis, R.R., Bell, D.S. and Klosterman, E.W. (1969) Effect of grazing management on the animal production from birdsfoot trefoil pastures. *Agronomy Journal* 61, 422–425.

Vos, G. and Jones, R.M. (1986) The role of stolons and rhizomes in legume persistence. In: *CSIRO Division of Tropical Crops and Pastures Annual Report 1985–86* CSIRO, Canberra, pp. 70–71.

Waghorn, G.C. and Shelton, I.D. (1992) The nutritive value of *Lotus* for sheep. *Proceedings of the New Zealand Society of Animal Production* 52, 89–92.

Waghorn, G.C., Ulyatt, M.J., John, A. and Fisher, M.T. (1987) The effect of condensed tannins on the site of digestion of amino acids and other nutrients in sheep fed on *Lotus corniculatus*. *British Journal of Nutrition* 57, 115–126.

Waghorn, G.C., Jones, W.T., Shelton, I.D. and McNabb, W.C. (1990) Condensed tannins and the nutritive value of pasture. *Proceedings of the New Zealand Grassland Association* 51, 171–176.

Wang, Y., Waghorn, G.C., Douglas, G.B., Barry, T.N. and Wilson, G.F. (1994) The effects of condensed tannin in *Lotus corniculatus* upon nutrient metabolism and upon body and wool growth in grazing sheep. *Proceedings of the New Zealand Society of Animal Production* 54, 219–222.

Wedderburn, M.E. (1986) Effect of applied nitrogen, increased inoculation, broadcast lime, and seed pelleting on establishment of *Lotus pedunculatus* cv. 'Grasslands Maku' in tussock grasslands. *New Zealand Journal of Experimental Agriculture* 14, 31–36.

Wedderburn, M.E. and Gwynne, D.C. (1981) Seasonality of rhizome and shoot production and nitrogen fixation in *Lotus uliginosus* under upland conditions in south-west Scotland. *Annals of Botany* 48, 5–13.

Wedderburn, M.E. and Lowther, W.L. (1985) Factors affecting establishment and spread of 'Grasslands Maku' lotus in tussock grasslands. *Proceedings of the New Zealand Grassland Association* 46, 97–101.

Wheeler, D.M. and Dodd, M.B. (1995) Effect of aluminium on yield and plant chemical concentrations of some temperate legumes. *Plant and Soil* 173, 133–145.

Wheeler, D.M., Edmeades, D.C., Christie, R.A. and Gardner, R. (1992) Effect of aluminium on the growth of 34 plant species: a summary of results obtained in low ionic strength solution culture. *Plant and Soil* 146 (12), 61–66.

White, S.K., McGraw, R.L. and Russelle, M.P. (1987) Effect of growth retardants on seed yield of birdsfoot trefoil. *Crop Science* 27, 608–610.

Wilson, G.P.M. (1993) *Lotus pedunculatus* Cav. (greater lotus) cv. Sharnae. *Australian Journal of Experimental Agriculture* 32, 794–795.

Winch, J.E., Watkin, E.M., Anderson, G.W. and Collins, T.L. (1969) The use of mixtures of granular dalapon, birdsfoot trefoil seed and fertilizer for roughland pasture renovation. *Journal of the British Grassland Society* 24, 302–307.

Wipfli, M.S., Wedberg, J.L. and Hogg, D.B. (1990a) Damage potential of three plant bug (Hemiptera, Heteroptera, Miridae) species to birdsfoot trefoil for seed in Wisconsin. *Journal of Economic Entomology* 83, 580–584.

Wipfli, M.S., Wedberg, J.L. and Hogg, D.B. (1990b) Cultural and chemical control strategies for three plant bug (Hemiptera, Heteroptera, Miridae) pests of birdsfoot trefoil in northern Wisconsin. *Journal of Economic Entomology* 83, 2086–2091.

Wyse, D.L. and McGraw, R.L. (1987) Control of white campion (*Silene alba*) in birdsfoot trefoil (*Lotus corniculatus*) with dinitroalanine herbicides. *Weed Technology* 1, 34–36.

7 Alsike Clover and Sainfoin

ALSIKE CLOVER

Introduction

Alsike clover (*Trifolium hybridum* L.) originated in northern Europe, its name coming from an area of Sweden where the species has grown for several centuries. The botanical name was allocated since it was wrongly considered a hybrid between red and white clovers. Alsike clover has been utilized in the temperate and subarctic zones of Europe, Asia and the Americas and in some upland regions of Australasia, but performs less well in warmer climates. It was introduced to Britain in the early decades of the nineteenth century and taken to North America in the mid-1800s, being first used in the midwestern states in the 1850s (Taylor, 1985).

During this century, it has been evaluated in nearly all temperate grassland regions, but has been used only in niche situations, where its adaptability helped it to contribute better than the other perennial legumes, or to complement them in multipurpose seed mixtures. It is a relatively short-lived species, persisting for 2 years or so in swards under normal management conditions, but, because it seeds freely, a longer perceived persistence could be due to regeneration from shed, ripe seed. It grows best in cool temperate conditions but is adapted to sites that are too wet, infertile or acid for red clover or lucerne (Townsend, 1985).

The Plant

Alsike clover has an upright growth form, similar to that of red clover, and develops many fine stems from a well-developed basal crown. Stems can grow

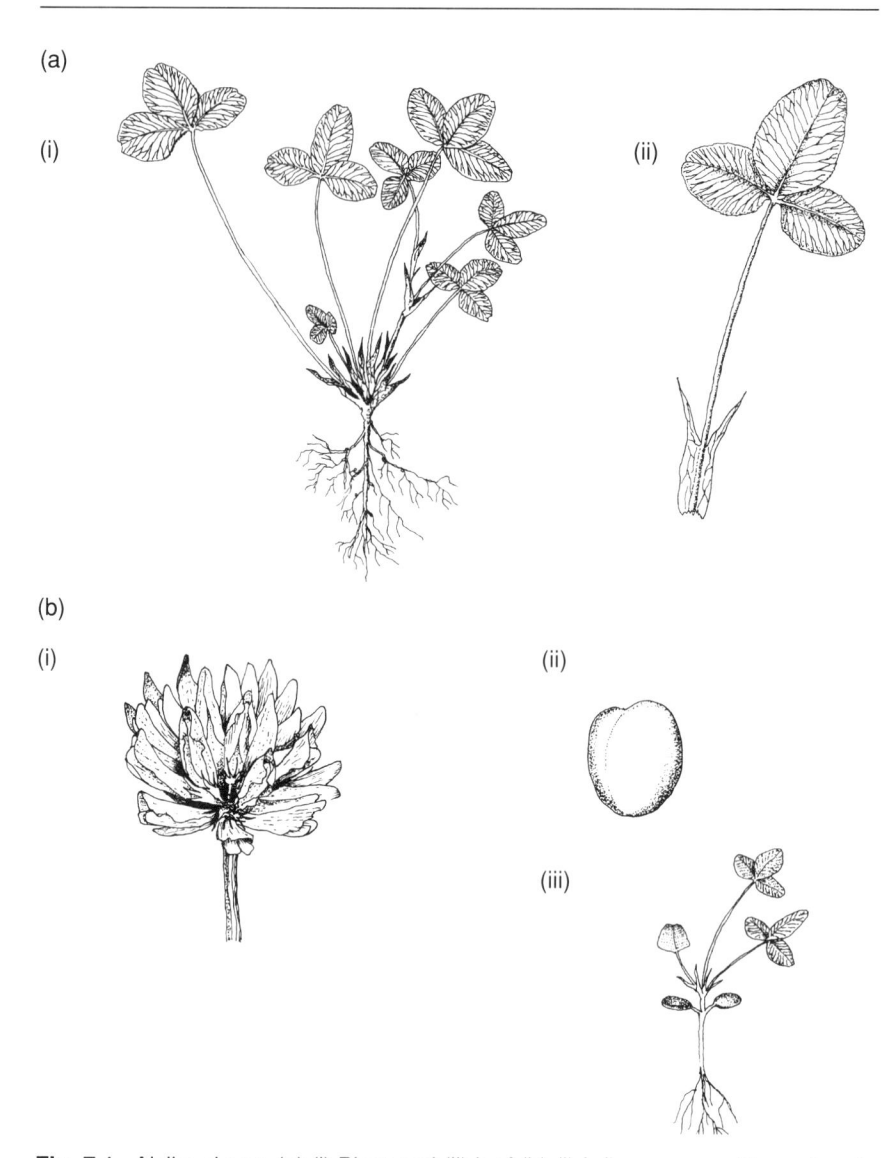

Fig. 7.1. Alsike clover. (a) (i) Plant and (ii) leaf.(b) (i) Inflorescence, (ii) seed and (iii) seedling.

to 50 cm in height, as they do not bear terminal flower-heads (Smetham, 1973), but they become prone to lodging. The plant has a branched tap-root system with numerous lateral branches. Leaves are trifoliate and glabrous, with oval leaflets, slightly serrated on their lower edges but without the leaf markings present in white and red clovers (Fig. 7.1). Stipules are 1–2.5 cm long, pointed and free for most of their length, with greenish venation.

Flowering is usually in midsummer, with globular flower-heads, 1–2.5 cm in diameter, borne on long axillary stalks. Flower-heads contain 30–50 florets. The flowers are pale pink or even whitish in colour, self-sterile and fertilized by honey-bees. The seeds are irregularly heart-shaped, smooth-surfaced and greenish-brown in colour, though darkening with age. Thousand-seed weight is approximately 0.7 g. No particular problems regarding seed quality are mentioned in the literature.

Breeding

Alsike clover is a self-incompatible, cross-pollinating legume, which is a natural diploid, with 16 chromosomes. Tetraploid forms ($2n = 32$) have been developed in Europe in the past and proved to be as fertile as the original diploids, which is unusual for artificially induced polyploids. Diploid ($2x$) and tetraploid ($4x$) plants have also been crossed in various combinations (Hyrkas et al., 1986).

Because the species has proved less valuable than red and white clovers, there has been less breeding effort. Nevertheless, in Scandinavian countries, Canada and elsewhere, improved selections have been released over the years, e.g. diploid Ermo Øtofte (Denmark), tetraploids Tetra and Frida (Sweden), diploids Aurora and Dawn (Canada) and diploid Grasslands G 50 (New Zealand). Diploid genetic stocks of alsike clover, C-33, C-34 and C-35, have recently been registered in the USA (Townsend, 1995). Attempts have also been made to cross alsike clover with Caucasian clover (*Trifolium ambiguum*) in the past, but the hybrid embryos aborted within 2 weeks after fertilization. Subsequent use of embryo-culture techniques has produced F_1 plants, but these were not fertile.

Seed production

Most of the world's seed supply of alsike clover is produced in North America. Seed-producing areas in Alberta, Canada, and the states of Idaho and Oregon in the USA provide highly satisfactory conditions. Honey-bees are the most important pollinators of the species. Seed yields of 400 kg ha^{-1} have been recorded and occasionally between 500 and 1000 kg ha^{-1} in specialist seed-growing areas, although the average North American seed yield is 160 kg ha^{-1} (Duke, 1981). Row spacing influences seed yield, with 15 cm spacing producing the highest seed yields in comparative trials (Pankiw et al., 1977), while wider spacings lowered yield; sowing rate also affects seed yield, optimum rates being 2–4.5 kg ha^{-1}. In trials during 1981–1989 in the Tula region of Russia, alsike clover gave 6-year average seed yields of 160 kg ha^{-1}, compared with 5-year average white-clover seed yields of 170–250 kg ha^{-1} (Korobov, 1990).

Agronomy and Management

Seed mixtures and establishment

Apart from seed crops, alsike clover is usually sown in seed mixtures with grasses and other legumes. For cutting systems, it is particularly compatible with red clover and with grasses such as timothy (*Phleum pratense*) and erect forms of cocksfoot (*Dactylis glomerata*). Between 2 and 5 kg ha^{-1} of alsike-clover seed used to be common in British mixtures, but the complex mixtures of the past have been superseded in the main by simpler mixtures, with fewer species, in which alsike is often omitted. A higher rate of alsike, 4–8 kg ha^{-1}, usually in mixture with red clover and a grass such as timothy, is recommended in North American mixtures (Townsend, 1985).

Because of its small seed size, shallow sowing, at 10–15 mm, is essential for successful establishment and, as with red clover, subsequent optimal performance is dependent on excellent establishment, since it does not spread vegetatively. Sowing may be in early spring or even in late summer in the cool moist regions where it performs best. Seed should be rhizobially inoculated with *Rhizobium leguminosarum* bv. *trifolii* (Burton, 1985). Seed can be broadcast on the soil surface and harrowed in, or else drilled, while aerial oversowing of alsike clover has been used to improve high country tussock grasslands (Keoghan and Allen, 1993).

Pests and diseases

Alsike clover is susceptible to many of the diseases and pests that attack red clover, but is resistant to scorch or northern anthracnose (*Kabatiella caulivora*) and southern anthracnose (*Colletotrichum trifolii*) (Townsend, 1985). Sooty blotch disease (*Cymadothea trifolii*) is common on alsike clover in temperate regions, and the species is also susceptible to rust (*Uromyces trifolii*) and powdery mildew (*Erysiphe trifolii*), although plants resistant to the latter disease have been noted. *Sclerotinia* root and crown rot (*Sclerotinia trifoliorum*) have been reported on alsike clover in high-altitude irrigated swards in the USA, and the species is very prone to *Sclerotinia* infection in Norway (Vestad, 1973). Effects of virus diseases seem less clear. Alsike mosaic-virus disease is reported to be widespread in the species in North America but has little apparent effect on performance. Although none of the viruses recorded in alsike clover seem to influence stand density or performance in Canada (McLaughlin and Boykin, 1988), severe damage from virus infection has been observed elsewhere.

There is a paucity of knowledge concerning the significance of pests, but aphids, for example, attack alsike clover, with varying impact (Townsend, 1985). Resistance to stem eelworm (*Ditylenchus dipsaci*) has been reported (Vestad, 1973), although it is a host for the root-lesion nematode (*Pratylenchus penetrans*) (Thies *et al.*, 1995).

Production

Basically, the agronomic and management requirements of alsike clover are similar to those of red clover. Alsike clover tolerates both wet and acidic soils, but it does respond to improved soil fertility. Annual production data are sparse, but dry matter (DM) yields up to 10.7 and 7.7 t ha^{-1} for first and second harvest years, respectively, are reported from high-altitude irrigated swards in western USA, with yield advantage in the second year to timothy/alsike-clover stands, compared with alsike-clover monocultures (Townsend, 1962). From multisite trials in northern Sweden, DM yields from timothy/alsike-clover mixtures – 5–6 t ha^{-1} in the first harvest year, declining to 4.5–5.5 t ha^{-1} in year 3 – were lower than those which contained red clover or red clover and alsike clover, while, in addition, alsike-clover contribution was inferior on both light- and heavy-textured soils and at soil pH levels from 4.9 to 6.4, although the inferiority was least at the lowest pH levels (Wallgren *et al.*, 1995). A range of alsike-clover germplasm was superior in yield and persistency to red and white clovers in the cold, infertile high country of New Zealand (Widdup and Ryan, 1994).

Feeding value

The feeding value of alsike clover is generally considered as being similar to that of red clover, although there is a lack of recent definitive evidence. In common with other legumes, it will be protein- and mineral-rich and, on account of its growth habit, digestibility will decline with advancing plant maturity, in a similar manner to red clover. Bloat can be a problem on clover-rich swards but can be combated by well documented measures, e.g. the use of surfactants for grazing animals. It has been recorded that the herbage of grazed alsike clover may cause photosensitization in horses (Ostergaard, 1992).

Utilization

Used alone or with a companion grass, e.g. timothy, good-quality hay or silage is attainable, especially if harvested before the forage is at an advanced stage of growth. The use of companion species reduces the incidence of lodging in hay crops (Pederson, 1995). As the clover's flowering is irregular, a reasonable seed crop may be obtained, in addition to a main conservation cut (Duke, 1981). Alsike clover is highly suitable for hay production at irrigated, high-altitude sites in western USA (Townsend, 1985). It has also proved of value as a stockpiled standing crop for use in autumn and winter (Widdup and Ryan, 1994). Alsike clover's tendency to be short-lived can be mitigated by managing it at some stage of the season to encourage a flowering/seed-setting phase, which allows some regeneration from shed seed.

References – Alsike Clover

Burton, J.C. (1985) *Rhizobium* relationships. In: Taylor, N.L. (ed.) *Clover Science and Technology*. ASA/CSSA/SSSA, Madison, Wisconsin, pp.161–185.

Duke, J.A. (1981) *Handbook of Legumes of World Economic Importance*. Plenum Press, New York.

Hyrkas, K., Kivinen, M. and Tigerstedt, P.A.D. (1986) Interspecific hybridisation of red clover (*Trifolium pratense* L.) with alsike clover (*Trifolium hybridum* L.) using *in vitro* embryo rescue. In: Horn, W., Jensen, C.J., Odenbach, W. and Schieder, O. (eds) *Proceedings of a Eucarpia International Symposium, Berlin (West), Germany*. Walter de Gruyter, Berlin, pp. 469–471.

Keoghan, J.K. and Allen, B.E. (1993) Pasture species for tussock grassland landscapes and farming systems. In: Floate, M. (ed.) *Guide to Tussock Grassland Farming*. AgResearch, New Zealand, pp. 39–54.

Korobov, P.P.A.D. (1990) *Agronomy of White Clover and Alsike Clover*. Report of the Tula Agricultural Research Institute, Tula, USSR (in Russian).

McLaughlin, M.R. and Boykin, D.L. (1988) Virus diseases of seven species of forage legumes in the southeastern United States. *Plant Disease* 72, 539–542.

Ostergaard, H.S.O. (1992) Photosensitization in horses – alsike clover can cause photosensitization. *Dansk-Veterinaertidsskrift* 75, (2) 49.

Pankiw, P., Bonin, S.G. and Lieverse, J.A.C. (1977) Effects of row spacing and seeding rates on seed yield in red clover, alsike clover and birdsfoot trefoil. *Canadian Journal of Plant Science* 57, 413–418.

Pedersen, G.A. (1995) White clover and other perennial clovers. In: Barnes, R.F., Miller, D.A. and Nelson, C.J. (eds) *Forages*, 5th edn, Vol. 1, *An Introduction to Grassland Agriculture*. Iowa State University Press, Ames, Iowa, pp. 227–236.

Smetham, M.L. (1973) Pasture legume species and strains. In: Langer, R.H.M. (ed.) *Pastures and Plants* A.H. & A.W. Reed, Wellington, pp. 85–128.

Taylor, N.L. (1985) Clovers around the world. In: Taylor, N.L. (ed.) *Clover Science and Technology*. ASA/CSSA/SSSA, Madison, Wisconsin, pp. 1–6

Thies, J.A., Peterson, A.D. and Barnes, D.K. (1995) Host suitability of forage grasses and legumes for root lesion nematode (*Pratylenchus penetrans*). *Crop Science* 35, 1647–1651.

Townsend, C.E. (1962) Performance of alsike clover varieties in a high-altitude meadow. *Crop Science* 2, 80–81.

Townsend, C.E. (1985) Miscellaneous perennial clovers. In: Taylor, N.L. (ed.) *Clover Science and Technology*. ASA/CSSA/SSSA, Madison, Wisconsin, pp. 563–567.

Townsend, C.E. (1995) Registration of C-33, C-34 and C-35 genetic root stocks of diploid alsike clover. *Crop Science* 35, 1519.

Vestad, R. (1973) Variety trials with alsike clover [Norwegian with English summary]. *Forskning-og-Forsoek-i-Landbruket* 24, 601–614.

Wallgren, B., Nilsdottir-Linde, N., Svanäng, K., Halling, M. and Magnét, B. (1995) Results from official variety trials in southern Sweden. In: *Special Publication 61*. Swedish University of Agricultural Sciences, Uppsala, pp. 8–13.

Widdup, K.H. and Ryan, D.L. (1994) Development of G 50 alsike clover for the South Island high country. *Proceedings of the New Zealand Grassland Association* 56, 107–111.

SAINFOIN

Introduction

Sainfoin (*Onobrychis viciifolia* Scop.), also referred to in the past as St Foin, cock's head or holy grass, is a perennial legume that has been grown in temperate Asia and Europe for several hundred years, although its use has declined in modern times. The name, sainfoin, is derived from French, meaning safe or healthy hay, a reference to its historical and beneficial use in feeding sick animals, and also probably to its non-bloating characteristic.

It performs well on calcareous soils, with a soil pH of 6.0 and above, and its erect growth habit makes it suited to cutting for conservation. Sainfoin is drought-resistant, being occasionally superior to lucerne in this respect, though lower-yielding (Rogers, 1976). It was evaluated in the USA in the early 1900s, but it has not become a major forage species, although there was renewed interest in the 1960s, resulting from the threat of alfalfa weevil (*Hypera* spp.) to lucerne on irrigated swards, and also from the need for dryland forage legumes (Hoveland and Townsend, 1985). It is now grown on the dry calcareous soils of western USA and Canada (Miller and Hoveland, 1995). Its potential as a source of high-quality herbage for grazing in the southern Great Plains of the USA has also been demonstrated (Karnezos *et al.*, 1994). Sainfoin is one of the perennial legumes utilized for grazing and hay in Mediterranean environments, e.g. in southern Italy (Martiniello and Ciola, 1994). Apart from an occasional interest in the species as a non-bloating legume, there is no commercial use of this legume in Australasia (Charlton, 1983).

The Plant

Sainfoin has numerous erect, hollow stems arising from basal buds on a branched crown and grows to a height of 40–80 cm. After cutting, branches grow out from axillary buds on the stem nodes below the point of cutting. The pinnate leaves comprise 10–28 leaflets, borne in pairs on long petioles and with a terminal leaflet (Fig. 7.2). It has a high weight per unit of leaf, when compared with lucerne (Sheehy and Popple, 1981). The stipules are broad and pointed. As many as 80 pink flowers are borne on densely flowering racemes, carried on axillary stalks. Pollination is mainly by honey-bees and the species is cross-pollinated. The seed pods, which are bilaterally compressed, each carry a single plattened, kidney-shaped seed. Individual seeds range in colour from olive-brown to brown or black and have a 1000-seed weight of around 15 g when hulled and 20 g when unhulled. Unlike many other forage legumes, the longevity of sainfoin seed is relatively short, extending only 5–8 years. The plant has a deep tap root, with a number of main

(a)

Fig 7.2. Sainfoin. (a) Plant. (b) (i) Leaf and (ii) inflorescence. (c) Seeds.

Alsike clover and sainfoin 281

Fig. 7.2 continued

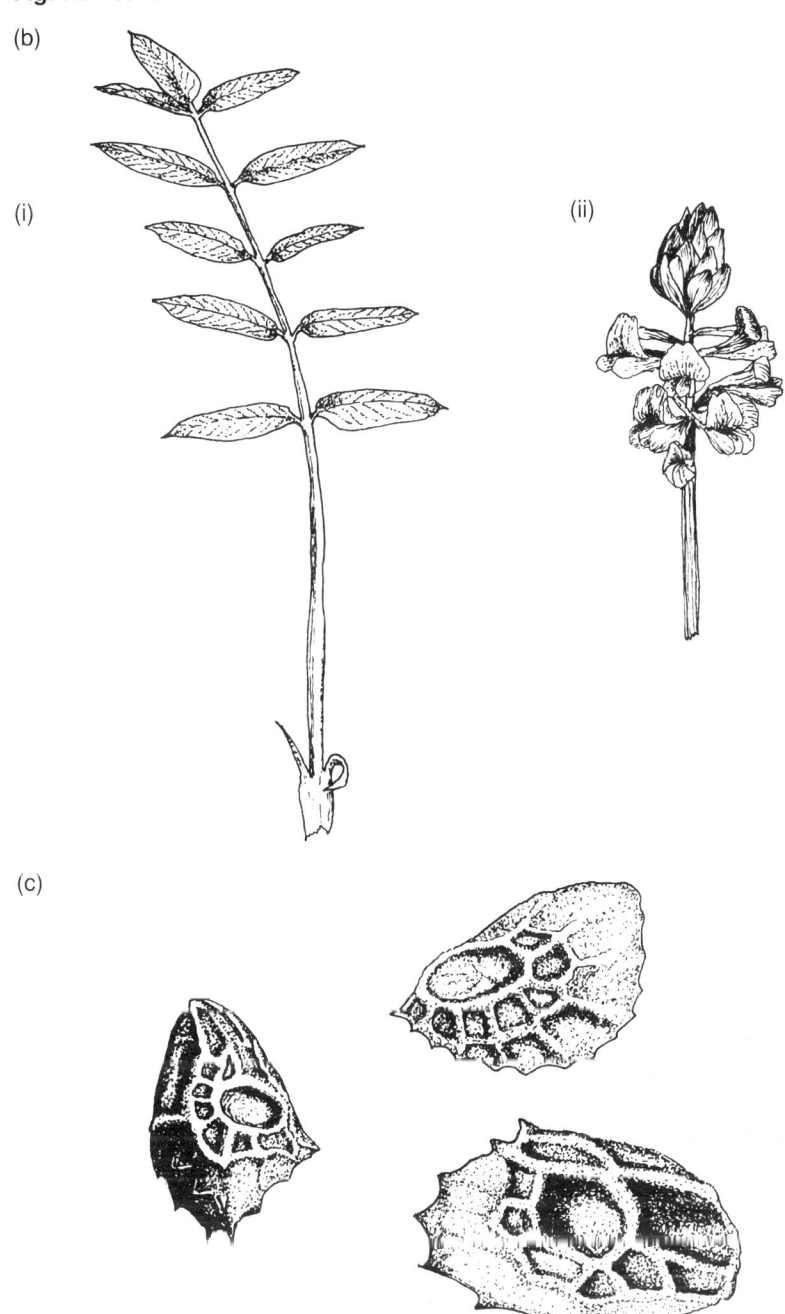

(b)

(i)　　　　　　　　　　　　(ii)

(c)

branches and numerous lateral roots, on which most of the nitrogen (N_2)-fixing nodules are found.

Physiology

Sainfoin is similar to lucerne with respect to root carbohydrate reserves affecting persistence of the stand. Typically, total non-structural carbohydrate (TNC) concentration in the root declines over the winter until late spring, when, if defoliation or level of utilization is not too severe, levels build up again through to late autumn (Mowrey and Matches, 1991). Sainfoin production is usually poor after periods of high temperature. A combination of high temperatures and defoliation results in the inability of photosynthesis and carbohydrate reserves to support high metabolic rates during high temperatures, and so a high proportion of plants die subsequent to a period of high temperature (Kallenbach et al., 1996). Sainfoin growth is unaffected by the range of day lengths likely to be encountered during the growing season, although long days induce flowering and increase production of fine roots (Kallenbach et al., 1995).

While the photosynthetic capacity of individual leaves is similar for lucerne and sainfoin, photosynthesis by a sainfoin canopy is lower than for a corresponding stage in regrowth of lucerne, due to the latter having a higher leaf-area index (LAI) and a more erect growth habit when regrowth is advanced (Sheehy and Popple, 1981). These factors result in a more efficient utilization of incident irradiance by lucerne than by sainfoin canopies.

Nitrogen fixation

Rhizobial inoculation is necessary when sainfoin seed is sown on land that has not grown forage legumes for a number of years. Sainfoin is easily infected with rhizobia from nodules of some other legumes, especially those from the same tribe (*Hedysarae*) and related tribes (e.g. *Galegae*). As a consequence of the latter, arctic rhizobia extracted from *Astragulus* and *Oxytropus* are infective towards sainfoin and can be more competitive in infecting roots of sainfoin at soil temperatures of 9–12°C than samples extracted from sainfoin nodules (Prevost and Bromfield, 1991). There is, therefore, an opportunity to select for *Rhizobium* for sainfoin introduced into areas with cold springs.

Low persistence in sainfoin has been attributed to nitrogen (N) deficiency due to low N_2 fixation, even when roots have abundant nodules. The efficiency of N_2 fixation in sainfoin is poor, in comparison with that for lucerne or red and white clover. About 10 mol carbon dioxide (CO_2) is required to fix 1 mol of N_2 in these three legumes, whereas about 20 are required to fix 1 mol of N_2 by sainfoin (Witty et al., 1983). Also, sainfoin invests less of its assimilate in leaf production and LAI than lucerne, and so energy availability may be a constraint on N_2 fixation in sainfoin. Nevertheless, in single-plant studies, sainfoin can fix as much N per unit of plant weight as lucerne (Hume et al., 1985).

As sainfoin yields in established crops are generally lower than those of lucerne, N_2 fixation would be expected to be correspondingly lower for sainfoin per unit ground area.

Plant Types

In the UK, there are two main types of sainfoin: common or single-cut, e.g. Cotswold Common and cv. Emyr, and giant or double-cut, e.g. English Giant. Common sainfoin is less vigorous during establishment, and so the stems do not elongate and flower much in the establishment year. Peak hay or silage yields are reached in the second or third harvest year and the stands can persist for 20–30 years. In contrast, giant sainfoin establishes quickly and develops flowers even in the year of sowing, it attains peak yield the following year and two conservation cuts may be taken annually. However, though higher-yielding, it is relatively short-lived and may persist for only 2 years, although local ecotypes may persist longer. Following selection, modern sainfoin cultivars do not necessarily fit the above categories closely and can be managed more flexibly. The cultivars Melrose and Nova were developed in Canada, cvs Eski, Remont and Renumex in the USA and cvs Zeus and Vala in Italy.

A seed crop may be taken in lieu of the main hay cut for common sainfoin types or after a hay cut for giant sainfoin types, which flower more than once in a season. Threshed seed yields may reach 1450 kg ha^{-1} (Hoveland and Townsend, 1985).

Agronomy and Management

Seed mixtures and establishment

Sainfoin is sown in monoculture at 50 kg ha^{-1} for hulled seed and double this rate for unhulled seed. Germination is quicker with the former, while the latter requires better moisture conditions so that the husks are softened. A non competitive grass species, such as meadow fescue, at 6 kg ha^{-1}, or timothy, at 2 kg ha^{-1}, makes a suitable companion grass, and its presence improves sward density and reduces the degree of weed ingress. While established monocultures effectively compete with weeds, they are susceptible to weed invasion during their establishment phase. Sainfoin seed is best drilled at 20–30 mm depth and, while it may be undersown in a cereal crop, direct sowing is more likely to ensure a successful establishment.

The rate of N_2 fixation in establishing sainfoin is low compared with some other legumes, as a consequence of a low LAI per unit of shoot growth and the inefficiency of the N_2-fixing process in sainfoin. Also, growth and N_2 fixation of sainfoin seedlings are beneficially influenced by large seed size (Cash and Ditterline, 1996). Another factor that benefits vigour of establishment is freedom from fungal attack, e.g. from *Fusarium* or *Cladosporium*. Damage can be minimized by dressing seeds with thiophanate methyl (Nan, 1995). Under experimental conditions, *Pythium* attack is reduced if roots are colonized by vesicular arbuscular mycorrhiza (Hwang *et al.*, 1993).

In south-west Canadian dryland, with 358 mm annual precipitation, sowing alternating rows of Russian wild-rye (*Psathyrostachys juncea*) and sainfoin, rather than mixing and seeding the mixture in the same rows, enhanced the persistence of sainfoin from 2 to 4 years, because of reduced interspecific competition (Kilcher, 1982). At a wetter site and over a 5-year period, sainfoin sown in alternate rows with crested wheatgrass *(Agropyron desertorum)* made a greater contribution to total herbage yield than when it was sown in rows in mixture with the grass, but this advantage was not apparent when the grass companion was Russian wild-rye (Hanna *et al.*, 1977). Lack of winter-hardiness has been a factor in reducing sainfoin persistence in south-west Canada (Jefferson *et al.*, 1994).

The addition of a legume, such as white clover, to sainfoin or sainfoin/grass mixtures can be advantageous, since the clover is a beneficial component when the swards are cut for conservation and then used for aftermath grazing. A sainfoin/birdsfoot trefoil mixture was stable when used for hay and hay-stockpile managements (Cooper, 1973), but, with sainfoin/lucerne mixtures, the composition shifted to lucerne dominance, particularly when both species were drilled together in rows and when drought stress occurred (Jefferson *et al.*, 1994).

Weed control

Because of their slow growth in the establishment phase, sainfoin swards are susceptible to weed invasion, unless there is good seed-bed preparation, including control of perennial weeds before cultivation. The use of a post-emergence sainfoin-safe herbicide, such as an MCPA/MCPB blend, is effective against seedling weeds, but the sainfoin seedlings must be adequately developed in order to tolerate the herbicide. In monocultures, carbetamide treatment in winter will maintain freedom from weeds. In Nebraska, USA, sainfoin yields at the first cut were increased by application of imazethapyr and weeds were controlled effectively (Wilson, 1994). Control of broad-leaved weeds with metribuzin, hexazinone or chlorsulfuron increases sainfoin yield significantly (Moyer *et al.*, 1990).

Pests and diseases

In the main, sainfoin is not susceptible to many pests or diseases, but it may be badly affected by clover rot (*S. trifoliorum*). The foliage is also severely attacked on occasion by powdery mildew (*E. trifolii*) and sometimes by *Verticillium* wilt (*Verticillium albo-atrum*).

The three most important pests identified in the USA are the *Rhizobium*-nodule-eating weevil (*Sitona scissifons*), the lygus bugs (*Lygus elisus, Lygus hesperus*) and the sainfoin bruchid (*Bruchudius unicolor*) (Gaudet *et al.*, 1980).

Production

Production varies according to soil fertility, but few fertilizer trials have been conducted in recent times. In common with other forage legumes, adequate phosphorus (P) and potassium (K) fertilization is required to sustain repeated crops of hay or silage. However, it is tolerant of soils with a low P status (Miller and Hoveland, 1995). Liming is essential to maintain soil pH above pH 6.0. Yields of 7–15 t DM ha^{-1} have been reported (Anon. 1982), although the upper level was representative of experimental plots, rather than farm practice. This range is lower than that from lucerne, which is usually superior to sainfoin in drought resistance and yield (Rogers, 1976). Under Canadian dryland conditions, the yield of sainfoin is about 20% less than that of lucerne (Goplen et al., 1991). A combination of lower LAI for a corresponding stage of growth, a less erect canopy structure, less efficient N_2-fixation system and lower water-use efficiency for part of the growing season contributes to sainfoin's inferiority to lucerne (Sheehy et al., 1984).

Loss of plants in very dry summers reduces the subsequent production of sainfoin, but timely irrigation will maintain stand density. Sainfoin's yield potential is highest in spring and so, if suffering from early-season drought, irrigation could be cost-effective, since water-use efficiency of sainfoin at that time of year is as high as that of lucerne, i.e. 15–18 kg DM mm^{-1} evapotranspiration (Bolger and Matches, 1990).

Feeding value

Sainfoin is protein-rich, but of moderate digestibility, when compared with white or red clover. At comparable stages of growth under grazing, sainfoin had lower crude protein and digestibility than lucerne (Karnezos et al., 1994). The decline of its digestibility with advancing maturity is not marked until after flowering, because the digestibility of the stem is similar to that of the leaf (Wilman and Asiedu, 1983). Sainfoin herbage is mineral-rich in comparison with grasses, but its calcium (Ca) and sodium (Na) contents are generally much lower than in other forage legumes (Spedding and Diekmahns, 1972).

Probably the main feature of its nutritive value is the presence of condensed tannins (CT) in the leaves, as a result of which the forage does not cause bloat in ruminants and, secondly, the protein is protected in the rumen, thereby leading to reduced ruminal fermentation and a better supply of amino acids for absorption in the small intestine than species without CT (Waghorn et al., 1990). Condensed tannins applied to a range of rumen microflora species were bound to the cell coats of some, resulting in changes in their morphology and reducing their protease activity (Jones et al., 1994). Compared with lucerne, about 50% more non ammonium N reaches the small intestine relative to N intake in sainfoin diets (Meissner et al., 1993). A sainfoin-only diet gave a significantly higher duodenal amino acid supply than a red-clover-only diet, but sainfoin/red-clover mixtures, containing 400 or 600 g sainfoin kg^{-1} total DM, did not show any interactive effects, implying that the CT in sainfoin had no beneficial effects on red-clover proteins (Beever and Siddons, 1986). Yearling

Angora goats grazing sainfoin gained more weight and produced more mohair of equal quality to goats grazing lucerne, a result attributed to the CT in sainfoin (Hart and Sahlu, 1993). In an assessment of several forage species fed *ad libitum*, sainfoin ranked 61% better than perennial ryegrass and second only to white clover for promoting lamb growth (Ulyatt, 1981).

Utilization

Sainfoin in monoculture or in mixture with a grass is essentially suited to a mainly cutting rather than grazing system. One or more cuts of hay or silage, together with aftermath grazing, is the normal management, the number of cuts possible being dependent on the sainfoin type or cultivar, since modern cultivars have been selected, *inter alia*, for improved response to defoliation and for longevity.

Stage of growth at cutting is the major determinant of the potential quality of hay or silage, and crops should be cut at midflowering for hay and early flowering for silage. The principles of good haymaking must be followed to maintain the quality and, in common with lucerne and other legumes cut for conservation, leaf shatter and loss during handling have to be avoided (Nash, 1985). Cold- or warm-air blowing reduces field-exposure time and the costs may be justified, since sainfoin may be sold as a high-value cash crop for racehorses. Similarly, with modern techniques of silage making, including wilting, short chopping and the use of an effective additive, silage with high nutritive value and voluntary intake can be ensured.

Aftermath growth in late season is nutritious and highly acceptable to grazing livestock, but overgrazing has to be avoided. Somewhat similar to lucerne, the plants benefit from a late-autumn rest before grazing off at a time when no further growth, which would deplete root carbohydrate reserves, is likely. Because of its erect growth habit and regrowth pattern from axillary buds on the stubble, it is most suited to cutting, but it may be tolerant of extensive grazing systems (Chassagne and Chambon, 1993). Severe grazing to almost 90% utilization has a deleterious effect on sainfoin persistence, due in part to a reduction in root TNC (Mowrey and Matches, 1991). However, stocking to achieve 75% removal of standing herbage in sainfoin swards does not reduce persistence more than lenient stocking at 50% herbage utilization (Mowrey and Volesky, 1993). Under rotational grazing, sainfoin and birdsfoot trefoil swards made better daily liveweight gains than a lucerne sward (Marten *et al.*, 1987).

References – Sainfoin

Anon. (1982) The future of sainfoin in British agriculture. In: *Proceedings of Meeting at Grassland Research Institute, Hurley*, Grassland Research Institute, Hurley, pp. 2–15.

Beever, D.E. and Siddons, R.C. (1986) Digestion and metabolism in the grazing ruminant. In: Milligan, L.P., Grovum, W.L. and Dobson, A. (eds) *Control of Digestion and*

Metabolism in Ruminants. Proceedings of the VI International Symposium on Ruminant Physiology, Banff, Canada. Prentice Hall, Englewood Cliffs, New Jersey, pp. 479–499.

Bolger, T.P. and Matches, A.G. (1990) Water-use efficiency and yield of sainfoin and alfalfa. *Crop Science* 30, 143–148.

Cash, S.D. and Ditterline, R.L. (1996) Seed size effects on growth and N_2-fixation of juvenile sainfoin. *Field Crops Research* 46, 145–151.

Charlton, J.F.L. (1983) Lotus and other legumes. In: Wratt, G.S. and Smith, H.C. (eds) *Plant Breeding in New Zealand*. Butterworths/DSIR, Wellington, pp. 253–262.

Chassagne, J. and Chambon, J. (1993) Le sainfoin: une légumineuse pour les sols de Causse. Application à la région agricole de Gramat et de Lomogne. *Fourrages* 134, 177–181.

Cooper, C.S. (1973) Sainfoin–birdsfoot trefoil mixtures for hay, hay–pasture, and hay–stockpile management regimes. *Agronomy Journal* 65, 752–754.

Gaudet, D.A., Sands, D.C., Mathre, D.E. and Ditterline, R.L. (1980) The role of bacteria in the root and crown rot complex of irrigated sainfoin in Montana. *Phytopathology* 70, 161–167.

Goplen, B.P., Richards, K.W. and Moyer, J.R. (1991) *Sainfoin for Western Canada*. O.N. Publication 1470/E, Agriculture Canada, Ottawa.

Hanna, M.R., Kozub, G.C. and Smoliak, S. (1977) Forage production of sainfoin and alfalfa on dryland in mixed- and alternate-row seeding with three grasses. *Canadian Journal of Plant Science* 60, 1481–1483.

Hart, S.P. and Sahlu, T. (1993) Mohair production and body-weight gains of yearling Angora goats grazing forages with different tannin levels. In: Baker, M.J. (ed.) *Proceedings of the XVII International Grassland Congress, New Zealand and Australia*, Vol. 1. New Zealand Grassland Association et al., Palmerston North, pp. 575–576.

Hoveland, C.S. and Townsend, C.E. (1985) Other legumes. In: Heath, M.E., Barnes, R.F. and Metcalfe, D.S. (eds) *Forages: The Science of Grassland Agriculture*, 4th edn. Iowa State University, Ames, Iowa pp. 146–153.

Hume, L.J., Withers, N.J. and Rhoades, D.A. (1985) Nitrogen fixation in sainfoin. 2. Effectiveness of the nitrogen-fixing system. *New Zealand Journal of Agricultural Research* 28, 337–348.

Hwang, S.F., Chakravarty, P. and Prevost, D. (1993) Effects of rhizobia, metalaxyl and VA mycorrhizal fungi on growth, nitrogen fixation and development of *Pythium* root rot of sainfoin. *Plant Diseases* 77, 1093–1098.

Jefferson, P.G., Lawrence, T., Irvine, R.B. and Kielly, G.A. (1994) Evaluation of sainfoin–alfalfa mixtures for forage production and compatibility at a semi-arid location in southern Saskatchewan. *Canadian Journal of Plant Science* 74, 785–791.

Jones, G.A., McAllister, T.A., Muir, A.D. and Cheng, K.J. (1994) Effects of sainfoin (*Onobrychis viciifolia* Scop.) condensed tannins on growth and proteolysis by four strains of rumenal bacteria. *Applied and Environmental Microbiology* 60, 1374–1378.

Kallenbach, R.L., Matches, A.G. and Mahan, J.R. (1995) Daylength influence on the growth and metabolism of sainfoin. *Crop Science* 35, 831–835.

Kallenbach, R.L., Matches, A.G. and Mahan, J.R. (1996) Sainfoin regrowth declines as metabolic rate increases with temperature. *Crop Science* 36, 91–97.

Karnezos, T.P., Matches, A.G. and Brown, C.P. (1994) Spring lamb production on alfalfa, sainfoin and wheatgrass pastures. *Agronomy Journal* 86, 497–502.

Kilcher, M.R. (1982) Persistence of sainfoin (*Onobrychis viciifolia* Scop.) in the semi-arid prairie region of southwestern Saskatchewan. *Canadian Journal of Plant Science* 62, 1049–1051.

Marten, G.C., Ehle, F.R. and Ristau, E.A. (1987) Performance and photosensitization of cattle related to forage quality of four legumes. *Crop Science* 27 138–145.

Martiniello, P. and Ciola, A. (1994) The effect of agronomic factors on seed and forage production in perennial legumes sainfoin (*Onobrychis viciifolia* Scop.) and French honeysuckle (*Hedysarum coronarium* L.). *Grass and Forage Science* 49, 121–129.

Meissner, H.H., Smuts, M., Vanniekerk, W.A. and Achinpongboateng, O. (1993) Rumen ammonia concentrations, and non-ammonia nitrogen passage to and apparent absorption from the small intestine of sheep ingesting subtropical, temperate and tannin-containing forages. *South African Journal of Animal Science* 23, 92–97.

Miller, D.A. and Hoveland, C.S. (1995) Other temperate legumes. In: Barnes, R.F., Miller, D.A. and Nelson, C.J. (eds) *Forages*, 5th edn, Vol.1, *An Introduction to Grassland Agriculture*. Iowa State University Press, Ames, Iowa, pp. 273–281.

Mowrey, D.P. and Matches, A.G. (1991) Persistence of sainfoin under different grazing regimes. *Agronomy Journal* 83, 714–716.

Mowrey, D.P. and Volesky, J.D. (1993) Feasibility of grazing sainfoin on the southern great plains. *Journal of Range Management* 46, 539–542.

Moyer, J.R., Hironaka, R., Kozub, G.C. and Bergen P. (1990) Effect of herbicide treatments on dandelion, alfalfa and sainfoin yields and quality. *Canadian Journal of Plant Science* 70, 1105–1113.

Nan, Z.B. (1995) Fungicide seed treatments of sainfoin control seedborne and root-invading fungi. *New Zealand Journal of Agricultural Research* 38, 413–420.

Nash, M.J. (1985) *Crop Production and Storage in Cool Temperate Climates*. Pergamon Press, Oxford.

Prevost, D. and Bromfield, E.S.P. (1991) Effect of low root temperature on symbiotic nitrogen fixation and competitive nodulation on *Onobrychis viciifolia* (sainfoin) by strains of Arctic and temperate rhizobia. *Biology and Fertility of Soils* 12, 161–164.

Rogers, H.H. (1976) Forage legumes. In: *Plant Breeding Institute Annual Report, 1975*. Plant Breeding Institute, Cambridge, pp.22–57.

Sheehy, J.E .and Popple, S.C. (1981) Photosynthesis, water relations, temperature and canopy structure as factors influencing the growth of sainfoin (*Onobrychis viciifolia* Scop.) and lucerne (*Medicago sativa* L.). *Annals of Botany* 48, 113–128.

Sheehy, J.E., Minchin, F.R. and McNeill, A. (1984) Physiological principles governing the growth and development of lucerne, sainfoin and red clover. In: Thomson, D.J. (ed.) *Forage Legumes*. Occasional Symposium No.16, British Grassland Society, Hurley, pp. 112–125.

Spedding, C.R.W. and Diekmahns, E.C. (eds) (1972) *Grasses and Legumes in British Agriculture*. Bulletin 49, Commonwealth Bureau of Pastures and Field Crops, Commonwealth Agricultural Bureaux, Farnham Royal.

Ulyatt, M.J. (1981) The feeding value of herbage: can it be improved? *New Zealand Journal of Agricultural Science* 15, 200–205.

Waghorn, G.C., Jones, W.T., Shelton, I.D. and McNabb, W.C. (1990) Condensed tannins and the nutritive value of herbage. *Proceedings of the New Zealand Grassland Association* 51, 171–176.

Wilman, D. and Asiedu, F.H.K. (1983) Growth, nutritive value and selection by sheep of sainfoin, red clover, lucerne and hybrid ryegrass. *Journal of Agricultural Science, Cambridge* 100, 115–126.

Wilson, R.G. (1994) Effect of imazethapyr on legumes and the effect of legumes on weeds. *Weed Technology* 8, 536–540.

Witty, J.F., Minchin, F.R. and Sheehy, J.E. (1983) Carbon costs of nitrogenase activity in legume root nodules determined using acetylene reduction. *Journal of Experimental Botany* 34, 951–963.

Serradella, Sulla and Tagasaste 8

PINK AND YELLOW SERRADELLAS

Introduction

Serradella is the common name for legumes of the genus *Ornithopus*, a group of summer annuals native to south-western Europe. However, they act as winter or cool-season annuals where they are grown in climatically mild regions, such as in parts of southern Australasia. This chapter concentrates on their role in this region, because of the number of developments that have been made there. Two species are in use as forage plants.

Pink or French serradella (*Ornithopus sativus* Brot.) is an erect form, which has no wild counterpart (Gladstones and McKeown, 1977). It is cultivated for forage in some parts of Europe, at high altitudes in Kenya, in the winter-rainfall zone of South Africa's Cape Province and also, to a lesser extent, in regions of Australia where mild, moist winters favour cool-season annuals. Its molecular nitrogen (N_2)-fixing ability is also exploited for tree growth when used as understorey for grazing in agroforestry situations in New Zealand.

Yellow serradella (*Ornithopus compressus* L.) occurs widely in natural pastures in countries bordering the Mediterranean, on non-calcareous soils that receive annual rainfall of more than 400 mm and at altitudes up to 1500 m. Since its introduction to Australia about 1950, it has performed and persisted well in several states and is the serradella species most widely used there. It grows on all soil types on which subterranean clover is grown but also on sandy, gritty soils, where the clover fails to persist (Gladstones and McKeown, 1977). Since it is relatively late-flowering, successful stands of yellow serradella are confined to regions receiving more than 500 mm annual rainfall.

To date, other serradella species are much more restricted, both ecologically and in their use, but they may provide germplasm of value for cross-breeding programmes in future. Common birdsfoot (*Ornithopus perpusillus*) is a European species found on sandy, stony soils. It is slender, like pink serradella, but is not regarded as a pasture plant. Moroccan serradella (*Ornithopus isthmocarpus*), regarded by some to be a subspecies of *O. sativus*, is found on the coastal lowlands of western Morocco, in southern Spain and in Portugal, but is not exploited as a forage plant. Slender serradella (*Ornithopus pinnatus*) is a low-yielding species confined to very infertile soils in areas where yellow serradella is common.

Since the use of pink and yellow serradella species as forage plants is usually in the same regions and their management is generally similar, they are jointly covered in the text below, but with distinction made when relevant.

The Plant

Pink serradella is pubescent, many-branched and semi-vine-like, but nevertheless forms a compact, dense sward. Initially, two large, flat cotyledons emerge, but subsequent growth is erect and, once it reaches half a metre or so in height, its foliage tends to lodge, becoming prostrate and spreading. Leaves are finely serrated and pinnate, with nine to 18 pairs of hairy leaflets, which are lanceolate or elliptical to ovate in shape, up to 1 cm long, and greyish-green in colour (Fig. 8.1).

Yellow serradella is similar in foliage features, except that it has a prostrate growth habit and a greener appearance in bulk. However, it has a deeper root system, enabling it to draw moisture and nutrients from greater depths than pink serradella or, for example, subterranean clover.

The inflorescence of pink serradella is an umbel of three to seven flowers on a long axillary stem, carried out beyond the leaf. Flowers are small, with a white or pink corolla. Flowering begins in early spring, once day length increases compared with previous winter conditions. The seeds develop in pods, which grow up to 3 cm long and are segmented, with constrictions between the segments. When mature, the segments, oblong and light brownish-red in colour and each tightly containing a seed, break off from the pods, when they fall to the soil surface. The 1000-seed weight of unhulled seeds is *c.* 4 g. The seeds are soft and nearly all will readily germinate.

Yellow serradella has similar flowering and seeding features to pink, but the flowers are yellow and the pods and seeds are hard, so that its seeds remain viable for long periods in the soil.

Breeding

The main serradella species are diploid, with a basic chromosome number of 14. Tetraploid forms have been developed, but their reduced ability to seed is a

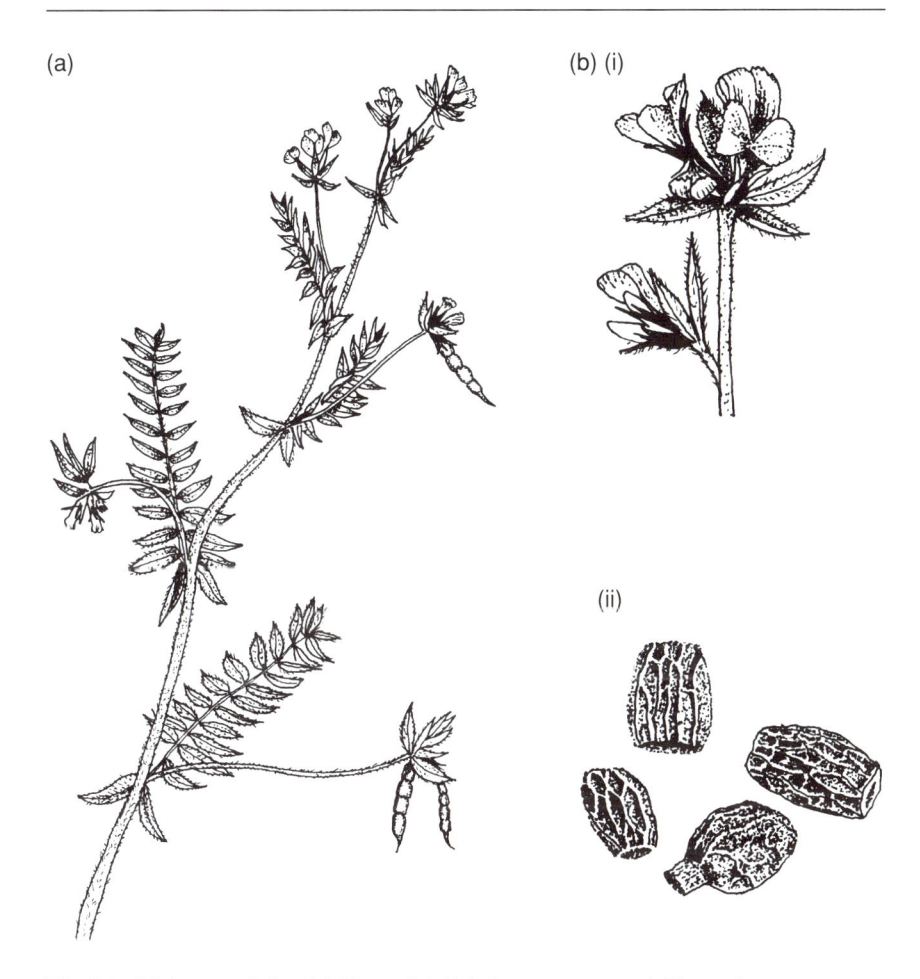

Fig 8.1. Pink serradella. (a) Plant. (b) (i) Inflorescence and (ii) seeds.

major disadvantage, since forage production after the first year is dependent on natural reseeding. Pink serradella cultivars include Grasslands Koha, from New Zealand (de Lautour and Rumball, 1986), Cadiz, an early-flowering Australian release, and Carnota and Tuy, from Spain.

Most available cultivars of yellow serradella have been developed in Australia. The first selected cultivar was Pitman, named after the farmer who first noticed the species growing on his farm in the state of Western Australia, and seed was released commercially in 1955. However, it is hard-seeded and has hard seed pods too, which makes it difficult to establish and regenerate in many situations. It is also late-flowering, which restricts it to rainfall areas of over 500 mm annually. Since then, an improvement programme has led to the release of several cultivars, e.g. Uniserra and Tauro, which are early flowering, thus making them suitable for low rainfall regions (Clark, 1983).

Attempts have been made to hybridize *O. sativus* and *O. compressus* and a single chimeral F1 hybrid was obtained by ovule–embryo culture (Williams *et al.*, 1975). Although the pollen germinability was very low, F_2–F_5 generations were obtained and plants selected for combinations of the valuable agronomic features of both parents. Selected plants of advanced generations had normal growth and fertility above 90%, enabling a selection, cv. Grasslands Spectra, which combined the good productivity and seed hullability of pink serradella with the more prostrate habit and greater persistence, through increased hard-seededness, of yellow serradella.

Seed production

When grown for commercial seed production, the seed is harvested by a conventional crop harvester before pod fall and segmentation occur, but some of the ripest seeds of highest germination capacity are usually lost while the immature seed pods, which remain attached to the plants, are harvested efficiently. Many of the seeds of yellow serradella remain in an unhulled state and are not scarified by the harvesting process (Bolland and Gladstones, 1987). To surmount these problems, some growers delay harvesting until late summer, after pod fall has occurred, and use suction harvesters to harvest the seed.

A mechanical dehuller, which has improved germination of yellow serradella seed from an initial 20% to 50%, has been developed (Sanders, 1996). However, immersion of seed in 98% sulphuric acid reduces hard-seededness to only a few per cent (Fu *et al.*, 1996).

Agronomy and Management

Seed mixtures

Since not all seeds of serradella are germinable, seed rates vary from 5 kg ha^{-1} to 15–20 kg ha^{-1}, the latter rate being used for hard-seeded yellow serradella. As more serradella cultivars become available, the mixing of cultivars of different maturity dates in one sward is being used to enhance sward performance. The inclusion of subterranean clover or rose clover (*Trifolium hirtum*) in mixtures with serradella is practised on occasion. A mixture of pink and yellow serradellas has been advocated in south-western Australia to surmount the hard-seededness feature of yellow serradella, using seed rates intermediate to those above; the pink serradella grows rapidly and dominates the new pasture during the first 2 years, thus allowing the slower establishing yellow serradella to increase in presence and become the long-term constituent (Clark, 1986).

Establishment

In late summer or early autumn, serradella seedlings emerge and establish vigorously, provided there is sufficient rainfall. Following forage production and utilization until early spring, flowering and seed setting are encouraged and the summer drought forces the plants to dry off. Sporadic rain in late summer then leads to germination of the self-sown seed, which starts the growth cycle again.

When inoculation of seed is required, as when sowing on a soil for the first time, standard procedures for legume inoculation should be followed. Rhizobia (*Bradyrhizobium*) which nodulate serradella effectively are known to persist well in sandy soils, and strains of inoculants of *Bradyrhizobium* spp. (*lupinus*) have been selected for effectiveness with serradella in specific regions (Ballard, 1996). Colonization in the rhizosphere of hulled seeds by *Bradyrhizobium* is higher than in dehulled seeds (Bowman *et al.*, 1995). Rhizobial inoculation of serradella species is unnecessary on soils that have already grown serradella or lupins (*Lupinus* spp.) in the past.

Fertilizer application is needed to correct significant nutrient deficiencies, and copper (Cu), zinc (Zn) and molybdenum (Mo), along with phosphorus (P) are applied, as necessary, in Western Australia although yellow serradella requires lesser quantities of these nutrients than other annual legumes (Cariss and Quinlivan, 1967; Clark, 1983). For the first few years after establishment, the pasture should be top-dressed with a phosphatic fertilizer such as superphosphate, at 20–40 kg phosphate (P_2O_5) ha^{-1} annually. The P requirement of yellow serradella is lower than for subterranean clover on light acid soils (Bolland, 1991), the requirement for P being reduced if the seed has a high P content (Bolland and Baker, 1990). It can also tolerate soils with low potassium (K) status better than subterrranean clover (Gladstones and McKeown, 1977).

In the extensively farmed situations where serradella is grown, it is not common practice to apply crop-protection chemicals during establishment, and pest and disease problems are minimal. Heavy grazing in early winter is used to control weeds, if necessary, but grazing pressure requires to be reduced in late winter/early spring to allow good seed set (Clark, 1986).

Feeding value

Pink serradella has high nutritive value (Table 8.1), and early reports indicated that it had no apparent toxins or oestrogenic substances (Gladstones and Barrett-Lennard, 1964).

Crude protein and digestibility typically decline with advancing plant maturity, during which there is a reduction in the leaf : stem ratio, although the rates of decline are slower than for legumes such as lucerne or red clover, possibly because of the indeterminate type of flowering (Lloveras and Iglesias, 1997).

Table 8.2 shows the nutritive value of pink-serradella silage in relation to silage from other forage species.

Utilization

The prime objective is to utilize forage during the late autumn–spring growth period, and then allow the sward to flower and set seed for the next season. Many of the management guidelines were developed by farmers (Bolland and

Table 8.1. Herbage quality of pink serradella and subterranean clover (from Taylor et al., 1977).

		Three-cut system			
Species	Quality	1	2	3	Single cut
Pink serradella	Crude protein (g kg^{-1} DM)	250	270	200	190
	Digestibility (%)	80	79	73	73
Subterranean clover	Crude protein (g kg^{-1} DM)	280	310	260	140
	Digestibility (%)	83	82	77	76

Table 8.2. Nutritive value of silages from different forage species in relation to nutrient concentrations required in the daily diet of livestock (from Taylor and Hughes, 1978).

	Energy content (MJ ME kg^{-1} DM)	Crude protein	Digestible crude protein	Ca	P
		(g kg^{-1} DM)			
Forages					
Pink serradella	9.5	189	120	12.2	3.1
Red clover	9.4	149	89	16.1	2.2
Maize	11.1	70	40	3.5	1.8
Oats	9.4	90	50	17.6	2.9
Animal requirements					
Lactating dairy cow (20–30 kg milk day^{-1})	10.1	50	114	4.7	3.5
Dry cow (maintenance)	8.4	85	51	3.4	2.6
Steers, 200 kg (liveweight gain, 0.75 kg day^{-1})	11.0	111	71	3.6	2.8

ME, metabolizable energy.

Gladstones, 1987). Once established, serradella is grazed in systems similar to those for subterranean clover and will carry fairly similar stocking rates, but it can also be cut for silage (Taylor and Hughes, 1978). Annual dry-matter (DM) yields are very variable, being dependent on several management factors, including site and defoliation frequency, but yields of 10–11 t ha^{-1} for pink serradella have been reported (Taylor et al., 1977; de Lautour and Rumball, 1986).

In general, frequent and severe defoliation delays flowering in both ser-

Table 8.3. Effect of grazing frequency on serradella re-establishment (unpublished data from J.F.L. Charlton).

	Pink serradella		Yellow serradella	
Frequency of grazing	Buried seed* (kg ha^{-1})	No. of seedlings (m^{-2})	Buried seed* (kg ha^{-1})	No. of seedlings (m^{-2})
Once	344	3193	371	537
Twice	46	192	788	1257
Thrice	64	112	511	561
Grazed then cut	53	41	522	337

*In top 3 cm of soil.

radella species, but seeding, and therefore subsequent autumn seedling density, is more adversely affected in pink than in yellow serradella by such grazing (Gladstones and Barrett-Lennard, 1964; Gladstones and Devitt, 1971). Nevertheless, seed production of yellow serradella was reduced by 50% or more by grazing in late spring (Conlan et al., 1994). Table 8.3 exemplifies the point and, by the third year of the trial illustrated, the quantity of buried seed was reduced to 9–18 kg ha^{-1} for pink serradella, in contrast to 165–644 kg ha^{-1} from yellow serradella, the highest quantities occurring in the less frequently grazed treatments. Frequent cutting of pink serradella reduced DM yields relative to infrequent cutting, as did severe relative to lax cutting intensity while frequent compared with infrequent grazing reduced yields of both pink and yellow serradella species (Taylor et al., 1977).

In the mild maritime winters of north-west Spain, pink serradella, sown in early autumn, provides more than 4 t DM ha^{-1} as high-quality forage by spring of the following year and can be treated as an annual catch crop, complementing maize, which is sown in late spring in a double-cropping system (Iglesias and Lloveras, 1997). Pink serradella is also used in that area on some farms as an understorey in vineyards, as it grows during the winter when the vines are dormant, providing forage, controlling weeds and supplying nitrogen (N) for the vines (Lloveras, 1987).

References – Serradella

Ballard, R.A. (1996) Assessment of strains of *Bradyrhizobium* spp. (*lupinus*) for serradella (*Ornithopus* spp.). *Australian Journal of Experimental Agriculture*, 36 63–70.
Bolland, M.D.A. (1991) Response of defoliated swards of subterranean clover and yellow serradella to superphosphate applications. *Australian Journal of Experimental Agriculture* 31, 777–783.

Bolland, M.D.A. and Baker, M. (1990) Effect of seed source and seed P concentration on the yield response of yellow serradella to superphosphate application. *Australian Journal of Experimental Agriculture* 30, 811–815.

Bolland, M.D.A. and Gladstones, J.S. (1987) Serradella (*Ornithopus* spp.) as a pasture legume in Australia. *Journal of the Australian Institute of Agricultural Science* 53, 5–10.

Bowman, A.M., Hebb, D.M., Munnich, D., Rumney, G.L. and Brockwell, J. (1995) Field persistence of *Bradyrhizobium* spp. (*lupinus*) inoculant for serradella (*Ornithopus compressus* L.). *Australian Journal of Experimental Agriculture* 3, 357–365.

Carris, H.G. and Quinlivan, B.J. (1967) Serradella. *Western Australia Journal of Agriculture* 8, 226–234.

Clark, S.G. (1983) *A Review of the Genus Ornithopus – Promising Pasture Legumes for Victoria*. Technical Report Series No. 83, Department of Agriculture, Government of Victoria.

Clark, S.G. (1986) *Yellow Serradella as a Pasture Legume for Deep Sand*. AgNote Series No. 3615, Department of Agriculture, Government of Victoria.

Conlan, D.J., Dear, B.S. and Coombes, N.E. (1994) Effects of grazing intensity and number of grazings on herbage production and seed yields of *Trifolium subterraneum*, *Medicago murex* and *Ornithopus compressus*. *Australian Journal of Experimental Agriculture* 34, 181–188.

de Lautour, G. and Rumball, W. (1986) Grasslands Koha pink serradella (*Ornithopus sativus* Brot.). *New Zealand Journal of Experimental Agriculture* 14, 464–467.

Fu, S.M., Hampton, J.G., Hill, M.J. and Hill, K.A. (1996) Breaking hard seed of yellow and slender serradella (*Ornithopus compressus* and *Ornithopus pinnatus*) by sulphuric acid scarification. *Seed Science and Technology* 24, 1–6.

Gladstones, J.S. and Barrett-Lennard, R.A. (1964) Serradella – a promising pasture legume in Western Australia. *Journal of the Australian Institute of Agricultural Science* 30, 258–262.

Gladstones, J.S. and Devitt, A.C. (l971) Breeding and testing early-flowering strains of yellow-flowering serradella (*Ornithopus compressus*). *Australian Journal of Experimental Agriculture and Animal Husbandry* 11, 431–439.

Gladstones, J.S. and McKeown, N.R. (1977) *Serradella – a Pasture Legume for Sandy Soils*. Bulletin 4030, Western Australian Department of Agriculture, Perth.

Iglesias, I. and Lloveras, J. (1997) Dry matter yield and nutritive value of cool season legumes for double cropping systems. (In preparation.)

Lloveras, J. (1987) Traditional cropping systems in northwestern Spain (Galicia). *Agricultural Systems* 23, 259–275.

Lloveras, J. and Iglesias, I. (1997) Accumulation of dry mass and evolution of nutritive value in serradella. (In preparation.)

Sanders, K.F. (1996) A dehuller for improving the germination of serradella seed. *Journal of Agricultural Engineering Research* 64, 187–195.

Taylor, A.O. and Hughes, K.A. (1978) Conservation-based forage crop systems for major or complete replacement of pasture. *Proceedings of the Agronomy Society of New Zealand* 8, 161–166.

Taylor, A.O., Hughes, K.A., Haslemore, R.M. and Holland, R. (1977) Influence of maturity and frequency of harvest on the nutritive quality of cool season forage legumes. *Proceedings of the Agronomy Society of New Zealand* 7, 45–49.

Williams, W.W., de Lautour, G. and Stiefel, W. (1975) Potential of serradella as a winter annual forage legume on sandy coastal soil. *New Zealand Journal of Experimental Agriculture* 3, 339–342.

SULLA

Introduction

Sulla (*Hedysarum coronarium* L.), also called Italian sainfoin, French honeysuckle or sweet vetch, is a short-lived, perennial legume, originating in the western Mediterranean basin and North Africa (Duke, 1981). It is one of the species of the genus *Hedysarum*, which comprises almost 100 species, from annuals to perennials and herbs to shrubs, many of which occur naturally in the Mediterranean region and some of which have some value as forage in arid and semiarid regions (De Olives, 1967; Kernick, 1978). Sulla forage is mainly used for silage or hay, as cut and carried green feed and sometimes as a green manure (Krishna, 1993). It is the main legume in southern Italy, where *c.* 250,000 ha are cultivated for grazing during the autumn–spring period and for hay during early summer (Martiniello and Ciola, 1994). It has been evaluated in parts of Spain (Gutiérrez Más, 1986), Portugal and North Africa. Sulla has also been introduced for forage or for land-conservation purposes, e.g. erosion control, in steep hill pastures or revegetating eroded areas, in Australasia and the USA.

The Plant

Sulla foliage growth is erect, 30–150 cm in height, and its succulent, pinnate leaves have oval to round leaflets arranged in pairs, with 7–15 pairs per leaf and a single terminal leaflet (Fig. 8.2). Leaves tend to be glabrous above and hairy below. The plant has a strong tap root, with numerous secondary roots. Flowering usually occurs from early summer to autumn and is stimulated by availability of soil moisture (Watson, 1982). Flower-heads are racemes of up to 35 florets, usually bright crimson in colour but ranging from pink to violet shades. After flowering, the red-tinged stems tend to become rather woody.

Seed pods have three to eight elliptical segments per pod, which are constricted between each seed and which split into brown unhulled seeds, with a rough, reticulated surface and with a 1000-seed weight of 11.0 g; the hulled seeds are creamy white to brown in colour, flattened, with an almost circular profile, and a 1000-seed weight of 5.3 g (Charlton, 1992).

Breeding

Sulla is a natural diploid species, with $2n = 16$ chromosomes. There are many local selections in the Mediterranean countries, Italian examples being Grimaldi, selected for improved cold tolerance, and Sparacia, which has an upright growth habit, making it better adapted to forage conservation than

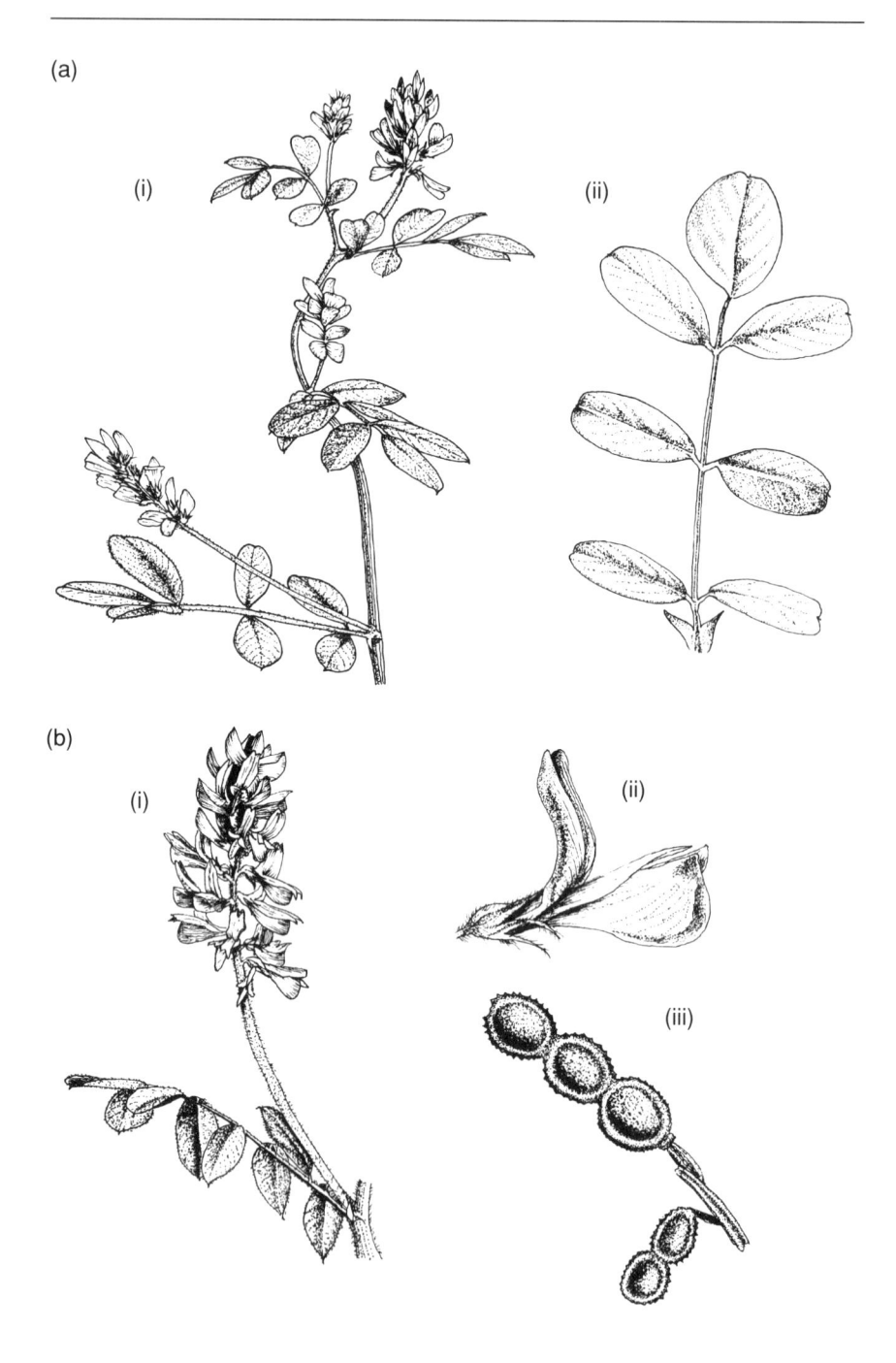

Fig. 8.2. Sulla. (a) (i) Aerial parts and (ii) leaf. (b) (i) Inflorescence, (ii) floret and (iii) seed pod.

the more prostrate types used to control soil erosion. There is scope to improve on existing ecotypes and cultivars available; for example, in Italy superior genotypes have been identified as having potential in breeding programmes to improve forage and seed production under Mediterranean conditions (Martiniello and Ciola, 1994). Grasslands Aokau and Necton are cultivars bred in New Zealand. The former, selected for soil conservation purposes (Lambrechtsen and Douglas, 1986), contains condensed tannins (CT), which eliminate the risk of bloat when grazed by ruminant stock.

Seed production

For satisfactory seed yields, crops should be closed to grazing before the start of spring. The crop is pollinated by honey-bees and seed maturity takes approximately 8 weeks. The optimum time for harvesting is when more than half the seeds are brown and the remainder purplish-red. Rolling the crop in early morning after light rain or dew, mowing and then leaving the wind-rows for 3–5 days before threshing is a satisfactory method. Seed yields can range from 200 kg ha^{-1} in the first year to over 500 kg ha^{-1} in subsequent years (Foote, 1988), although, under irrigation, seed yields exceeded 1500 kg ha^{-1} (Martiniello and Ciola, 1994). The proportion of hard seed at harvest can be high, especially when there is hot, dry weather during seed ripening. Treating seed with hot water, 60–75°C for a minute, breaks the dormancy of hard seed.

Agronomy and Management

Establishment

Sulla is mainly sown alone but can be sown with a cereal or in mixture with other legumes. For dehulled seed 5–10 kg ha^{-1} is recommended, and for hulled seed 15–20 kg ha^{-1} (Foote, 1988). Seed may be drilled or broadcast and harrowed as long as sowing is shallow. It thrives on soils with a pH greater than 6.0–6.5 with free forms of calcium (Ca) (Gutiérrez Más, 1986), but is likely to fail on acid or waterlogged soils. Thus, in many soils, liming may be necessary prior to seeding and, in common with most legumes, adequate available soil P is a prerequisite.

Sulla requires a specific strain of *Rhizobium hedisari* for effective nodulation. The rhizobia that inoculate sulla are not in any of the common cross-inoculation groups, and even isolates from other species of *Hedysarum* are not effective (Krishna, 1993). The widely used peat based inoculation method works well, 20% gum-arabic solution being an excellent protective additive.

For successful establishment, sulla requires good control of invasive broad-leaved weeds, using approved herbicides that are effective in controlling grass weeds (Watson, 1982).

Sulla appears to be resistant to aphids but powdery-mildew disease (*Erysiphe trifolii*) has adversely affected its performance in Italian situations.

Feeding value
The forage is of high nutritive value, particularly the leaflets and reproductive parts, compared with stems and petioles, and in the autumn/winter period, compared with late spring (Pinto et al., 1993). The CT in sulla clearly add to its potential as a forage. Apart from preventing bloat in ruminants, CT in plants protect protein in the rumen from degradation to ammonia and thus increased quantities of amino acids are subsequently absorbed in the abomasum (Waghorn et al., 1989). However, herbage intake and digestibility of CT-containing legumes are lowered if the level of CT in the forage is too high (Barry, 1989). Inactivating CT in sulla with polyethylene glycol increased digestibility by 2 percentage units (Stienezen et al., 1996).

The liveweight gain in lambs fed sulla was greater than that in lambs fed sulla with tannin effects removed (Barry, 1989). However, higher rates of voluntary intake and body growth in lambs grazing sulla, compared with lambs grazing grass/white-clover swards, were attributed to a very high readily fermentable : structural carbohydrate ratio (Terrill et al., 1992). Condensed tannins in sulla protected grazing lambs against the deleterious effects of intestinal nematodes (*Trichostrongylus colubriformis*), so that liveweight gains were high, whereas lambs grazing lucerne lost weight and condition, although protein protection in the rumen may also have been a factor (Niezen et al., 1995); faecal contamination of wool in sheep and its serious consequence, fly-strike, caused by maggots of the blowfly (*Lucilia serricata* and *Lucilia cuprina*), were also reduced by intake of CT.

Utilization
Cutting before the onset of flowering ensures that forage of high nutritional quality is obtained, since the quality peaks before the onset of flowering and, thereafter, stems become fibrous. Wilting and short chopping are advantageous when making silage. In cutting trials in New Zealand, DM yields ranged from 12 to 18 t ha^{-1} (Rys et al., 1988).

The forage is highly acceptable to sheep and cattle. For grazing, a rotational system is best, since the regrowths come primarily from axillary buds on the lower parts of the stem. A recommendation for established swards is to graze when the foliage reaches 30–50 cm in height, and there is then 3–5 t DM ha^{-1} present. In New Zealand, this sward state is achieved in 6–8 weeks in spring and every 8–10 weeks during summer and autumn, and annual yields of 14–16 t DM ha^{-1} have been recorded (Foote, 1988). In grazing trials with sheep, DM yields ranged from 7.4 to 18.0 t ha^{-1} and there was no difference in yield of hard- and lax-grazed swards, but grazing in late autumn significantly decreased plant density and yield (Krishna et al., 1990). Regrowth comes mainly from leaf axils or scars located on the plant crowns, rather than from axillary buds in the stubble remaining after grazing, and so damage to the crowns, by excess treading for example, should be minimized (Krishna and Kemp, 1993). Sulla pastures yielding 6–7 t DM ha^{-1} under semiarid conditions in Sardinia, Italy, were effectively exploited by grazing with dairy sheep (Sulas et al., 1995).

References – Sulla

Barry, T.N. (1989) Condensed tannins: their role in ruminant protein and carbohydrate digestion and possible effects upon the rumen ecosystem. In: Nolan, J.V., Long, R.A. and Demeyer, D.I. (eds) *The Role of Protozoa and Fungi in Ruminant Digestion*. University of New England Press, Armidale, New England, pp. 153–169.

Charlton, J.F.L. (1992) Some basic concepts of pasture seed mixtures. *Proceedings of the New Zealand Grassland Association* 53, 37–40.

De Olives, G. (1967) *La Zulla*. Serie Técnica No. 23, Publicaciónes de Capacitación Agraria, Madrid, 47 pp.

Duke, J.A. (1981) *Handbook of Legumes of World Economic Importance*. Plenum Press, New York and London.

Foote, A.S. (1988) Local cultivar adaptation for Mediterranean's sulla. *New Zealand Journal of Agriculture* 153, 25–27.

Gutiérrez Más, J.C. (1986) Aproximación a la ecologia, de la zulla (*Hedysarum coronarium* L.) Distribución geográfica de poblaciones naturales en la provincia de Cádiz. Posibilidades de expansion en la Península Ibérica. *Pastos* 16, 1–16

Kernick, M.D. (1978) *Hedysarum* spp. In: *Ecological Management of Arid and Semi-arid Rangelands in Africa and the Near and Middle East (EMSAR-II)*, Vol. IV, *Indigenous Arid and Semi-arid Forage Plants of North Africa, the Near and Middle East*. Technical Data Sheet No. 25, FAO, Rome, pp. 597–619.

Krishna, H. (1993) Sulla (*Hedysarum coronarium* L.): an agronomic evaluation. PhD thesis, Massey University, Palmerston North, New Zealand.

Krishna, H. and Kemp, P.D. (1993) Regrowth sites of sulla (*Hedysarum coronarium* L.) after defoliation. In: Baker, M.J. (ed.) *Proceedings of the XVII International Grassland Congress, New Zealand and Australia*, Vol. 1. New Zealand Grassland Association *et al*., Palmerston North, pp. 149–150.

Krishna, H., Kemp, P.D. and Newton, S.D. (1990) 'Necton' sulla – A preliminary agronomic evaluation. *Proceedings of the New Zealand Grassland Association* 52, 157–159.

Lambrechtsen, N.C. and Douglas, G.B. (1986) Management and uses of *Hedysarum coronarium* (sulla). In: *Water and Soil Miscellaneous Publication* No. 94, pp. 263–266.

Martiniello, P. and Ciola, A. (1994) The effect of agronomic factors on seed and forage production in perennial legumes sainfoin (*Onobrychis viciifolia* Scop.) and French honeysuckle (*Hedysarum coronarium* L.). *Grass and Forage Science* 49, 121–129.

Niezen, J.H., Waghorn, T.S., Charleston, W.A.G. and Waghorn, G.C. (1995) Growth and gastrointestinal nematode parasitism in lambs grazing either lucerne (*Medicago sativa*) or sulla (*Hedysarum coronarium*) which contains condensed tannins. *Journal of Agricultural Science, Cambridge* 125, 281–289.

Pinto, P.A., Barrados, G.T. and Tenreiro, P.C. (1993) Growth analysis and chemical composition of sulla (*Hedysarum coronarium* L.). In: *Proceedings of the XVII International Grassland Congress, New Zealand and Australia*. New Zealand Grassland Association *et al*., Palmerston North, pp. 587–589.

Rys, G.T., Smith, N. and Slay, M.W. (1988) Alternative forage species in Hawkes Bay. *Proceedings of the Agronomy Society of New Zealand* 18, 75–80.

Stienezen, M., Waghorn, G.C. and Douglas, G.B. (1996) Digestibility and effects of condensed tannins on digestion of sulla (*Hedysarum coronarium*) when fed to sheep. *New Zealand Journal of Agricultural Research* 39, 215–221.

Sulas, L., Porqueddu, C., Roggero, P.P. and Bullitta, P. (1995) The role and potential of sulla (*Hedysarum coronarium* L.) in the Mediterranean dairy sheep farming system. In: *Proceedings of the Fifth International Rangeland Congress, Salt Lake City, USA*, pp. 543–544.

Terrill, T.H., Douglas, G.B., Foote, A.G., Purchas, R.W., Wilson, G.F., and Barry, T.N. (1992) Effect of condensed tannins upon body growth, wool growth and rumen metabolism in sheep grazing sulla (*Hedysarum coronarium*). *Journal of Agricultural Science, Cambridge* 119, 265–273.

Waghorn, G.C., Jones, W.T., Shelton, I.D. and McNabb, W.C. (1989) Condensed tannins and the nutritive value of pasture. *Proceedings of the New Zealand Grassland Association* 51, 171–175.

Watson, M.J. (1982) *Hedysarum coronarium* – a legume with potential for soil conservation and forage. *New Zealand Journal of Agricultural Science* 16, 189–193.

TAGASASTE

Introduction

Tagasaste (*Chamaecytisus palmensis* (Christ.) Hutch.) is a perennial leguminous shrub or small tree that has been developed for forage production in specialist situations, but particularly those with hot, dry summers. The plant has become popular in recent decades in the Mediterranean region and parts of Australasia and Africa. It has also been known as tree lucerne, false tree lucerne and lucerne tree, but tagasaste, its original name, is preferred, since it is botanically dissimilar to lucerne (*Medicago sativa*). The genus *Chamaecytisus* contains approximately 30 species, all of which are European trees or shrubs. Two *Chamaecytisus* species originated in the Canary Islands, namely, *C. palmensis* and *C. proliferus*. The former is endemic to the island of La Palma, whereas the second species, known locally as escabon, grows on several islands. They were identified as potential forage legumes for wider use after the First World War, although *C. palmensis* (sometimes referred to as *C. proliferus* subspecies *palmensis*) had been long recognized as a forage source.

Tagasaste seeds were sent to the Botanic Gardens in Adelaide, South Australia, from the Royal Botanic Gardens, Kew, England, in 1879 (Dann and Trimmer, 1986). Subsequent reports praised its drought tolerance and fodder potential. Its seed was distributed to the British colonies from Kew in 1879 for forage-plant trials, although it is reported being introduced as a hedge plant to New Zealand, where it is now widely naturalized (Webb, 1982). Since then, it has become naturalized in milder temperate regions in several countries. The first notable proponent of tagasaste was Dr Victor Perez, a Canary Islander, who recommended that plants be spaced 2–3m apart and cut two or three times a year, enabling them to last 10–20 years. Spanish reports tell of island farmers taking two or three cuts per year from tagasaste stands to feed cattle, sheep, goats and horses (Davies, 1982).

Commercial development of tagasaste expanded in Australasia more recently, when Western Australian farmers began using it on deep sandy soils within the past 20 years (Wiley *et al.*, 1994). This particular soil type had failed to support productive pasture or annual legumes, but a management system was developed that successfully incorporated tagasaste, initially by direct grazing stands for a month in late summer/early autumn, and then using mechanical cutting to make its forage available throughout the year. Sir James McCusker, one of the leading farmers in this region, encouraged commercial tagasaste use by funding the Martindale Research Project in 1985, and much valuable information on using this legume became available from this project.

The Plant

Tagasaste grows to a height of 5 m and almost to the same diameter (Dann and Trimmer, 1986). Plant growth habit varies from prostrate to upright, but the forms can be manipulated by cutting management in the early life of the plant. The plant is very deep-rooting, having roots up to 8 m long (Davies, 1982). It often has long, drooping branches, with dull, bluish-green trifoliate leaves and minute stipules (Fig. 8.3). White leguminous-type flowers are produced during late winter to early spring and are both self- and cross-pollinated. Bumble-bees are the main pollinators. Black, flattened seed pods develop, growing up to 5 cm long and each containing about ten seeds. The seeds are flattened, oval and black, with a 1000-seed weight of 22 g. Tagasaste has the reproductive characteristics of shorter-lived species, but exerts a high reproductive effort, compared with other leguminous tree species.

Although no cultivars are listed, tagasaste exhibits wide phenotypic variation in features such as flowering time, growth habit, vigour, leaf size, leaf hairiness, frost tolerance and taste (Wright and Menzies, 1985). In a study of variation in 65 accessions, there was wide variation both within and between populations which would be potentially valuable in breeding programmes (Woodfield and Forde, 1987).

Agronomy and Management

Establishment

Tagasaste is established from seed or rooted seedlings, although stands from planted seedlings are the favoured method and are more productive in the first 2 years than stands grown from seed. The seed is usually very hard, with low germination capacity when unscarified, but commercial scarifiers, such as

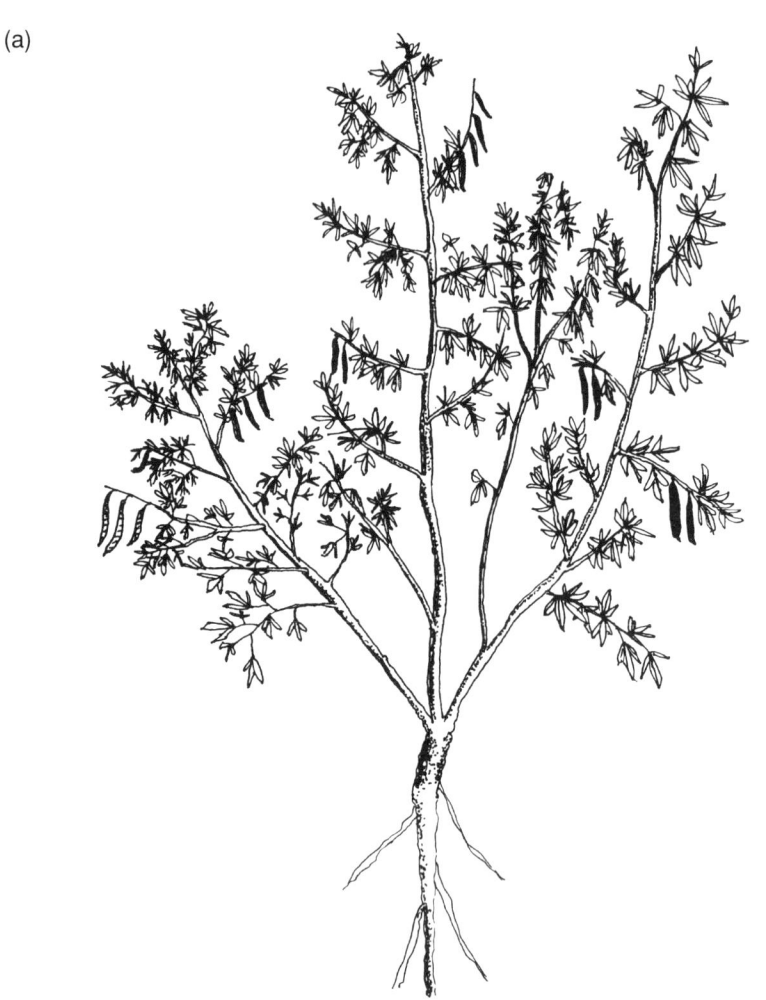

Fig. 8.3. Tagasaste. (a) Tree. (b) (i) Foliage and (ii) inflorescences. (c) (i) Pods and (ii) seeds.

those used for treatment of subterranean-clover seed, can be successfully used. Another successful method is hot-water treatment, whereby seed are dropped into boiling water, which is then immediately removed from its heat source; left to soak, germinable seeds start to swell and can be separated from dormant, hard seeds by sieving (Dann and Trimmer, 1986). Seeds must also be inoculated with effective *Bradyrhizobium/Rhizobium lotii* before sowing, to ensure successful establishment.

In Chile, N_2 fixation accounted for 86% of the 49 g of N taken up per plant over 2 years (Ovalle *et al.*, 1996); at 5000 plants ha^{-1}, this represented an annual fixation rate of over 100 kg N ha^{-1}. *Rhizobium* isolated from the root nodules of tagasaste showed infection features similar to *Bradyrhizobium* in some instances and *Rhizobium lotii* in others (Gault *et al.*, 1994).

Fig. 8.3 continued

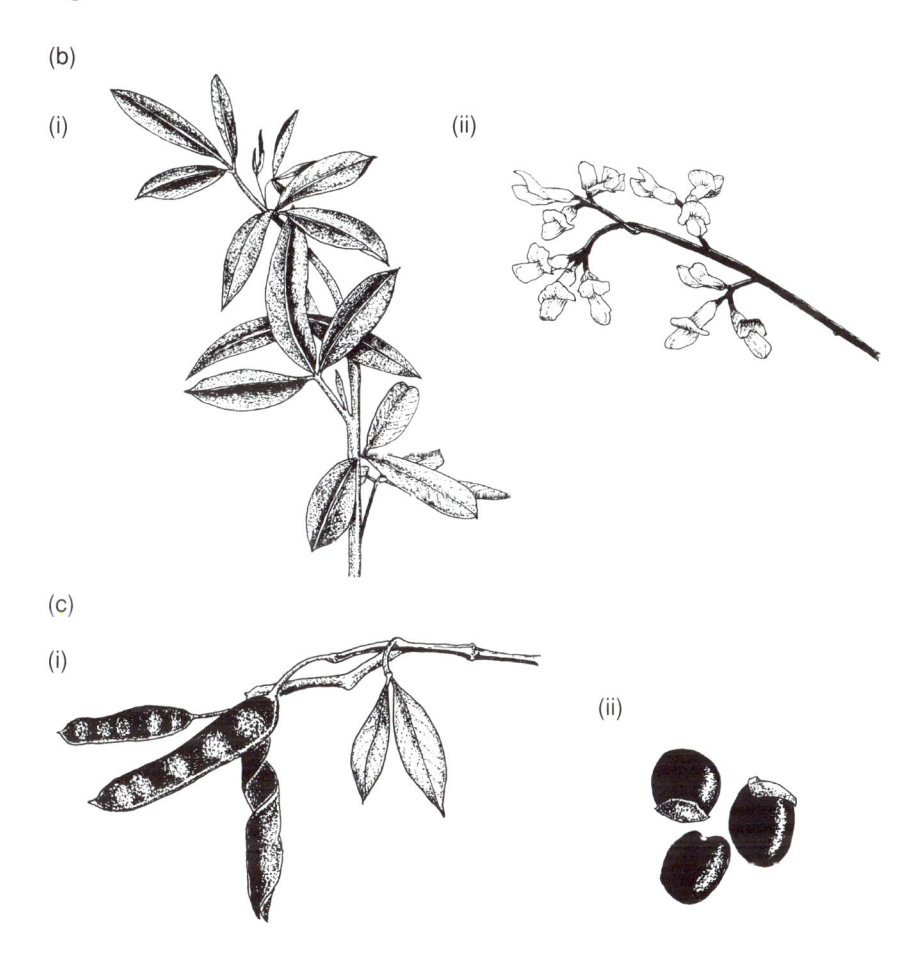

(b)

(i)   (ii)

(c)

(i)   (ii)

Although tagasaste is normally autumn-sown, some stands have grown successfully after sowing in winter in Australia (Wiley *et al.*, 1994). Any pasture growing between sown rows helps to prevent soil erosion during stand establishment. Experimental plantings have been made at plant densities of 2000 ha^{-1}, using rows set 5 m apart (Oldham, 1993), and this allows ready observation of stock. Densities as high as 10,000 plants ha^{-1} have been suggested, but establishment costs increase with closer spacing and yet forage yields do not when row spacing is closer than 6 m (Wiley and Maughan, 1993). The optimum row spacing seems to be between 6 and 10 m, with plants every 2 m within rows (Wiley *et al.*, 1994). However, in Australia, a plantation with 20,000 plants ha^{-1} produced 40% more DM, including pasture, than one at 2500 ha^{-1}, although individual tree biomass was lowest at the highest density (McGowan and Mathews, 1994). Corners of fields are also planted as copses for browsing. Plants can be safely grazed or trimmed at 11 months after sowing or when 25 cm high, the aim being to encourage the development of a multistemmed plant tolerant to grazing.

Insect pests tend to cause many tagasaste-stand failures during the first year, especially from direct seeding. Thus, young stands should be checked at least once weekly during the first few weeks and sprayed with pesticide, as necessary. Young plants are very acceptable and wild animals, such as rabbits, which can seriously damage establishing stands, must be controlled. Tree guards, protective fencing and repellents are available for use during establishment. If this is done during the first 2 years, spacing between trees should be sufficient to permit inter-row cropping or haymaking. Bark stripping is a minor problem, although some animals can cause severe local stripping and locust attack can accelerate such damage (Oldham, 1993).

Forage production

Tagasaste grows well in mild, temperate climates but can be frost-tender. It has been grown successfully in areas with as little as 300 mm annual rainfall and it grows well in deep sands, where its roots can penetrate beyond 10 m (Wiley *et al.*, 1994). It prefers deep, free-draining light soils, even though acidic, but does not withstand waterlogging. General fertilizer requirements are similar to those for other forage legumes, and minor element deficiencies must be corrected wherever they occur. The plants respond well to P, with application of 40–80 kg P_2O_5 ha^{-1} being common in Western Australia (Dann and Trimmer, 1986).

Yields are typically influenced by a host of factors and thus are very variable. Annual DM production of 13–18 t ha^{-1} may be achievable, but the edible DM (EDM) of mainly leaf and fine stem is likely to be 50–60% of this total (Townsend and Radcliffe, 1987; Douglas *et al.*, 1996).

Feeding value

Although tagasaste has been classed as of moderate feeding value, it is an economic alternative feed for sheep and cattle, instead of hand-fed grain or hay, in dryland Australia (Oldham and Mattinson, 1990). Tagasaste usually has a lower DM digestibility than pasture, e.g. 73% vs. 67% *in vivo* digestibility, and also lower concentrations of minerals (McGowan *et al.*, 1995). The forage quality varies with plant part (Table 8.4; Dann and Trimmer, 1986) and management, but quality can be maintained throughout the year (Douglas *et al.*, 1996).

Grazing tagasaste only once yearly gives forage adequate for sheep maintenance and perhaps wool growth, but not rapid liveweight gain. Tagasaste foliage contains significant concentrations of phenolic compounds, which, at concentrations of 70 g kg^{-1} DM, are associated with good acceptability and a daily voluntary intake of 1 kg $head^{-1}$ by young merino sheep (Borens and Poppi, 1990); in contrast, tagasaste foliage with total phenolics of over 170 g kg^{-1} DM are rejected by sheep. Such levels can arise soon after insect attack and seem to be an induced pest defence mechanism (Oldham, 1993).

Table 8.4. Protein and digestibility of tagasaste plant parts (from Wiley et al., 1994).

Plant part	Crude protein (g kg^{-1} DM)	Digestibility (%)
Leaf	200–300	70–80
Fine stem	90	50–60
Large stem	60	40–50
Wood	30	Unknown
Bark	130	Unknown

Utilization

Utilization normally involves a combination of rotational grazing and cutting, since continuous sheep grazing kills the trees. One grazing system that has been developed involves an optimal 30-day grazing period with a maximum of 45 days before a rest period (Oldham, 1993). Once grazing starts, regular observation is advisable and a grazing 'line' soon develops, usually at about 1.2 m above ground level. The tree rows are then cut to about 0.6 m height, in order to ensure subsequent access for stock and the sheep are removed, once the cut material has been eaten.

Intakes as low as 200 g head^{-1} day^{-1} of supplementary freshly cut tagasaste foliage were sufficient for maintenance of ewe weaners grazing dried-out annual ryegrass/subterranean-clover pasture during autumn (Hemsley et al., 1987); the sheep gained about 60 g liveweight head^{-1} compared with others who lost weight grazing the dry pasture, although fed a lupin-grain supplement. Tagasaste-supplemented sheep also grew 11 g wool head^{-1} daily, compared with 6 g head^{-1} on the control dry pasture, and gave 30% more wool for the year. In other work, ewe weaners grazing tagasaste during summer consistently grew more clean wool than sheep grazing dry pastures or crop stubble, but effects on other wool characteristics varied depending upon class of sheep (Oldham, 1993). An annual carrying capacity of 3000 sheep-grazing days (SGD) ha^{-1} has been reported, while cutting produced annual forage yields of 3 t EDM ha^{-1} (Southern, 1988). Sheep with restricted access of 1–3 days per week to tagasaste during autumn, at stocking rates of more than 100 ha^{-1}, maintained liveweight and wool tensile strength, and this management increased tagasaste annual carrying capacity to 6000 ha^{-1}.

Other Uses

Although forage production is the main use for tagasaste, its adaptability to a wide range of environments may make it suitable for several other purposes. *Inter alia*, erosion control, shelter belts for grazing stock or for protecting horticultural crops, firewood in some countries and understorey grazing in the early phase of agroforestry are possibilities (Davies, 1982)

References

Borens, F.M.P. and Poppi, D.P. (1990) The nutritive value for ruminants of tagasaste (*Chamaecytisus palmensis*), a leguminous shrub. *Animal Feed Science and Technology* 28, 275–292.

Dann, P. and Trimmer, B. (1986) Tagasaste: a tree legume for fodder and other uses. *New Zealand Journal of Agricultural Science* 20, 142–151.

Davies, D.J.G. (1982) Tree lucerne: an historical perspective and discussion of potential uses in new Zealand. In: *Proceedings of a Workshop, Lincoln*, pp. 6–16.

Douglas, G.B., Bulloch, B.T. and Foote, A.G. (1996) Cutting management of willows (*Salix* spp.) and leguminous shrubs for forage during summer. *New Zealand Journal of Agricultural Research* 39, 175–184.

Gault, R.R., Pilka, A., Hebb, D.M. and Brockwell, J. (1994) Nodulation studies on legumes exotic to Australia: symbiotic relationships between *Chamaecytisus palmensis* (tagasaste) and *Lotus* spp. *Australian Journal of Experimental Agriculture* 34, 385–394.

Hemsley, J.A., Peter, D., Downs, P.A., Smith, J. and Buscall, D. (1987) *Nutritional Comparison of Lupin Grain vs. Tagasaste for Growth and Wool Production by Merino Ewe Weaners Grazing Mature Annual Pasture.* Martindale Research Project – Scientific Review April 1987, School of Agriculture, The University of Western Australia, Nedlands.

McGowan, A.A. and Mathews, G.L. (1994) Effect of interrow spacing on the production of tagasaste and associated pastures. *Australian Journal of Experimental Agriculture* 34, 487–490.

McGowan, A.A., Mathews, G.L. and Moate, P.J. (1995) Mineral balance of sheep fed pasture, tagasaste or tagasaste with a mineral supplement. *Australian Journal of Experimental Agriculture* 31, 51–54.

Oldham, C.M. (1993) Tagasaste (*Chamaecytisus palmensis*) – a fodder shrub and alternative legume forage. In: Michalk, D.L., Craig, A.D. and Collins, W.J. (eds) *Proceedings of the Second National Alternative Pasture Legume Workshop, Coonawarra, Southern Australia.* South Australia Research and Development Institute, Adelaide, pp. 88–109.

Oldham, C.M. and Mattinson, B.C. (1990) Advances in research on tagasaste. In Scott, P.R. (ed.) *Integration of Trees into the Agricultural Landscape.* Technical Report 102, Division of Resource Management, Department of Agriculture Western Australia, Perth, pp. 46–62.

Ovalle, C., Longeri, L., Aronson, J., Herrera, A. and Avendano, J. (1996) N_2-fixation, nodule efficiency and biomass accumulation after two years in three Chilean legume trees and tagasaste (*Chaemycytisus proliferus* ssp. *palmensis*). *Plant and Soil* 179, 131–140.

Southern, P.J. (1988) *Fertiliser Requirements for Tagasaste.* Productivity Focus 7/No. 3, CSBP & Farmers, Perth, Australia.

Townsend, R.J. and Radcliffe, J.E. (1987) Establishment and management of tagasaste. *Proceedings of the New Zealand Grassland Association* 48, 109–113.

Webb, C.J. (1982) Tree lucerne: its taxonomic status and naturalisation in New Zealand. In: Logan, L.A. (ed.) *Proceedings of a Workshop, Lincoln.* Crop Research Division, DSIR, Christchurch, pp. 2–5.

Wiley, T. and Maughan, C. (1993) Recent results on a number of management factors thought to affect the productivity of paddocks of tagasaste. In: Oldham, C.M. and

Allen, G. (eds) *Advances in Research on Tagasaste*, No. 3. South Australian Department of Agriculture Press, Adelaide, pp. 37–52.

Wiley, T., Oldham, C.M., Allen, G. and Wiese, T. (1994) *Tagasaste.* Bulletin 4291, Department of Agriculture, Western Australia, 23pp.

Woodfield, D.R. and Forde, M.B. (1987) Genetic variability within tagasaste. *Proceedings of the New Zealand Grassland Association* 48, 103–108.

Wright, D. and Menzies, A. (1985) Phenotypic variability within tagasaste. In: Logan, L.A. and Radcliffe, J.E. (eds) *Fodder Trees – A Summary of Current Research in New Zealand.* Report No. 106, Crop Research Division, DSIR, Christchurch.

Prospects for Forage Legumes

9

It is obvious from the preceding chapters that forage legumes are playing an immensely important role in sustainable grassland agriculture in temperate areas of the world and this role will continue. Much is known about the science of growth, development and utilization of the major forage legumes but their potential is not usually fully realized in practice. Most research work on forage legumes is based on the study of components of systems and only when the findings from these studies are synthesized into practical animal production systems can the technical and economic benefits and disbenefits of legume-based swards be evaluated. Such work, together with decision-support roles by field advisory services, is particularly necessary in countries where nitrogen (N)-fertilized grass swards are the norm or in developing temperate countries or regions where grassland receiving low inputs could potentially be improved by the introduction of forage legumes. Models of specific animal production systems can also contribute to assessment of the impact of introducing or increasing forage-legume technology.

Research and development effort on most of the widely used forage-legume species has been, and still is, considerable, though rather small for some of the lesser used species. To take one example, the *Proceedings of the International Grassland Congress* in 1993 had 78 entries for white clover and 49 for lucerne but only four each for birdsfoot trefoil and sainfoin, while there were none for alsike clover and the serradellas. Nevertheless, the relative importance or popularity of individual legume species fluctuates from time to time with the emergence of new research findings, the release of improved cultivars or innovative developments.

It has been customary to evaluate animal production systems in terms of technical efficiency and profitability. However, consideration now has to be given to environmental impact in evaluation of any new system. For example, excessive use of inorganic fertilizers is no longer environmentally acceptable in many countries, irrespective of cost–benefit, due to the increased public awareness of their contribution to contamination of underground and surface waters. Increasing floral and faunal diversity, and controls on the use of inorganic fertilizers and application of organic manures are being supported by financial incentive from governments in countries such as those in the European Union (EU) to encourage their wider adoption. It will be evident from the preceding chapters that forage legumes can play an important role in these measures.

Another EU policy that has implications for the use of forage legumes is one that encourages limits to be placed on stock numbers per unit area. The amount of feed required per unit area is likely to fall within the limit of what can be produced by forage legumes, and the higher nutritive value of forage is likely to increase production per animal, the higher output per head at least partly compensating for the lost production from reduced stock numbers.

Animal products from legume-based swards are perceived by consumers in the world market as being more 'natural' than equivalent products from intensely managed, highly N-fertilized grassland, although, to the majority of consumers, product price will still be the overriding factor. The proposed substitution of forage legumes for fertilizer N, thereby saving non-renewable energy entailed in the manufacture of the fertilizer, has been an ongoing issue for the last 25 years or so in countries that have adopted the fertilizer-N route. However, the reality is that energy has been plentiful and fertilizer N has been cost-effective, and so a resurgence in legume use is unlikely unless this scenario changes. In instances where the profitability of N-fertilizer-based systems is similar to that from legume-based systems, the perceived greater reliability of the fertilizer-N system usually persuades the farmer to opt for it rather than rely on the system that depends on biologically produced N.

For plant growth and development, N acquisition and supply are second only to photosynthesis in terms of importance, and biological molecular nitrogen (N_2) fixation accounts for 65% of the N currently used in world agriculture (Vance, 1997). A high proportion of this N comes from forage legumes in the world's grasslands. Taking New Zealand's white-clover-based grasslands as an example, the estimated quantity of N fixed is 1.57 Mt over 13.5 Mha of grassland and, adding to this clover's contribution to animal diets seed sales and honey production, the value of white clover to the country's economy is over $US3 billion (thousand million) (Caradus et al., 1995). Improving the efficiency of the legume plant–*Rhizobium* symbiosis and extending the use of forage legumes to less developed grasslands would be of enormous benefit, not only in terms of agricultural productivity but also in reducing the use of non-renewable energy for fertilizer-N manufacture. Simultaneously with such developments, management practices to minimize N emissions to the soil and aerial environments would be necessary.

The role of forage legumes could expand in the future if projected elevated atmospheric carbon dioxide and consequential global, or at least regional, warming come to pass; this seems highly probable, as a result of the carbon emissions from human activities. Atmospheric warming will be particularly beneficial to the vast high-latitude, cold-winter zones of the northern hemisphere. The effects of elevation of atmospheric temperature and carbon dioxide, in conjunction with resultant changes in precipitation and evapotranspiration, require consideration when modelling the consequences of global warming or when planning forage-legume research for future eventualities.

The projected increase in world population, particularly in the less developed regions, makes it inevitable that existing or potentially fertile grasslands will be used to produce crops for direct human consumption, while peripheral poorer land will be used to satisfy the demand for ruminant products. Already, in some densely populated countries, ruminant production is largely confined to upland and mountainous natural grasslands in need of the N_2 fixation and herbage-quality advantages that introduced forage legumes can confer. Some of the pioneer and lesser legumes may be particularly suitable for these areas, but, assuming the overall soil fertility can be improved economically and climatic conditions permit, such pioneer legumes could eventually be replaced by the more productive major legume species. However, breeding of the latter for stress factors, such as tolerance to cold or drought and to soil acidity or salinity, would clearly need to be expanded since there are technical and economic limitations to manipulating the environment.

The inevitable instability of forage legume–grass associations, resulting from the ready, but variable, availability of legume N in the soil and the higher efficiency of utilization of soil mineral N by grass, is a characteristic that limits the attractiveness of grass–legume mixtures for long-term stands. The challenge for research is to devise managements that control the competitiveness of the grass (or legume, in some circumstances) so that a long-term balance can be sustained. Ironically, it is likely that success in improving the N_2-fixing potential of the legume–*Rhizobium* association may exacerbate instability of the grass/legume association, due to the increased amount of N cycling within the system, especially in grazed swards. The variability in legume-N availability also limits the use of forage legumes as green manures. As with all organic manures, the content of available nutrients is variable. Research is required to improve production of N available to the succeeding crop, so that total N availability can be controlled.

There is an enormous pool of genetic variability in forage legumes and, while this has been tapped to breed improved cultivars, particularly in recent decades, there are still ample opportunities for further improvement by conventional and innovative breeding techniques. Recent advances and the current status in the application of biotechnologies to forage-legume improvement have been reviewed (McKersie and Brown, 1997). As pointed out, different biotechnologies are being used to identify, create, preserve or transfer genetic variability, particularly in the fields of N_2 fixation, forage quality and tolerance or resistance to biotic and abiotic stresses.

Methods now exist for genes to be transferred to most of the major forage legumes, e.g. the use of *Agrobacterium tumefaciens* to mediate the introduction of cloned genes into white clover, lucerne and subterranean clover and

Agrobacterium rhizogenes for birdsfoot trefoil and sainfoin (White, 1997). However, it is important to identify the single-gene traits which are most desirable for better forage-legume growth and utilization and which can potentially be incorporated into forage-legume genomes by gene-transfer technology. Also, a range of fertile hybrids between white clover and Caucasian clover have been successfully produced, with the combined traits of the parental species in varying proportions (Hussain and Williams, 1997). It is only a matter of time until attempts are successful in incorporating bloat resistance into white clover, red clover and lucerne, via condensed tannins (CT) from the non-bloat-inducing sainfoin, the *Lotus* species or sulla (Morris and Robbins, 1997). The advantages of CT in improving protein digestion/amino acid absorption are also well documented. When the first-fruits of biotechnology will emerge commercially is a matter of conjecture – perhaps within the next 8–12 years, *inter alia* given the period for cultivar evaluation between the breeder's final selection and approved listing for commercial use, a period that varies from country to country. However, public confidence in biotechnological advances pertaining to food production generally has still to be fully won over.

In conclusion, it is evident that worldwide there is now greater appreciation of the role and potential of forage legumes. Their usage on a world scale is likely to intensify, certainly in the long if not the short term, as new challenges and opportunities in grassland agriculture present themselves. Through the use of biotechnology, significant advances in creating a range of cultivars with improved characteristics is forthcoming. As eulogized by Gladstones (1975), 'Legumes make possible an ecologically sound non-exploitive and yet productive agricultural system, with which a hopefully stabilized [human] population can live in permanent balance or better.' The authors of this book trust that they have contributed towards this goal.

References

Caradus, J.R., Woodfield, D.R. and Stewart, A.V. (1995) Overview and vision for white clover. In: Woodfield, D.R. (ed.) *White Clover: New Zealand's Competitive Edge.* Grassland Research and Practice Series No. 6, New Zealand Grassland Association, Palmerston North, pp. 1–6.

Gladstones, J.S. (1975) Legumes and Australian agriculture. *Journal of the Australian Institute of Agricultural Sciences* 41, 227–239.

Hussain, S.W. and Williams, W.M. (1998) Development of a fertile genetic bridge between *Trifolium ambiguum* M. Bieb. and *T. repens*. *Theoretical and Applied Genetics* (in press).

McKersie, B.D. and Brown, D.C.W. (eds) (1997) *Biotechnology and the Improvement of Forage Legumes*. CAB International, Wallingford.

Morris, P. and Robbins, M.P. (1997) Manipulating condensed tannins in forage legumes. In: McKersie, B.D. and Brown, D.C.W. (eds) *Biotechnology and the Improvement of Forage Legumes*. CAB International, Wallingford, pp. 147–173.

Vance, C.P. (1997) Nitrogen fixation capacity. In: McKersie, B.D. and Brown, D.C.W. (eds) *Biotechnology and the Improvement of Forage Legumes*. CAB International, Wallingford, pp. 375–407.

White, D.W.R. (1997) Potential of biotechnology to alter pasture yield and quality. In: Welch, R.A.S., Burns, D.J.W., Davis, S.R., Popay, A.I. and Prosser, C.J. (eds) *Milk Composition, Production and Biotechnology*. CAB International, Wallingford, pp. 441–454.

Index

Alsike clover
   adaptation 273
   bloat in livestock 277
   breeding 275, 277
   conservation 277
   cultivars 275
   diseases 276
   establishment
     rhizobial inoculation 276
     Rhizobium leguminarosum bv. *trifolii* 276
     seed mixtures 276
     sowing methods 276
   feeding value 277
   flowering 275, 277
   grazing 277
   morphology 273–275
   nutritive value 277
   origins 273
   persistence 273
   pests 276
   production 277
   seed production 275, 277
   soil
     conditions 273, 277
     pH 277
   viruses 276
   world distribution 273

Birdsfoot trefoil
   adaptation 7, 245–246
   animal intake 261
   animal performance 261–262
   antibloating characteristic 8, 261
   breeding
     aims 255
     cultivars 255, 263
     ecotypes 254
     herbicide resistance 258
     ploidy 254
   conservation 262
   cutting management 262–263
   diseases 259
   drought tolerance 246, 260
   establishment
     direct drilling 259
     monocultures 257
     oversowing roughland 259
     seed mixtures 257, 262–263
     seed rates 257
     seed treatments 252
     sowing depth 257
   fertilizer use 258
   grazing
     autumn stockpiling 262
     management 263
   honey production 246
   morphology
     growth habit 247
     leaves 247
     seed 251, 256
   narrow-leaf birdsfoot trefoil
     adaptation 246
     breeding 254
     distribution 246
     growth habit 247
     seed rate 257
   nitrogen fixation
     amounts 4, 254
     nodulation 253
     rhizobial inoculation 253
     temperature effects 254

Birdsfoot trefoil *cont.*
   nutritive value
      amino acid absorption 8, 267
      condensed tannins (CT) 6, 8, 260–261
      digestibility 260
      protein protection 8, 261
   origins 245
   persistence 263
   pests
      nematode inhibition 8, 262
      nematodes 259
   physiology
      critical mineral concentrations 253
      flowering 245–251
      germination 252
      light effects 252
      plant nutrition 252–253
      regrowth 262
      seedling emergence 252
      temperature effects 252
      total non-structural carbohydrates (TNC) 262
   seed production
      chemical growth regulation 256
      hard-seededness 251
      pests 258–259
      pod dehiscence 251
      production 256
      weed control 256
      yields 256
   soil
      drainage conditions 246
      salinity 246
   weed control 258
   world distribution 7, 245–246

Chaemaecytisus palmensis see Tagasaste
Chaemaecytisus proliferus see Tagasaste

Greater lotus
   adaptation 245–246
   agroforestry 246
   animal performance 261–262
   breeding
      aims 255
      cultivars 255
      ploidy 254
   conservation 262
   establishment
      natural reseeding 260
      oversowing 257
      seed treatment 252
      sowing rate 257
      sowing time 257
   grazing management 263
   honey production 246
   morphology
      growth habit 247
      leaves 247
      rhizomes 247
      seed 251–252, 256
   nitrogen fixation
      amounts 254
      Bradyrhizobium 253
      mineral nutrient effects 254
      rhizobial inoculation 253
      Rhizobium lotii 254
   nutritive value
      amino acid absorption 8, 261
      condensed tannins (CT) 6, 8, 260–261
      digestibility 260
      protein protection 8
   origins 245
   persistence 263
   pests
      grass grub (*Costelybra zealandica*) 259
      nematode inhibition 8, 262
   physiology
      aluminium tolerance 253
      critical mineral concentrations 253
      flowering 248–249
      germination 252
      light effects 252
      plant nutrition 253
      temperature effects 252
   seed production
      chemical growth regulation 256
      hard-seededness 251–252
      pod dehiscence 251
      production 256
      yields 256
   soil acidity 253
   weed control 258
   world distribution 7–8, 245–246

Hedysarum coronarium see Sulla

Lotus corniculatus see Birdsfoot trefoil
Lotus tenuis see Birdsfoot trefoil, narrow-leaf birdsfoot trefoil
Lotus uliginosus see Greater lotus
Lucerne
   alternative uses 161
   ancestry 107

Index

animal intake
　comparison with grass 7, 159–160
　factors affecting 157–158
animal performance
　beef cattle 159
　milk production 160–161
　sheep 161
antiquality factors
　bloat in livestock 6, 131, 155
　oestrogens 147, 155–156, 159
　saponins 155
breeding
　aims 123, 127, 130
　cultivar differences 113–114, 119, 123, 160
　cultivar registration 132–133
　cultivar types 7, 128, 131
　genetic engineering 7, 128, 131, 314
　history 132
　lucerne species 107–108
　methods 7, 127–128, 133
　rate of improvement 128
　somatic hybrids 131
chemical composition (Table 3.7) 152
　critical mineral concentrations 125–126
　crude protein (CP) 118, 152–154, 157, 160
　diagnosis and recommendation integrated system (DSIR) 125
　mineral contents 116, 124–127, 151
　mineral toxicity 115, 125, 130
　nitrogen 113, 119, 125
　starch 113
　total non-structural carbohydrates (TNC) 112, 116–117, 119
　water soluble carbohydrates (WSC) 113, 116, 157
conservation
　big bale silage 156
　dehydration 155, 158, 159, 160
　hay 107, 153–154, 156
　leaf shatter 156
　silage 107, 153–155, 157, 159
defoliation
　effects 118
　frequency 118, 120, 135–136, 156
　severity 118–120
diseases (Table 3.5) 117
　Alternaria leaf spot 148
　bacterial wilt 118, 132, 148
　fungicide control 147
　Fusarium crown rot 147
　Fusarium wilt 132, 148
　Phytophthora root rot 133, 147, 149

Rhizoctonia crown rot 118
Sclerotinia crown rot 118, 129, 148
spring black stem 129
summer black stem 129
Verticillium wilt 129, 132, 133, 148
establishment
　companion grasses 123, 135–137, 139, 158–159
　cover crop 138–139
　direct drilling 138
　seed mixtures 135–138
　seed rates 137
　sowing date 138
　sowing depth 137
　sowing methods 139
fertilizer use
　lime 140
　nitrogen 139
　phosphorus 140, 148
　potassium 140, 148
　slurry 139
grazing
　continuous 107, 158
　intervals 158
　rotational 107, 158–159
　selective 159
　tolerance 129–130
herbage quality
　amino acids (Table 3.8) 153–154
　chemical composition 152
　digestibility 112, 152–153
　fibre 112, 118, 152, 154, 160
　metabolizable energy 160
　protein degradation 153–154, 160
irrigation 115, 134, 140
Medicago falcata 207–209, 128, 130
Medicago media 107–109, 128, 130
morphology
　axillary buds 114, 119
　crown buds 114, 119
　flowers 109
　roots 109, 111
　seed 109
nitrogen fixation
　amounts 4, 120
　defoliation effects 117, 122
　drought effects 111, 115
　fertilizer nitrogen effects 122, 139
　rhizobial inoculation 121, 138
　Rhizobium meliloti 120–121, 123
　root nodules 120, 122
　soil acidity 121
　vesicular arbuscular mycorrhiza (VAM) interaction 121

Lucerne cont.
    nitrogen transfer 123–124
    nutritive value 118, 119, 130–131, 153, 156
    origins 108
    persistence 118, 119, 146, 157, 158
    pests (Table 3.4) 144
        alfalfa seed chalcid 145
        alfalfa weevil 7, 143
        biological control 143
        blue aphid 133
        clover weevil 144, 147
        control methods 146
        gall midge 146
        leaf hoppers 144–145
        leatherjackets 145
        pea aphid 7, 133
        root lesion nematode 7, 129, 145
        spittlebugs 144
        spotted aphid 7, 133
        stem nematode 7, 129, 145
    physiology
        assimilate partitioning 116–117
        cold-hardiness 114, 130
        defoliation 117–120
        dormancy 112
        drought resistance 114–115
        flowering 112
        germination 111–112
        light effects 113–114
        light quality 136
        moisture 114, 130, 140
        photomorphogenesis 113
        photosynthesis 113, 115
        pollination 109, 134
        temperature effects 112–113
        water use efficiency 115
    production
        annual 7, 149–150
        models 151
        prediction 149–150
        seasonal 149–150
    seed production
        hard-seededness 133
        management 134–135
        production 133–135
        yields 134–135
    soil
        conditions 137
        pH 148
        salinity 130
    viruses 149
    weed control
        chemical 142–143
        cultural 141
    winter-hardiness 112–113, 129
    world distribution 6–7, 107–109

Medicago sativa see Lucerne

Onobrychis viciifolia see Sainfoin
Ornithopus compressus see Serradella
Ornithopus sativus see Serradella

Red clover
    adaptation 182
    animal intake 5, 210–211
    animal performance 5, 211
    antiquality factors
        bloat in livestock 206–207
        formononetin 194, 206
        oestrogenic effects 6, 206
    breeding 192–193
        aims 192–193, 201, 203
        cultivar types 186–187, 193–194
        genetic engineering 207, 316
        methods 192–193
        tetraploids 187–188, 192, 195, 205
    chemical composition
        critical mineral concentrations 188
        crude protein (CP) 196, 204–205
        mineral contents 187–188
        nitrogen 187
        starch 189–190
        total available carbohydrates (TAC) 183, 189
        total non-structural carbohydrates (TNC) 185–186
        water soluble carbohydrates (WSC) 205, 207–209
    conservation
        hay 207
        silage 207–209
        silage fermentation 208–209
    defoliation
        autumn 186, 190, 201, 205–206
        effects 188–190, 204
    diseases
        clover rot 201
        fungicide control 201–202
        Fusarium spp. 201–202
        northern anthracnose *see* Lucerne, diseases, scorch

powdery mildew (*Erysiphe trifolii*) 202
scorch (*Kabatiella caulivora*)
southern anthracnose (*Colletrichum trifolii*) 202
establishment
  companion grass 196–197
  cover crop 197–198
  direct drilling 198–199, 204
  direct sowing 198
  hard seed 182
  seed mixtures 195–197
  slot seeding 204
  sowing date 197
  undersowing 197–198
feeding value 211
fertilizer use
  lime 199
  nutrients 199–200
  slurry 199–200
grazing
  continuous 209–210
  rotational 209–210
  tolerance 193
green manure 182
herbage quality
  chemical composition 204–206, 211
  digestibility 204–205
irrigation 187
morphology
  axillary buds 183
  roots 183
  seed 183
  stems 183
nitrogen fixation
  amounts 4, 190
  rhizobial inoculation 190–191, 198
  Rhizobium leguminosarum bv. *trifolii* 198
  root nodules 188, 190–191
  vesicular arbuscular mycorrhiza (VAM) 187–188
nitrogen transfer 191–192
novelty uses 182
nutritive value 204–206
origins 181
persistence 193, 200–203
pests
  clover cyst nematode (*Heterodera trifolii*) 199, 203
  clover-stem eelworm (*Ditylenchus dipsaci*) 199, 202–203
  insecticide control 201–202
  root-knot nematodes (*Meloidogyne* spp.) 199, 203
  root-lesion nematodes (*Pratylenchus* spp.) 199–203
  slugs (*Deroceras reticulatum*) 203
  weevils (*Sitona* spp.) 203
physiology
  carbon fluxes 189, 191
  cation exchange capacity 187
  cold-hardiness 185–186, 197–198
  flowering 185–187
  germination 182, 192–193
  light effects 186–187
  moisture 187
  photosynthesis 184
  pollination 194
  temperature effects 183–184
production
  annual production 197, 203–204
  seasonality 203
  wheel-tracking effects 207–208
seed production
  chemical growth regulation 194, 205
  harvesting 194
  yields 195
soil
  compaction 207–208
  conditions 199
  pH 188, 190–191, 199–200
sward renovation 198
viruses 202
weed control 200–202
winter-hardiness 185–186, 197–198
world distribution 7, 181

Sainfoin
  adaptation 8, 279
  amino acid absorption 8
  animal performance 285–286
  antibloating characteristic 8, 279, 285–286
  breeding 283
  chemical composition
    condensed tannins (CT) 8, 285
    crude protein 285
    minerals 285
    total non-structural carbohydrates (TNC) 279, 286
  conservation
    hay 279, 283, 286
    silage 283, 286
  cultivars 283
  diseases 283–284
  ecotypes 282–283

Sainfoin *cont.*
　　establishment
　　　companion grass 283–284
　　　rhizobial infection 282
　　　rhizobial inoculation 282
　　　seed mixtures 283–284
　　　sowing methods 276
　　feeding value 285–286
　　flowering 280
　　grazing 279
　　irrigation 279
　　morphology 279–281
　　nitrogen fixation 282–283
　　nutritive value 285–286
　　origins 279
　　persistence 282, 286
　　pests 284
　　physiology
　　　drought resistance 280–285
　　　light effects 282
　　　photosynthesis 282
　　　temperature effects 282
　　　winter-hardiness 284
　　production 283–285
　　seed production 283
　　soil
　　　conditions 279, 284
　　　pH 279, 284
　　weed control 283–284
　　winter-hardiness 284
　　world distribution 279
Serradella (pink and yellow)
　　adaptation 291–292
　　Bradyrhizobium spp. (*lupinus*) 295
　　breeding
　　　cultivars 294
　　　hybridization 294
　　conservation 296
　　establishment
　　　hard seed 293–294
　　　seed mixtures 294
　　　sowing rates 294
　　fertilizer use 295
　　grazing management 295–297
　　morphology 292
　　origins 291
　　production 296–297
　　rhizobial inoculation 295
　　seed production
　　　buried seed 297
　　　hard-seededness 293–294
　　　harvesting 294
　　　pods 292
　　　production 296–297
　　　treatment 294

　　species 291–292
　　uses 297
　　utilization 291–292
　　weed control 295
　　world distribution 291–292
Subterranean clover
　　adaptation 7, 225
　　animal performance 238
　　antiquality factors
　　　bloat in livestock 238
　　　formononetin 4, 236–237
　　breeding
　　　aims 232
　　　cultivars 232–233
　　　ecotypes 232
　　chemical composition 236
　　conservation 238–239
　　diseases 232, 235
　　establishment
　　　buried seed 229, 234
　　　companion species 233, 236
　　　false strike of seedlings 234
　　　hard seed 234
　　　oversowing 234
　　　seed rate 233
　　　sowing depth 231
　　fertilizer use
　　　lime 231, 235
　　　phosphorus 234–235
　　　trace elements 235
　　grazing
　　　establishing swards 236
　　　pressure 238
　　　systems 238
　　morphology
　　　growth habit 227
　　　leaves 227
　　　seed 229–231
　　　seed burrs 229–230
　　nitrogen fixation
　　　amounts 231
　　　nodulation 231
　　　rhizobial inoculation 231
　　　Rhizobium leguminosarum bv. *trifolii*
　　　　231
　　nutritive value 236–238
　　pests 235
　　physiology
　　　flowering 227–228
　　　germination rate 234
　　　moisture 231
　　　nitrogen 231
　　　self fertility 229, 232
　　　shade tolerance 226
　　　temperature 230, 234

production 236, 239
seed production
 burrs 229–230
 hard-seededness 230–231, 234
 in Australia 226, 233
soil
 fertility requirements 234–235
 pH 235
 waterlogging 235
subclover species 226–227
Trifolium brachycalcinum 226–227
Trifolium yanninicum 226–227
viruses 232, 235
weed control 233–234
world distribution 225, 226
Sulla
 animal performance 300
 anti-bloating characteristic 8, 301, 302
 breeding
  cultivars 299, 301
  ecotypes 299–301
 conservation
  hay 299
  silage 302
 disease 301
 establishment
  seed mixtures 301
  seed rates 301
  seed treatment 299
 feeding value 302
 grazing management 299, 302
 morphology 299
 nematode control 302
 nutritive value
  amino acid absorption 8
  condensed tannins (CT) 8, 301, 302
  protein protection 8, 302
 origins 299
 rhizobial inoculation 301
 Rhizobium hedisari 301
 seed production
  hard-seededness 301
  harvesting 301
  pods 299
  production 301
  yields 301
 uses 299
 weed control 301

Tagasaste
 adaptation 304–305, 308
 animal performance 308–309
 cutting management 304–305, 309
 ecotypes 305

establishment
 rooted seedlings 305
 seed treatment 305–306
 sowing time 307
 tree density 307–308
feeding value 308
fertilizer use 308
grazing management 309
intake 308–309
morphology 305
nitrogen fixation
 amount 306
 Bradyrhizobium lotii 306
 rhizobial inoculation 306
 Rhizobium lotii 306
origin 304
pests 308
production 308–309
seed production 305
soil type 305, 308
uses 309
utilization 309
world distribution 7, 304–305
Temperate forage legumes
 advantages
  condensed tannins (CT) 6, 8, 316
  feeding value 1, 4–5
  nitrogen fixation 1, 3–4
  soil structure 4
 disadvantages
  bloat in livestock 5–6
  phyto-oestrogens 6
 future use 3
 genetic engineering 315–316
 global warming 314–315
 history 1
 minor species 8
 nitrogen
  cycle 2
  environmental protection 2–3
  fixation 2, 4
  fossil fuel energy 2–3
 nutritive value 4–5
 prospects 313–316
 world distribution 6–8
Trifolium hybridum see Alsike clover
Trifolium pratense see Red clover
Trifolium repens see White clover
Trifolium subterranean see Subterranean clover

White clover
 alternative uses
  ground cover 79
  organic farming 78–79

White clover *cont.*
  alternative uses *cont.*
    source of honey 79
    understorey for cereals 79
  animal intake
    comparisons with grass 75–76
    factors affecting 75–76
    grass/white clover mixtures 76–78
    monocultures 75–76
  animal performance
    beef cattle 78
    sheep 76–77
  antiquality factors
    bloat in livestock 6, 72
    cyanogenesis 59, 72–73
  breeding
    aims 29, 40–41
    cultivar evaluation 42–43, 74
    cultivar types 17, 22–23, 38–40, 42, 68
    genetic engineering 40, 315–316
    history 38–39
    hybridization 40
    methods 39–40
    rate of improvement 41–42
  chemical composition
    critical mineral concentrations 34–38
    crude protein 42, 70, 78
    diagnosis and recommendation integrated system (DSIR) 37–38
    mineral contents 34–38
    mineral toxicity 37
    nitrogen 34–70
    starch 17, 28
    total available carbohydrates (TAC) 29
    total non-structural carbohydrates (TNC) 23
  conservation
    big bale silage 75
    hay 70
    integration with grazing 69–70
    monocultures 70
    silage 69–70
  defoliation
    effects 29
    frequency 63–64
    severity 63
  diseases
    black blotch (*Cymodothea trifolii*) 61
    burn *see* White clover, diseases, pepper spot
    clover root (*Sclerotinia trifoliorum*) 61
    fungicide control 61
    Fusarium spp. 61
    leaf spot (*Pseudopeziza trifolii*) 61
    pepper spot (*Leptosphaerulina trifolii*) 61
    sooty blotch *see* White clover, diseases, black blotch
  establishment
    companion grass 47–49
    cover crop 50
    direct drilling 50–52
    direct sowing 50
    natural reseeding 52
    seed mixtures 42, 47–49
    slot seeding 51–52
    sowing date 49–50
    sowing depth 49
    sowing rate 46–47
    undersowing 50
  feeding value 76–78
  fertilizer use
    autumn 55
    lime 52
    mineral nutrients 37, 52
    nitrogen 52–55
    slurry 56
    spring 54–55
    trace elements 56
  grazing
    animal species effects 5, 66–68
    autumn stockpiling 64
    continuous 66–67
    excretal return effects 67–68
    intensity 65
    rest interval 66
    rotational 67
    selective 66–67
    trampling effects 67
    winter grazing 65
    zero grazing 69
  herbage quality
    chemical composition 70–71
    digestibility 71
    target clover contents 3, 4–5, 71–72, 78
  irrigation 26, 70
  morphology
    leaves 17, 22
    petiole extension 17, 21–22, 25
    phenotypic plasticity 17, 66
    roots 16–17, 28
    stolon burial 19–20
    stolon development 17, 19, 27–29
    stolon persistence 28–29

nitrogen fixation
  amounts 2, 31–32, 312
  drought effects 27
  heavy metal effects 31
  rhizobial inoculation 30, 74
  Rhizobium leguminosarum bv. *trifolii* 30–31
  root nodules 17, 31
  seedlings 22
  shade effects 32
  soil acidity effects 30–31
  vesicular arbuscular mycorrhiza (VAM) 58
nitrogen transfer
  cycling 33–34, 57
  estimates 32–34
  losses 33–34, 57
  stress conditions 33
novelty uses 79
nutritive value 70–72
origins 15
persistence 19, 28, 29, 66
pests
  grass grub (*Costelybra zealandica*) 60
  insecticide control 58–60
  leatherjackets (*Tipula* spp.) 59
  lucerne flea (*Sminthurus viridis*) 60
  nematodes (*Pratylenchus* spp.) 60
  porina (*Wiseana cervinata*) 60
  red-legged earth mite (*Halotydeus destructor*) 60
  slugs (*Deroceras reticulatum*; *Limax* spp.) 59
  stem eelworm (*Sclerotinia trifoliorum*) 59–60
  weevils (*Sitona* spp.) 60
physiology
  carbon dioxide assimilation 24
  carbon dioxide elevation 29–30
  cation exchange capacity 34
  cold-hardiness 23
  defoliation 28–29
  flowering 20, 25–26
  germination 21
  global warming 29–30
  light effects 19, 23–26
  light quality 25, 27
  mineral nutrition 34–38
  moisture 26–27
  perennation 27–29
  photosynthesis 23–26
  pollination 20
  seedling growth 21–22
  temperature effects 22–23
  water stress 23, 26–27
production
  annual 52–53, 62–64
  autumn 55
  endophyte effects 48
  seasonal 53–55
  spring 23, 54–55
  winter 23
rangeland improvement
  constraints 74
  fertilizer use 74–75
  liming 74–75
  rough grazing 73–74
  two-pasture system 73–74
seed production
  certification 46
  chemical growth regulation 45
  hard-seededness 20–21
  harvesting 45–46
  world production 43
  yield components 45–46
  yields 43–44
soil
  compaction 67
  conditions 26, 74
  pH 30–31, 36–37, 74–75
sward renovation 50–52
viruses 62
weed control 57–58
winter-hardiness 23
world distribution 6, 15–16